Yearbook of Astronomy 2019

Front Cover: A Cassini image of Saturn's moon Mimas with the outer part of the planet's A ring and the narrow F ring in the background. Mimas is one of many moons that gravitationally interact with the rings to produce wave structures. See the article *The Cassini-Huygens Mission to the Saturn System* for a summary of results from this mission. (NASA/JPL-Caltech/Space Science Institute)

YEARBOOK OF ASTRONOMY 2019

Brian Jones

WHITE OWL
AN IMPRINT OF PEN & SWORD BOOKS LTD.
YORKSHIRE – PHILADELPHIA

First published in Great Britain in 2018 by
Pen & Sword WHITE OWL
An imprint of
Pen & Sword Books Ltd
Yorkshire – Philadelphia

ISBN 9781526737038

A CIP catalogue record for this book is available from the British Library.

Typeset in Dante By Mac Style

Printed and bound in India by Replika Press Pvt. Ltd.

Pen & Sword Books Ltd incorporates the Imprints of Pen & Sword Books Archaeology, Atlas, Aviation, Battleground, Discovery, Family History, History, Maritime, Military, Naval, Politics, Railways, Select, Transport, True Crime, Fiction, Frontline Books, Leo Cooper, Praetorian Press, Seaforth Publishing, Wharncliffe and White Owl.

For a complete list of Pen & Sword titles please contact

PEN & SWORD BOOKS LIMITED
47 Church Street, Barnsley, South Yorkshire, S70 2AS, England
E-mail: enquiries@pen-and-sword.co.uk
Website: www.pen-and-sword.co.uk

or

PEN AND SWORD BOOKS
1950 Lawrence Rd, Havertown, PA 19083, USA
E-mail: Uspen-and-sword@casematepublishers.com

Contents

Article Section

Miscellaneous

Editor's Foreword

The Yearbook of Astronomy 2019 is the latest edition of what has long been an indispensable publication, the annual appearance of which has been eagerly anticipated by astronomers, both amateur and professional, for well over half a century. As ever, the Yearbook is aimed at both the armchair astronomer and the active backyard observer. Within its pages you will find a rich blend of information, star charts and guides to the night sky coupled with an interesting mixture of articles which collectively embrace a wide range of topics, ranging from the history of astronomy to the latest results of astronomical research; space exploration to observational astronomy; and our own celestial neighbourhood out to the farthest reaches of space.

The *Monthly Star Charts* have been compiled by David Harper and show the night sky as seen throughout the year. Two sets of twelve charts have been provided, one set for observers in the Northern Hemisphere and one for those in the Southern Hemisphere. Between them, each pair of charts depicts the entire sky as two semi-circular half-sky views, one looking north and the other looking south.

In response to helpful feedback from readers of the 2018 edition, two new items have been added to the first section of the Yearbook. *The Planets in 2019* gives a concise summary of the observing conditions for each of the planets during the year and *Some Events in 2019* is a calendar of significant Solar System events occurring throughout 2019. Lists of *Phases of the Moon in 2019* and *Eclipses in 2019* are also provided. Further improvements and additions are planned for the 2020 edition of the Yearbook, including visibility diagrams for Mercury and Venus and finder charts for the outer planets. In the meantime, suggestions from readers are welcomed.

The *Monthly Sky Notes* have been compiled by Lynne Marie Stockman and give details of the positions and visibility of the planets throughout 2019. The sky notes now contain additional information to that provided in the 2018 edition, which it is hoped will enhance their usefulness to observers. Each section of the Monthly Sky Notes is accompanied by a short article, the range of which includes not only observing guides to specific constellations on view at the time but also items on a wide variety of astronomy-related topics.

The Monthly Sky Notes and Articles section of the book rounds off with *Comets in 2019*, *Minor Planets in 2019* and *Meteor Showers in 2019*, all three written and provided by Neil Norman and all three titles being fairly self-explanatory, describing as they do the occurrence and visibility of examples of these three classes of object during and throughout the year.

In his article *Astronomy in 2018* Rod Hine covers a range of topics, including a summary of the continuing interest in the TRAPPIST-1 star system and a close look at the ongoing mysteries of Dark Matter and Dark Energy and the latest attempts by scientists to explain them. This is followed by *Solar System Exploration in 2018* in which Peter Rea provides information on a wide range of space missions, including the eagerly anticipated BepiColombo mission to Mercury and the exciting asteroid sample return mission OSIRIS-Rex. Launched in September 2016 to study the carbon-rich asteroid 101955 Bennu, the objectives of OSIRIS-Rex include orbiting and mapping the asteroid prior to touching down on the surface to collect, and subsequently return, a small sample to Earth for analysis.

In 2019 we celebrate the 400th anniversary of the birth of Jeremiah Horrocks whose achievements during his short life suggest that he had the potential to be one of the greatest of astronomers. In his article *Anniversaries in 2019* Neil Haggath explores the short-lived but brilliant career of Jeremiah Horrocks, who was born in 1619 and whose life was tragically cut short by his death in 1641 at the age of just 22 years. Among the other anniversaries covered in Neil's article is what is perhaps the most obvious one of all, that of the 50th anniversary of the first landing of humans on the Moon, this by the Apollo 11 mission in 1969.

In his article *The Cassini-Huygens Mission to the Saturn System* Carl Murray provides us with a detailed overview of the magnificent achievements of this historic mission. Carl has been directly involved with the Cassini project since the 1990s, and his article presents us with an interesting and informative summary of what can only be described as a highly successful mission.

This is followed by *Science Fiction and the Future of Astronomy* by Mike Brotherton, who informs us that one of the primary jobs of science fiction is to try to forecast the future in an entertaining and perhaps adventurous way. Although not always successful, science fiction has tended to hit the mark reasonably often in a number of areas, including technology, forecasting the development of space travel, robots and ever smaller and more powerful computers. In this fascinating article, Mike looks to the future, and the opportunities astronomers may be presented with through the ongoing evolution of the telescopes and other instrumentation they use to explore the Universe.

Well known for his discovery of Phobos and Deimos, the two moons of Mars, in August 1877, the famous American astronomer Asaph Hall is the subject of *Asaph Hall: Man of Mars* by Neil Norman. The article provides us not only with the story of the discoveries of Phobos and Deimos, but also a great deal about the life and other achievements of Asaph Hall.

Following his article *Double and Multiple Stars* in the 2018 edition of the Yearbook, John McCue has now taken this a stage further. In *Getting the Measure of Double Stars*, John points out that the observation of double stars can be a rewarding pursuit, even in these days of light polluted skies. He then goes on to inform us of the different methods and techniques that can be used to observe and measure binary stars. The steady accumulation over time of these observations and measurements provides valuable information, which eventually allows the orbits of binary star components around each other to be refined and calculated. As the article clearly demonstrates, binary star observation is one of the areas of observation to which the backyard astronomer can make useful contributions.

Two important centenaries are then highlighted, the first of which is that of the founding of the International Astronomical Union (IAU), described in Susan Stubbs' informative article *100 Years of the International Astronomical Union* which, as well as covering the early years of the IAU, goes on to tell us a great deal about the role of the IAU in modern day astronomy.

This is followed by Neil Haggath's article *In Total Support of Einstein: Eddington's Eclipse, 1919* in which he describes Sir Arthur Eddington's expedition to Príncipe off the west coast of Africa to witness the solar eclipse of 29 May 1919, the results of which were used to verify the bending of light by the Sun's gravity, as predicted by Einstein's General Theory of Relativity.

In his article *The First Micro-Quasar* regular contributor David Harland tells the fascinating story behind the momentous discovery of a micro-quasar which, at the time it was first brought to light in 1978-1980, represented an entirely new class of astronomical object.

This is followed by our penultimate article *Father Lucian Kemble and the Kemble Asterisms* in which Steve Brown examines the life and achievements of Franciscan Friar and amateur astronomer Lucian Kemble, who first came to the attention of many through his discovery of the three attractive asterisms that now bear his name – Kemble 2, Kemble's Kite and the famous Kemble's Cascade, each of which is examined in detail.

The constellation of Cetus (the Whale) plays host to Mira, one of the best known variable stars in the entire sky. In our final article *Mira 'The Wonderful'* Roger Pickard explores not only the early history of observation of Mira, but also what subsequent and current research has told us about this famous star, and other objects of its type.

The final section of the book starts off with *Some Interesting Variable Stars to Observe in 2019* by Roger Pickard which contains useful information on variables, including the well known long-period variable star Mira, as well as predictions for timings of minimum brightness of the famous eclipsing binary Algol for 2019. *Some Interesting Double Stars* and *Some Interesting Nebulae, Star Clusters and Galaxies* present a selection of objects for you to seek out in the night sky. The lists included here are by no means definitive and may well omit your favourite celestial targets. If this is the case, please let us know and we will endeavour to include these in future editions of the Yearbook.

The book rounds off with a selection of *Astronomical Organizations*, which lists organizations and associations across the world through which you can further pursue your interest and participation in astronomy (if there are any that we have omitted please let us know) and *Our Contributors*, which contains brief background details of the numerous writers who have contributed to the Yearbook of Astronomy 2019. Finally, following the introduction of a *Glossary* to the Yearbook of Astronomy in the 2018 edition, this has been expanded for 2019 with the addition of many new entries of brief but informative explanations for many of the words, and much of the terminology, used in the Yearbook (as well as a few more).

Over time new topics and themes will be introduced into the Yearbook to allow it to keep pace with the increasing range of skills, techniques and observing methods now open to amateur astronomers, this in addition to articles relating to our rapidly-expanding knowledge of the Universe in which we live. There will be an interesting mix, some articles written at a level which will appeal to the casual reader and some of what may be loosely described as at a more academic level. The intention is to fully maintain and continually increase the usefulness and relevance of the Yearbook of Astronomy to the interests of the readership who are, without doubt, the most important aspect of the Yearbook and the reason it exists in the first place.

As ever, grateful thanks are extended to those individuals who have contributed a great deal of time and effort to the Yearbook of Astronomy 2019, not least of which is David Harper, who has provided updated versions of his excellent Monthly Star Charts. These were generated specifically for what has been described as the new generation of the Yearbook of Astronomy, and the charts add greatly to the overall value of the book to star gazers. Equally important are the efforts of Lynne Marie Stockman who has put together the Monthly Sky Notes. Their combined efforts have produced what can justifiably be described as the backbone of the Yearbook of Astronomy.

Also worthy of mention is Garry (Garfield) Blackmore who, as well as preparing many of the illustrations for publication, has produced the artwork for the smaller star charts included with a number of the articles. Thanks also go to Jonathan Wright, Lori Jones, Kate Bamforth, Janet Brookes and Paul Wilkinson of Pen & Sword Books Ltd without whose combined help, belief and confidence in the Yearbook of Astronomy, this much-loved and iconic publication may well have disappeared for ever.

Brian Jones – Editor
Bradford, West Riding of Yorkshire

February 2018

As many of you will be aware, the future of the Yearbook of Astronomy was under threat following the decision to make the 2016 edition the last. However, the series was rescued, both through the publication of a special 2017 edition (which successfully maintained the continuity of the Yearbook) and a successful search for a publisher to take this iconic publication on and to carry it to even greater heights as the Yearbook approaches its diamond jubilee in 2022.

The Yearbook of Astronomy 2017 was a limited edition, although copies are still available to purchase. It should be borne in mind that you would not be obtaining the 2017 edition as a current guide to the night sky, but as the landmark edition of the Yearbook of Astronomy which fulfilled its purpose of keeping the series alive, and which heralded in the new generation of this highly valued and treasured publication. You can order your copy of the 2017 edition at **www.starlight-nights.co.uk/subscriber-2017-yearbook-astronomy**

Preface

The information given in this edition of the Yearbook of Astronomy is in narrative form. The positions of the planets given in the Monthly Sky Notes often refer to the constellations in which they lie at the time. These can be found on the star charts which collectively show the whole sky via two charts depicting the northern and southern circumpolar stars and forty-eight charts depicting the main stars and constellations for each month of the year. The northern and southern circumpolar charts show the stars that are within 45° of the two celestial poles, while the monthly charts depict the stars and constellations that are visible throughout the year from Europe and North America or from Australia and New Zealand. The monthly charts overlap the circumpolar charts. Wherever you are on the Earth, you will be able to locate and identify the stars depicted on the appropriate areas of the chart(s).

There are numerous star atlases available that offer more detailed information, such as *Sky & Telescope's POCKET SKY ATLAS* and *Norton's STAR ATLAS and Reference Handbook* to name but a couple. In addition, more precise information relating to planetary positions and so on can be found in a number of publications, a good example of which is *The Handbook of the British Astronomical Association*, as well as many of the popular astronomy magazines such as the British monthly periodicals *Sky at Night* and *Astronomy Now* and the American monthly magazines *Astronomy* and *Sky & Telescope*.

About Time

Before the late 18th century, the biggest problem affecting mariners sailing the seas was finding their position. Latitude was easily determined by observing the altitude of the pole star above the northern horizon. Longitude, however, was far more difficult to measure. The inability of mariners to determine their longitude often led to them getting lost, and on many occasions shipwrecked. To address this problem King Charles II established the Royal Observatory at Greenwich in 1675 and from here, Astronomers Royal began the process of measuring and cataloguing the stars as they passed due south across the Greenwich meridian.

Now mariners only needed an accurate timepiece (the chronometer invented by Yorkshire-born clockmaker John Harrison) to display GMT (Greenwich Mean Time). Working out the local standard time onboard ship and subtracting this from GMT gave the ship's longitude (west or east) from the Greenwich meridian. Therefore mariners always knew where they were at sea and the longitude problem was solved.

Astronomers use a time scale called Universal Time (UT). This is equivalent to Greenwich Mean Time and is defined by the rotation of the Earth. The Yearbook of Astronomy gives all times in UT rather than in the local time for a particular city or country. Times are expressed using the 24-hour clock, with the day beginning at midnight, denoted by 00:00. Universal Time (UT) is related to local mean time by the formula:

Local Mean Time = UT – west longitude

In practice, small differences in longitude are ignored and the observer will use local clock time which will be the appropriate Standard (or Zone) Time. As the formula indicates, places in west longitude will have a Standard Time slow on UT, while those in east longitude will have a Standard Time fast on UT. As examples we have:

Standard Time in

New Zealand	UT +12 hours
Victoria, NSW	UT +10 hours
Western Australia	UT + 8 hours
South Africa	UT + 2 hours
British Isles	UT
Eastern Standard Time	UT −5 hours
Central Standard Time	UT −6 hours
Pacific Standard Time	UT −8 hours

During the periods when Summer Time (also called Daylight Saving Time) is in use, one hour must be added to Standard Time to obtain the appropriate Summer/Daylight Saving Time. For example, Pacific Daylight Time is UT −7 hours.

Using the Yearbook of Astronomy as an Observing Guide

Notes on the Monthly Star Charts

The star charts on the following pages show the night sky throughout the year. There are two sets of charts, one for use by observers in the Northern Hemisphere and one for those in the Southern Hemisphere. The first set is drawn for latitude 52°N and can be used by observers in Europe, Canada and most of the United States. The second set is drawn for latitude 35°S and show the stars as seen from Australia and New Zealand. Twelve pairs of charts are provided for each of these latitudes.

Each pair of charts shows the entire sky as two semi-circular half-sky views, one looking north and the other looking south. A given pair of charts can be used at different times of year. For example, chart 1 shows the night sky at midnight on 21 December, but also at 2am on 21 January, 4am on 21 February and so forth. The accompanying table will enable you to select the correct chart for a given month and time of night. The caption next to each chart also lists the dates and times of night for which it is valid.

The charts are intended to help you find the more prominent constellations and other objects of interest mentioned in the monthly observing notes. To avoid the charts becoming too crowded, only stars of magnitude 4.5 or brighter are shown. This corresponds to stars that are bright enough to be seen from any dark suburban garden on a night when the Moon is not too close to full phase.

Each constellation is depicted by joining selected stars with lines to form a pattern. There is no official standard for these patterns, so you may occasionally find different patterns used in other popular astronomy books for some of the constellations.

Any map projection from a sphere onto a flat page will by necessity contain some distortions. This is true of star charts as well as maps of the Earth. The distortion on the half-sky charts is greatest near the semi-circular boundary of each chart, where it may appear to stretch constellation patterns out of shape.

The charts also show selected deep-sky objects such as galaxies, nebulae and star clusters. Many of these objects are too faint to be seen with the naked eye, and you will need binoculars or a telescope to observe them. Please refer to the table of deep-sky objects for more information.

Selecting the Correct Charts

The table below shows which of the charts to use for particular dates and times throughout the year and will help you to select the correct pair of half-sky charts for any combination of month and time of night.

The Earth takes 23 hours 56 minutes (and 4 seconds) to rotate once around its axis with respect to the fixed stars. Because this is around four minutes shorter than a full 24 hours, the stars appear to rise and set about 4 minutes earlier on each successive day, or around an hour earlier each fortnight. Therefore, as well as showing the stars at 10pm (22h in 24-hour notation) on 21 January, chart 1 also depicts the sky at 9pm (21h) on 6 February, 8pm (20h) on 21 February and 7pm (19h) on 6 March.

The times listed do not include summer time (daylight saving time), so if summer time is in force you must subtract one hour to obtain standard time (GMT if you are in the United Kingdom) before referring to the chart. For example, to find the correct chart for mid-September in the northern hemisphere at 3am summer time, first of all subtract one hour to obtain 2am (2h) standard time. Then you can consult the table, where you will find that you should use chart 11.

The table does not indicate sunrise, sunset or twilight. In northern temperate latitudes, the sky is still light at 18h and 6h from April to September, and still light at 20h and 4h from May to August. In Australia and New Zealand, the sky is still light at 18h and 6h from October to March, and in twilight (with only bright stars visible) at 20h and 04h from November to January.

Local Time	18h	20h	22h	0h	2h	4h	6h
January	11	12	1	2	3	4	5
February	12	1	2	3	4	5	6
March	1	2	3	4	5	6	7
April	2	3	4	5	6	7	8
May	3	4	5	6	7	8	9
June	4	5	6	7	8	9	10
July	5	6	7	8	9	10	11
August	6	7	8	9	10	11	12
September	7	8	9	10	11	12	1
October	8	9	10	11	12	1	2
November	9	10	11	12	1	2	3
December	10	11	12	1	2	3	4

Legend to the Star Charts

STARS

Symbol Magnitude

0 or brighter
1
2
3
4
5

Double star
Variable star

DEEP-SKY OBJECTS

Symbol Type of object

Open star cluster
Globular star cluster
Nebula
Cluster with nebula
Planetary nebula
Galaxy

Magellanic Clouds

Star Names

There are over 200 stars with proper names, most of which are of Roman, Greek or Arabic origin although only a couple of dozen or so of these names are used regularly. Examples include Arcturus in Boötes, Castor and Pollux in Gemini and Rigel in Orion.

A system whereby Greek letters were assigned to stars was introduced by the German astronomer and celestial cartographer Johann Bayer in his star atlas Uranometria, published in 1603. Bayer's system is applied to the brighter stars within any particular constellation, which are given a letter from the Greek alphabet followed by the genitive case of the constellation in which the star is located. This genitive case is simply the Latin form meaning 'of' the constellation. Examples are the stars Alpha Boötis and Beta Centauri which translate literally as 'Alpha of Boötes' and 'Beta of the Centaur'.

As a general rule, the brightest star in a constellation is labelled Alpha (α), the second brightest Beta (β), and the third brightest Gamma (γ) and so on, although there are some constellations where the system falls down. An example is Gemini where the principal star (Pollux) is designated Beta Geminorum, the second brightest (Castor) being known as Alpha Geminorum.

There are only 24 letters in the Greek alphabet, the consequence of which was that the fainter naked eye stars needed an alternative system of classification. The system in popular use is that devised by the first Astronomer Royal John Flamsteed in which the stars in each constellation are listed numerically in order from west to east. Although many of the brighter stars within any particular constellation will have both Greek letters and Flamsteed numbers, the latter are generally used only when a star does not have a Greek letter.

The Greek Alphabet

α	Alpha	ι	Iota	ρ	Rho
β	Beta	κ	Kappa	σ	Sigma
γ	Gamma	λ	Lambda	τ	Tau
δ	Delta	μ	Mu	υ	Upsilon
ε	Epsilon	ν	Nu	φ	Phi
ζ	Zeta	ξ	Xi	χ	Chi
η	Eta	ο	Omicron	ψ	Psi
θ	Theta	π	Pi	ω	Omega

The Names of the Constellations

On clear, dark, moonless nights, the sky seems to teem with stars although in reality you can never see more than a couple of thousand or so at any one time when looking with the unaided eye. Each and every one of these stars belongs to a particular constellation, although the constellations that we see in the sky, and which grace the pages of star atlases, are nothing more than chance alignments. The stars that make up the constellations are often situated at vastly differing distances from us and only appear close to each other, and form the patterns that we see, because they lie in more or less the same direction as each other as seen from Earth.

A large number of the constellations are named after mythological characters, and were given their names thousands of years ago. However, those star groups lying close to the south celestial pole were discovered by Europeans only during the last few centuries, many of these by explorers and astronomers who mapped the stars during their journeys to lands under southern skies. This resulted in many of the newer constellations having modern-sounding names, such as Octans (the Octant) and Microscopium (the Microscope), both of which were devised by the French astronomer Nicolas Louis De La Caille during the early 1750s.

Over the centuries, many different suggestions for new constellations have been put forward by astronomers who, for one reason or another, felt the need to add new groupings to star charts and to fill gaps between the traditional constellations. Astronomers drew up their own charts of the sky, incorporating their new groups into them. A number of these new constellations had cumbersome names, notable examples including Officina Typographica (the Printing Shop) introduced by the German astronomer Johann Bode in 1801; Sceptrum Brandenburgicum (the Sceptre of Brandenburg) introduced by the German astronomer Gottfried Kirch in 1688; Taurus Poniatovii (Poniatowski's Bull) introduced by the Polish-Lithuanian astronomer Martin Odlanicky Poczobut in 1777; and Quadrans Muralis (the Mural Quadrant) devised by the French astronomer Joseph-Jerôme de Lalande in1795. Although these have long since been rejected, the latter has been immortalised by the annual Quadrantid meteor shower, the radiant of which lies in an area of sky formerly occupied by Quadrans Muralis.

During the 1920s the International Astronomical Union (IAU) systemised matters by adopting an official list of 88 accepted constellations, each with official spellings and abbreviations. Precise boundaries for each constellation were then drawn up so that every point in the sky belonged to a particular constellation.

The abbreviations devised by the IAU each have three letters which in the majority of cases are the first three letters of the constellation name, such as AND for Andromeda, EQU for Equuleus, HER for Hercules, ORI for Orion and so on. This trend is not strictly adhered to in cases where confusion may arise. This happens with the two constellations Leo (abbreviated LEO) and Leo Minor (abbreviated LMI). Similarly, because Triangulum (TRI) may be mistaken for Triangulum Australe, the latter is abbreviated TRA. Other instances occur with Sagitta (SGE) and Sagittarius (SGR) and with Canis Major (CMA) and Canis Minor (CMI) where the first two letters from the second names of the constellations are used. This is also the case with Corona Australis (CRA) and Corona Borealis (CRB) where the first letter of the second name of each constellation is incorporated. Finally, mention must be made of Crater (CRT) which has been abbreviated in such a way as to avoid confusion with the aforementioned CRA (Corona Australis).

The table shown on the following pages contains the name of each of the 88 constellations together with the translation and abbreviation of the constellation name. The constellations depicted on the monthly star charts are identified with their abbreviations rather than the full constellation names.

The Constellations

Andromeda	Andromeda	AND
Antlia	The Air Pump	ANT
Apus	The Bird of Paradise	APS
Aquarius	The Water Carrier	AQR
Aquila	The Eagle	AQL
Ara	The Altar	ARA
Aries	The Ram	ARI
Auriga	The Charioteer	AUR
Boötes	The Herdsman	BOO
Caelum	The Graving Tool	CAE

Camelopardalis	The Giraffe	CAM
Cancer	The Crab	CNC
Canes Venatici	The Hunting Dogs	CVN
Canis Major	The Great Dog	CMA
Canis Minor	The Little Dog	CMI
Capricornus	The Goat	CAP
Carina	The Keel	CAR
Cassiopeia	Cassiopeia	CAS
Centaurus	The Centaur	CEN
Cepheus	Cepheus	CEP
Cetus	The Whale	CET

Chamaeleon	The Chameleon	CHA
Circinus	The Pair of Compasses	CIR
Columba	The Dove	COL
Coma Berenices	Berenice's Hair	COM
Corona Australis	The Southern Crown	CRA
Corona Borealis	The Northern Crown	CRB
Corvus	The Crow	CRV
Crater	The Cup	CRT
Crux	The Cross	CRU
Cygnus	The Swan	CYG
Delphinus	The Dolphin	DEL
Dorado	The Goldfish	DOR
Draco	The Dragon	DRA
Equuleus	The Foal	EQU
Eridanus	The River	ERI
Fornax	The Furnace	FOR
Gemini	The Twins	GEM
Grus	The Crane	GRU
Hercules	Hercules	HER
Horologium	The Pendulum Clock	HOR
Hydra	The Water Snake	HYA
Hydrus	The Lesser Water Snake	HYI
Indus	The Indian	IND
Lacerta	The Lizard	LAC
Leo	The Lion	LEO
Leo Minor	The Lesser Lion	LMI
Lepus	The Hare	LEP
Libra	The Scales	LIB
Lupus	The Wolf	LUP
Lynx	The Lynx	LYN
Lyra	The Lyre	LYR
Mensa	The Table Mountain	MEN
Microscopium	The Microscope	MIC

Monoceros	The Unicorn	MON
Musca	The Fly	MUS
Norma	The Level	NOR
Octans	The Octant	OCT
Ophiuchus	The Serpent Bearer	OPH
Orion	Orion	ORI
Pavo	The Peacock	PAV
Pegasus	Pegasus	PEG
Perseus	Perseus	PER
Phoenix	The Phoenix	PHE
Pictor	The Painter's Easel	PIC
Pisces	The Fish	PSC
Piscis Austrinus	The Southern Fish	PSA
Puppis	The Stern	PUP
Pyxis	The Mariner's Compass	PYX
Reticulum	The Net	RET
Sagitta	The Arrow	SGE
Sagittarius	The Archer	SGR
Scorpius	The Scorpion	SCO
Sculptor	The Sculptor	SCL
Scutum	The Shield	SCT
Serpens Caput and Cauda	The Serpent	SER
Sextans	The Sextant	SEX
Taurus	The Bull	TAU
Telescopium	The Telescope	TEL
Triangulum	The Triangle	TRI
Triangulum Australe	The Southern Triangle	TRA
Tucana	The Toucan	TUC
Ursa Major	The Great Bear	UMA
Ursa Minor	The Little Bear	UMI
Vela	The Sail	VEL
Virgo	The Virgin	VIR
Volans	The Flying Fish	VOL
Vulpecula	The Fox	VUL

The Monthly Star Charts

Northern Hemisphere Star Charts

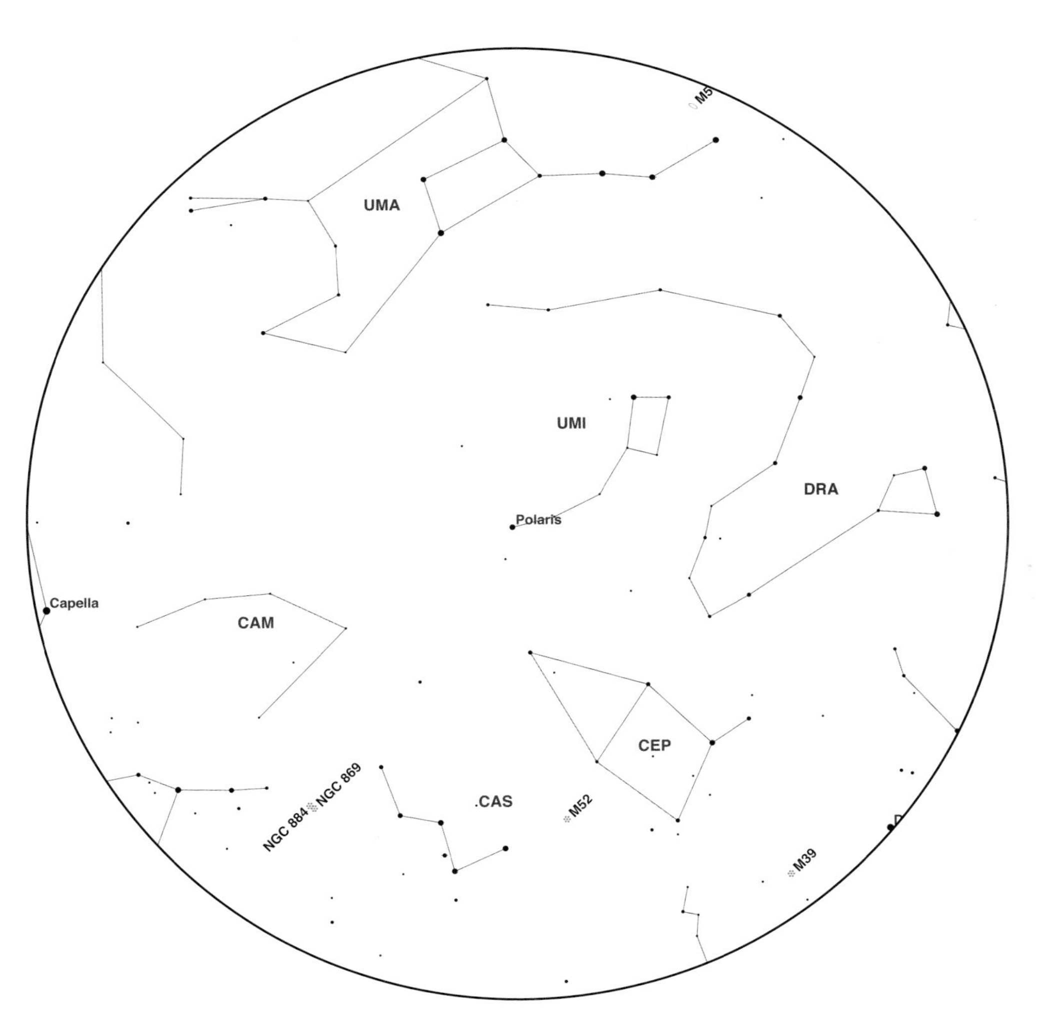

This chart shows stars lying at declinations between +45 and +90 degrees. These constellations are circumpolar for observers in Europe and North America.

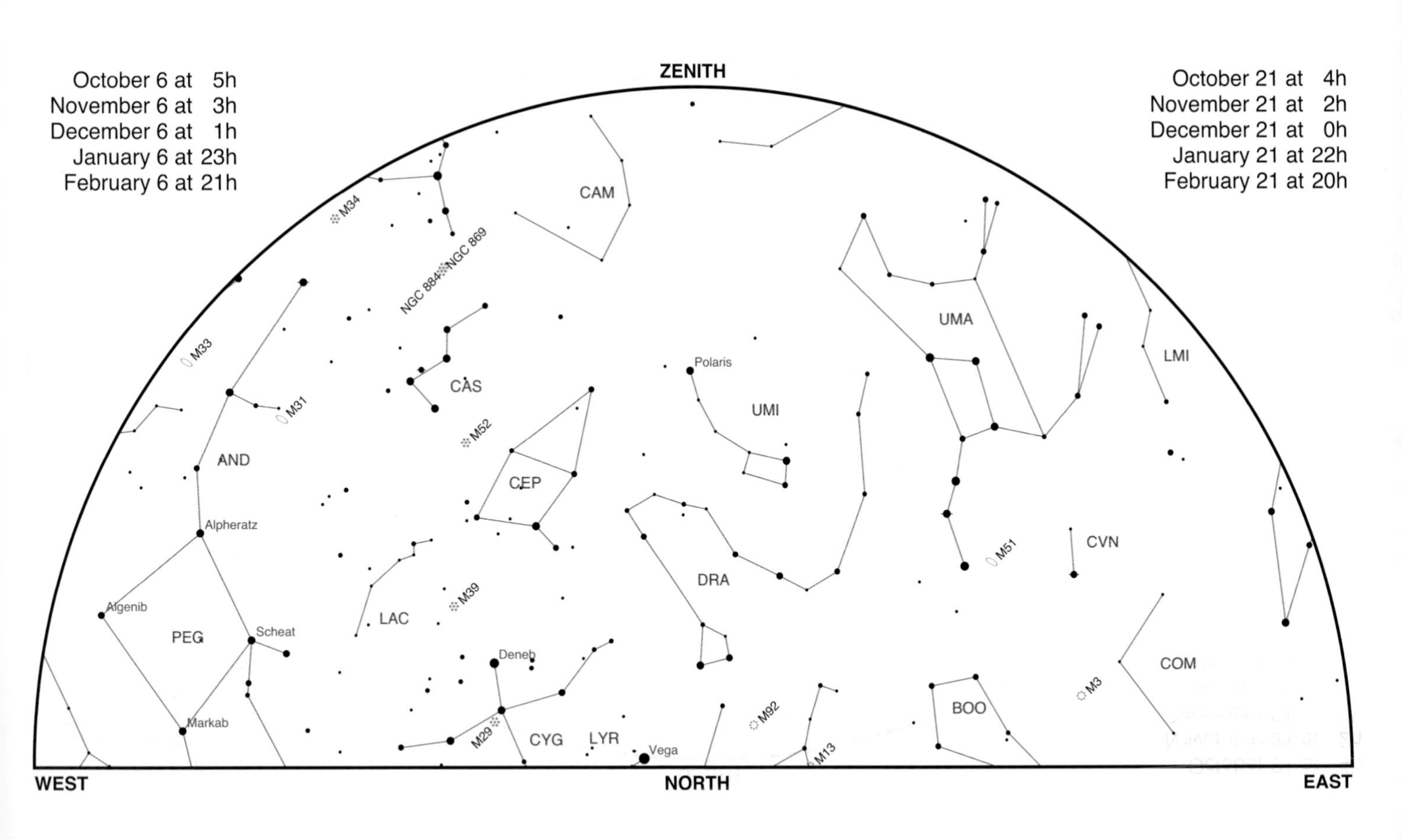
1N
October 6 at 5h
November 6 at 3h
December 6 at 1h
January 6 at 23h
February 6 at 21h
October 21 at 4h
November 21 at 2h
December 21 at 0h
January 21 at 22h
February 21 at 20h
ZENITH
WEST
NORTH
EAST
CAM
M34
NGC 884
NGC 869
UMA
LMI
M33
CAS
Polaris
UMI
M31
M52
AND
CEP
Alpheratz
DRA
M51
CVN
M39
Algenib
PEG
Scheat
LAC
Deneb
COM
M3
M92
BOO
Markab
M29
CYG
LYR
Vega
M13

1S

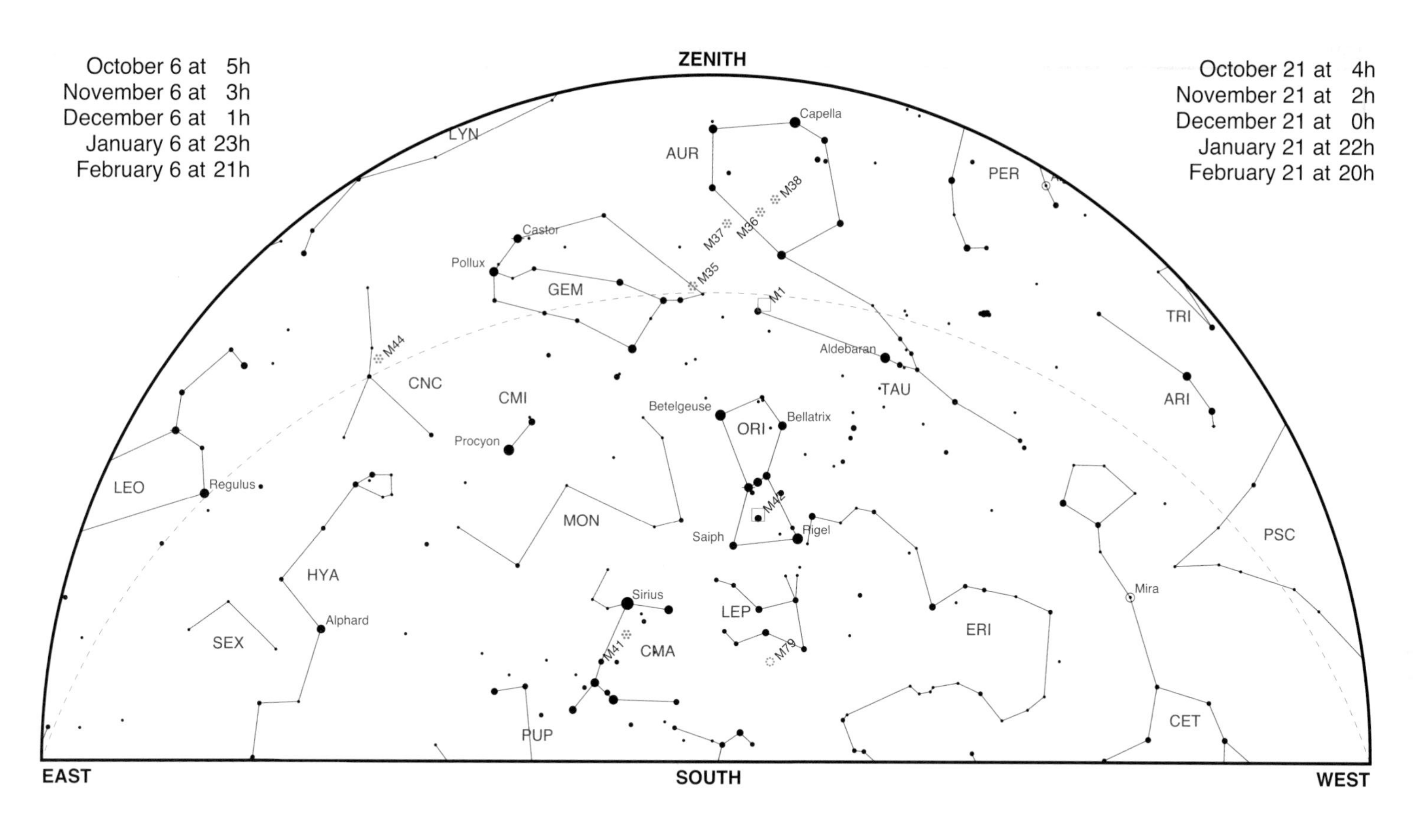
October 6 at 5h
November 6 at 3h
December 6 at 1h
January 6 at 23h
February 6 at 21h
ZENITH
October 21 at 4h
November 21 at 2h
December 21 at 0h
January 21 at 22h
February 21 at 20h
LYN
AUR
Capella
PER
M38
M36
M37
Castor
Pollux
GEM
M35
M1
TRI
M44
CNC
CMI
Aldebaran
TAU
ARI
Betelgeuse
Bellatrix
ORI
Procyon
LEO
Regulus
MON
M42
Saiph
Rigel
PSC
HYA
Sirius
LEP
Mira
ERI
SEX
Alphard
M41
CMA
M79
CET
PUP
EAST
SOUTH
WEST

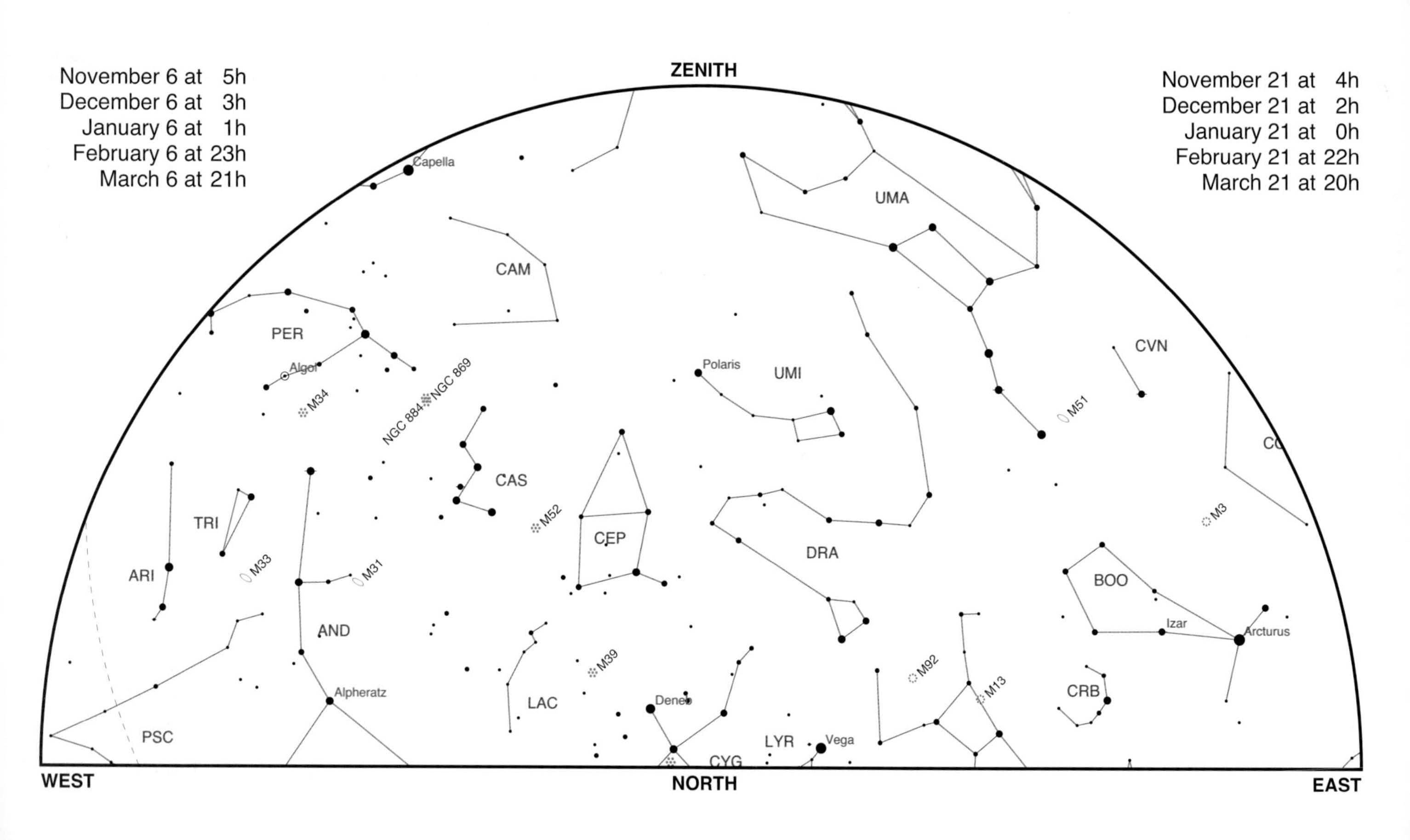
2N
November 6 at 5h
December 6 at 3h
January 6 at 1h
February 6 at 23h
March 6 at 21h
November 21 at 4h
December 21 at 2h
January 21 at 0h
February 21 at 22h
March 21 at 20h
ZENITH
WEST
NORTH
EAST
Capella
CAM
UMA
PER
Algol
M34
NGC 884
NGC 869
Polaris
UMI
CVN
M51
CAS
TRI
ARI
M33
M31
M52
CEP
DRA
M3
BOO
Izar
Arcturus
AND
Alpheratz
M39
LAC
Deneb
CYG
LYR
Vega
M92
M13
CRB
PSC

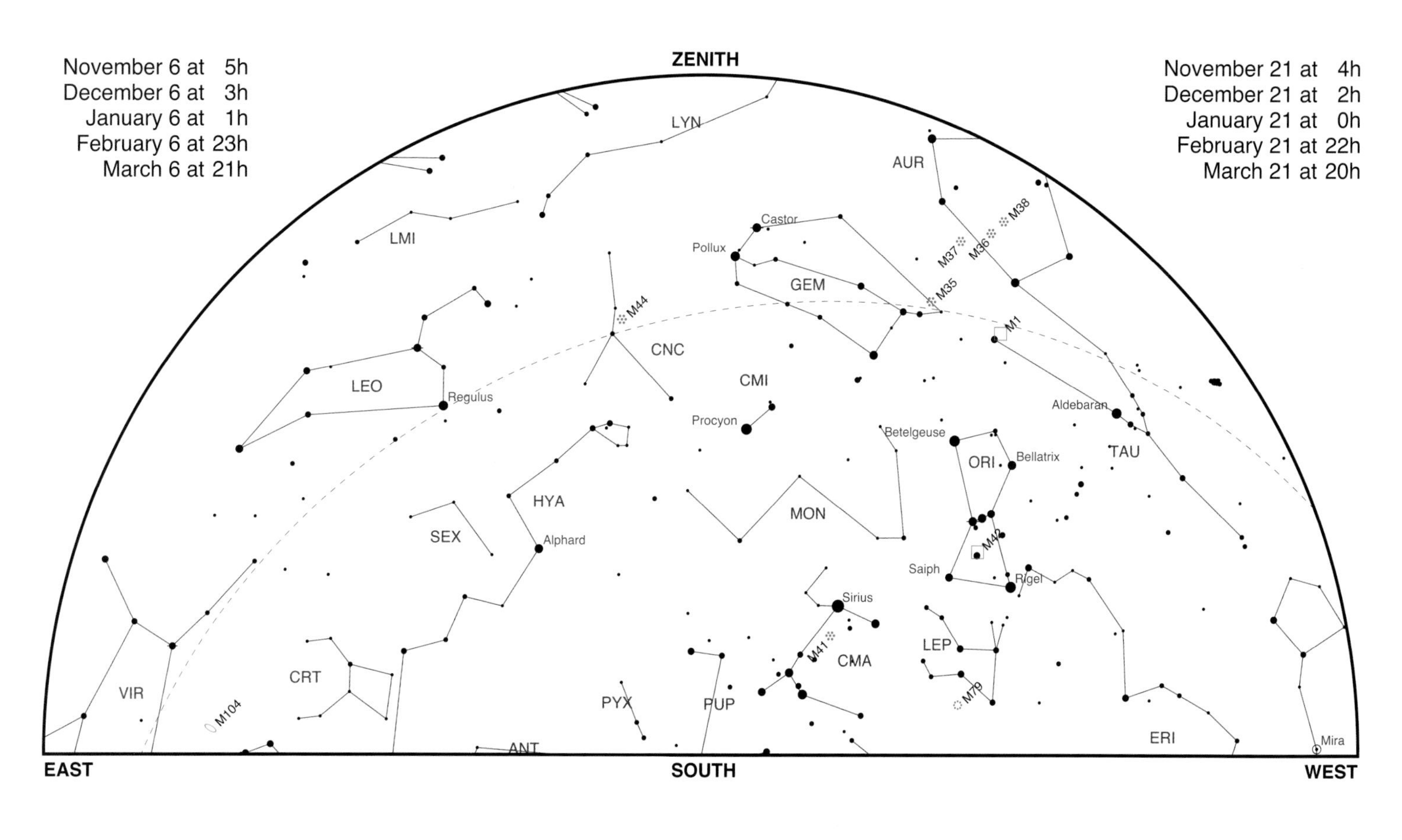
November 6 at 5h
December 6 at 3h
January 6 at 1h
February 6 at 23h
March 6 at 21h
ZENITH
November 21 at 4h
December 21 at 2h
January 21 at 0h
February 21 at 22h
March 21 at 20h
LYN
AUR
LMI
Castor
Pollux
GEM
M37
M36
M38
M35
M44
M1
CNC
LEO
CMI
Regulus
Procyon
Aldebaran
Betelgeuse
TAU
ORI
Bellatrix
HYA
MON
SEX
Alphard
M42
Saiph
Rigel
Sirius
LEP
M41
CMA
VIR
CRT
M104
PYX
PUP
M79
ERI
ANT
Mira
EAST
SOUTH
WEST
2S

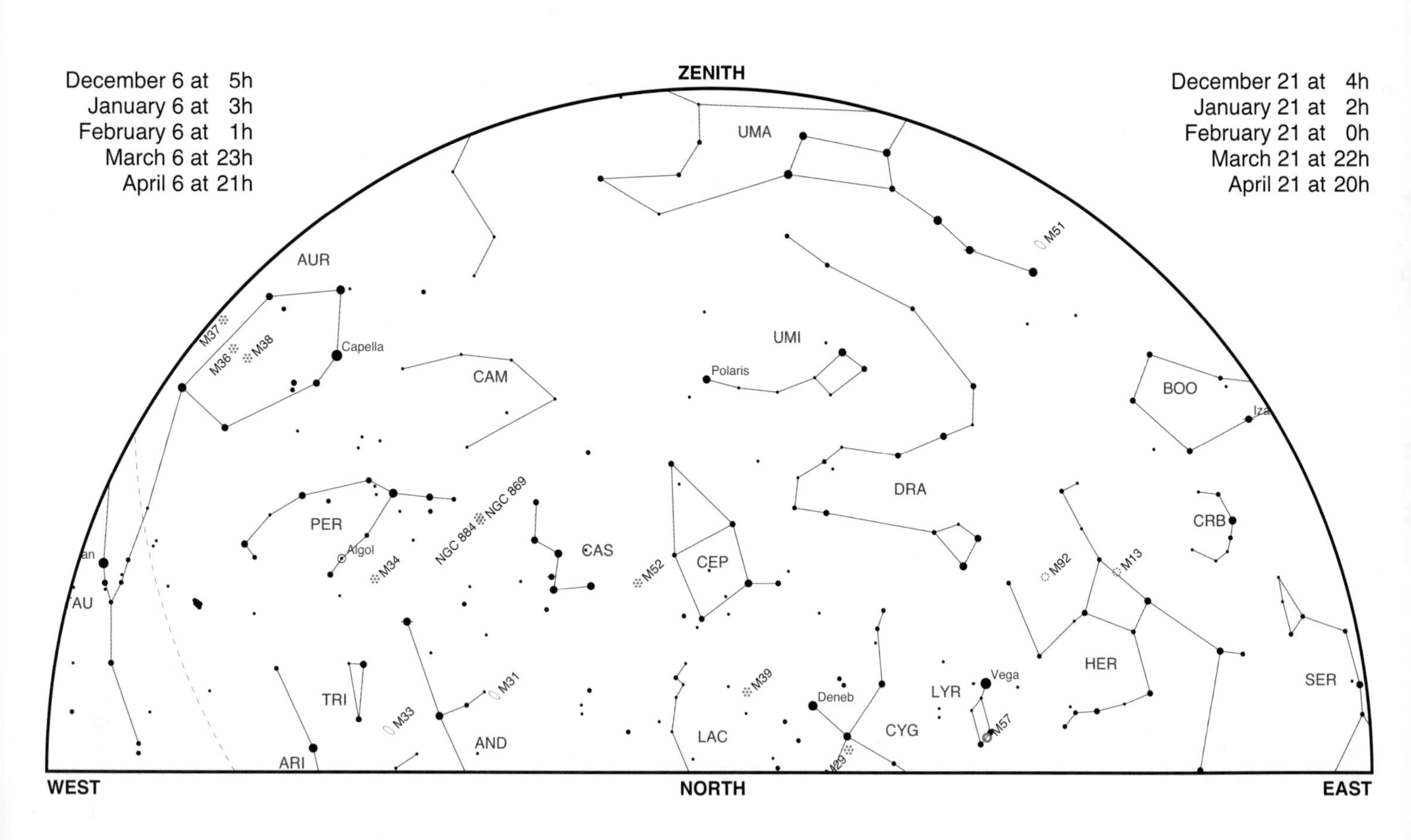
3N
December 6 at 5h
January 6 at 3h
February 6 at 1h
March 6 at 23h
April 6 at 21h
ZENITH
December 21 at 4h
January 21 at 2h
February 21 at 0h
March 21 at 22h
April 21 at 20h
UMA
M51
AUR
M37
M36
M38
Capella
CAM
UMI
Polaris
BOO
Iza
DRA
CRB
PER
Algol
M34
NGC 884
NGC 869
CAS
M52
CEP
M92
M13
AU
HER
TRI
M31
M33
AND
ARI
M39
LAC
Deneb
M29
CYG
LYR
Vega
M57
SER
WEST
NORTH
EAST

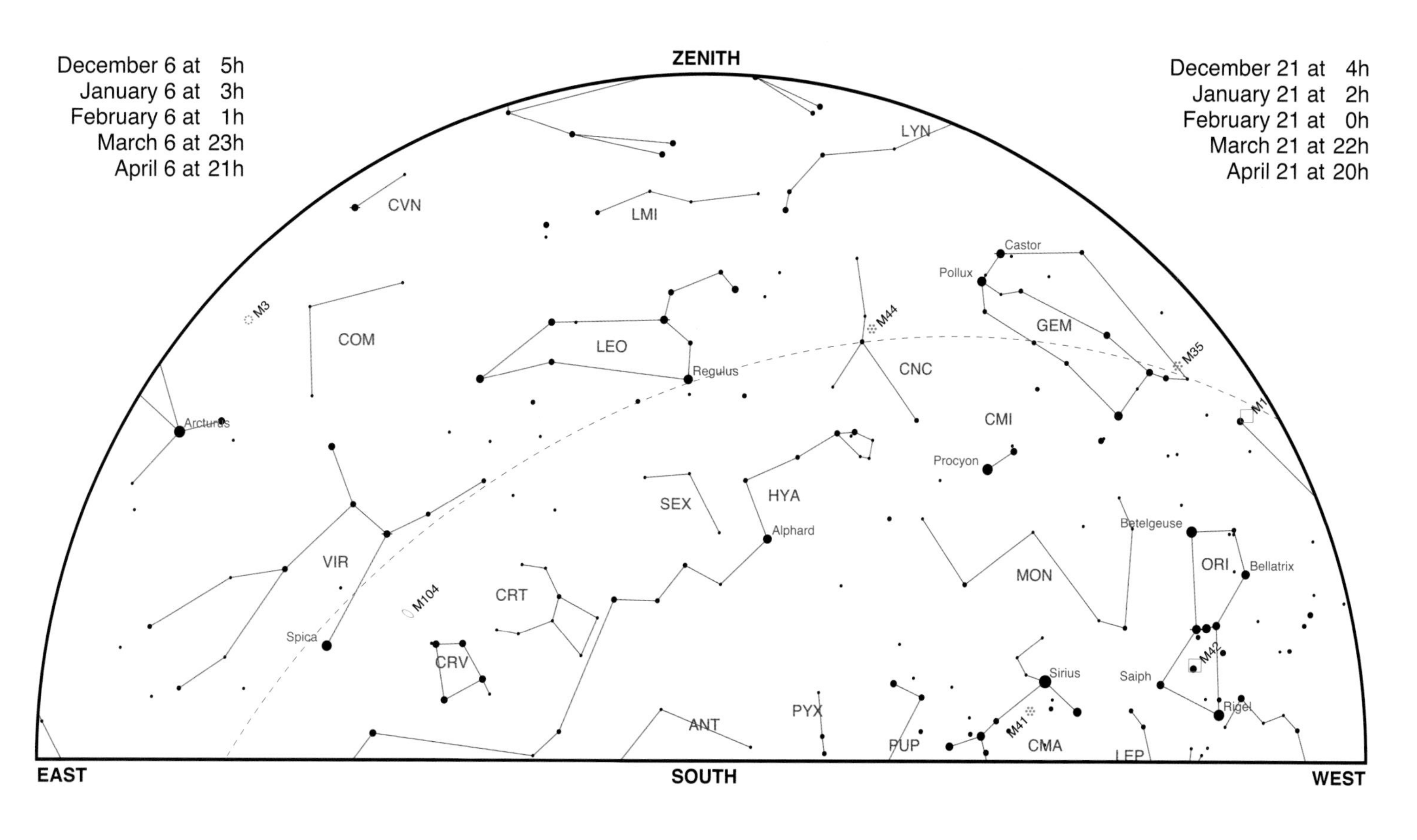
3S
ZENITH
December 6 at 5h
January 6 at 3h
February 6 at 1h
March 6 at 23h
April 6 at 21h
December 21 at 4h
January 21 at 2h
February 21 at 0h
March 21 at 22h
April 21 at 20h
EAST
SOUTH
WEST
LYN
CVN
LMI
Castor
Pollux
M3
COM
LEO
M44
GEM
M35
Regulus
CNC
CMI
M1
Arcturus
Procyon
SEX
HYA
Alphard
Betelgeuse
VIR
MON
ORI
Bellatrix
M104
CRT
Spica
CRV
M42
Sirius
Saiph
Rigel
ANT
PYX
M41
PUP
CMA
LEP

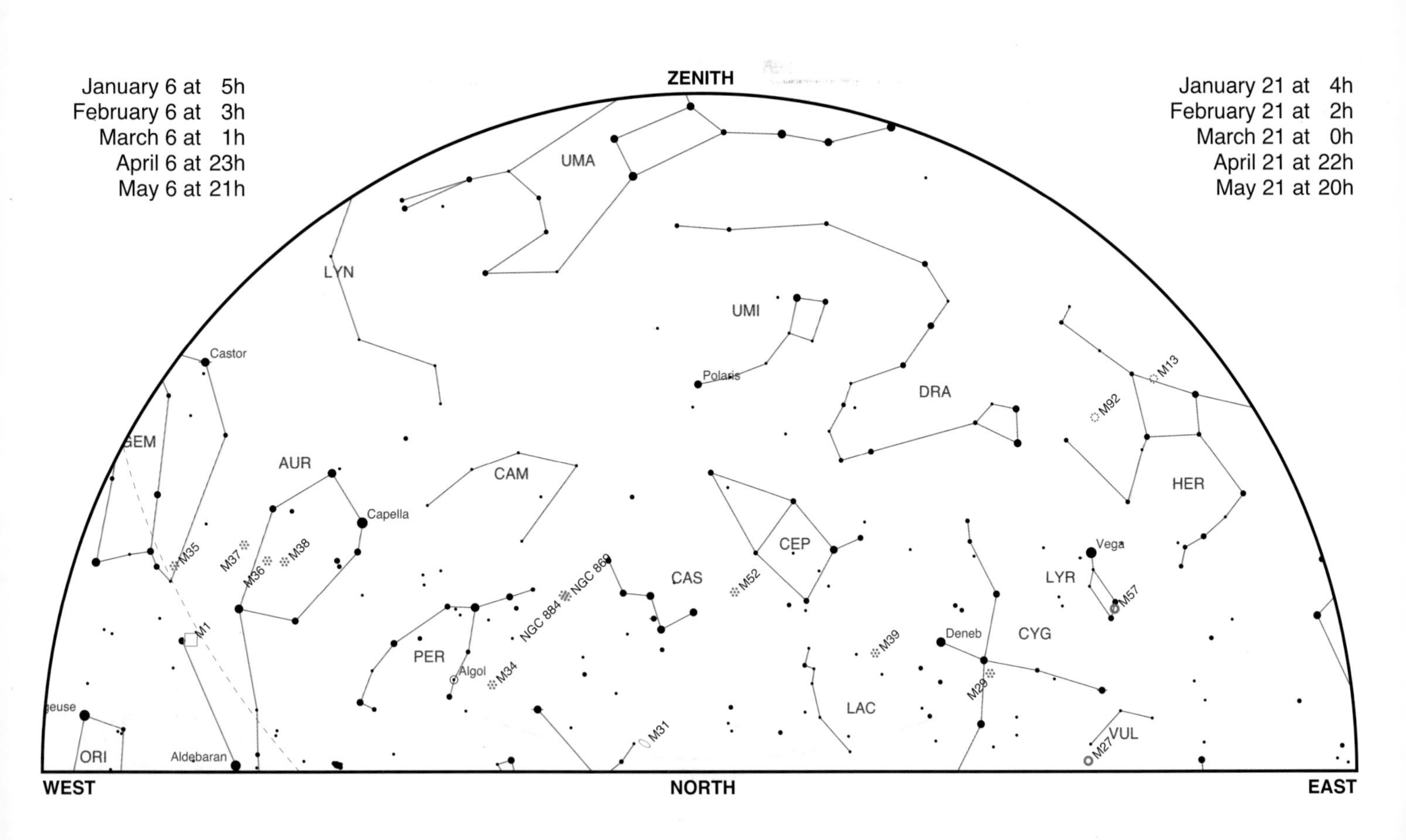
4N
January 6 at 5h
February 6 at 3h
March 6 at 1h
April 6 at 23h
May 6 at 21h
January 21 at 4h
February 21 at 2h
March 21 at 0h
April 21 at 22h
May 21 at 20h
ZENITH
WEST
NORTH
EAST
UMA
LYN
UMI
Polaris
DRA
HER
M13
M92
Castor
GEM
AUR
Capella
CAM
CEP
CAS
M52
M35
M37
M36
M38
NGC 884
NGC 869
PER
Algol
M34
M1
Aldebaran
ORI
M31
LAC
M39
Deneb
CYG
M29
LYR
Vega
M57
VUL
M27

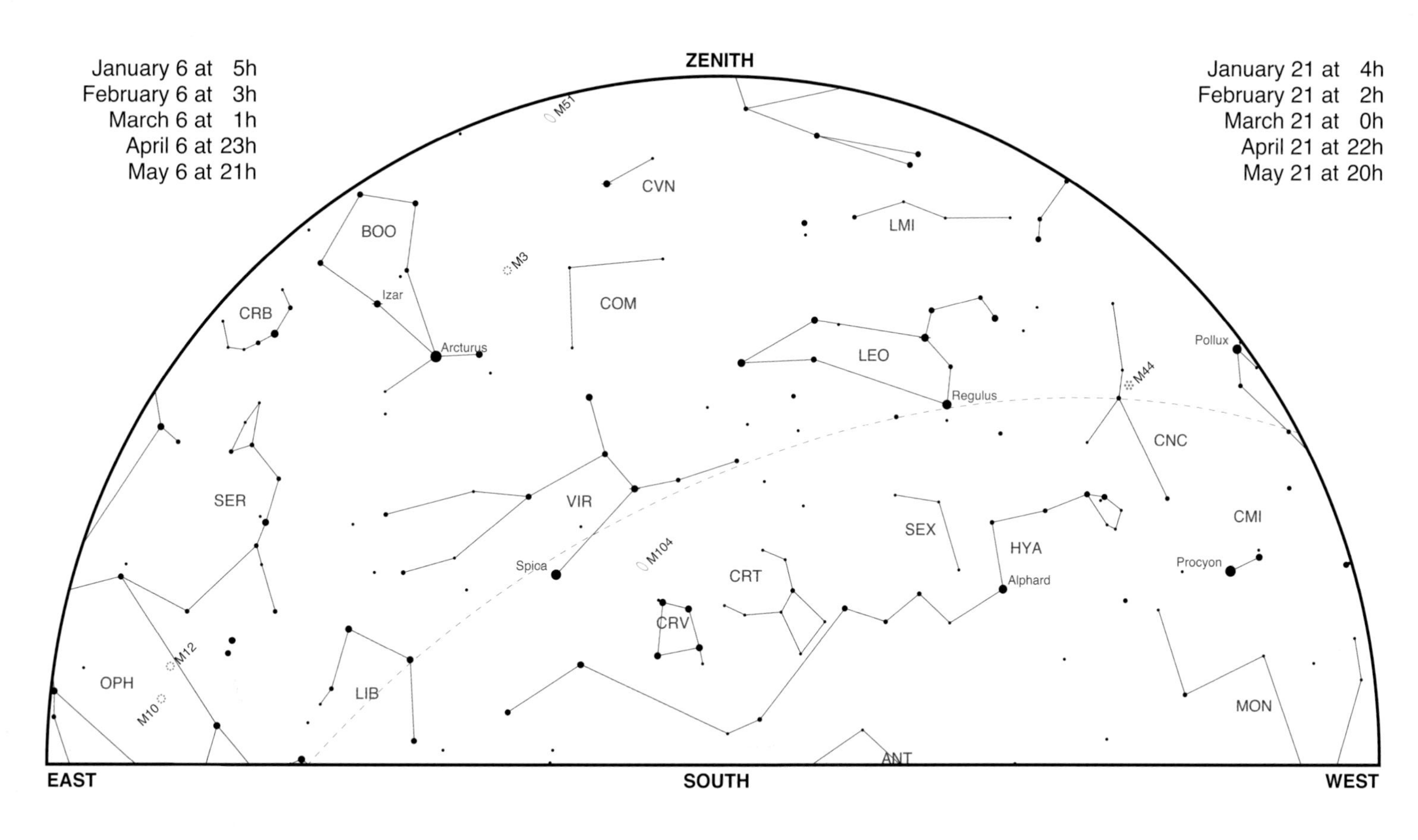
January 6 at 5h
February 6 at 3h
March 6 at 1h
April 6 at 23h
May 6 at 21h
ZENITH
January 21 at 4h
February 21 at 2h
March 21 at 0h
April 21 at 22h
May 21 at 20h
M51
CVN
LMI
BOO
M3
COM
Izar
CRB
Arcturus
LEO
Pollux
M44
Regulus
CNC
SER
VIR
SEX
CMI
M104
HYA
Spica
Procyon
CRT
Alphard
CRV
M12
OPH
LIB
MON
M10
ANT
EAST
SOUTH
WEST
4S

5N

January 6 at	7h
February 6 at	5h
March 6 at	3h
April 6 at	1h
May 6 at	23h

January 21 at	6h
February 21 at	4h
March 21 at	2h
April 21 at	0h
May 21 at	22h

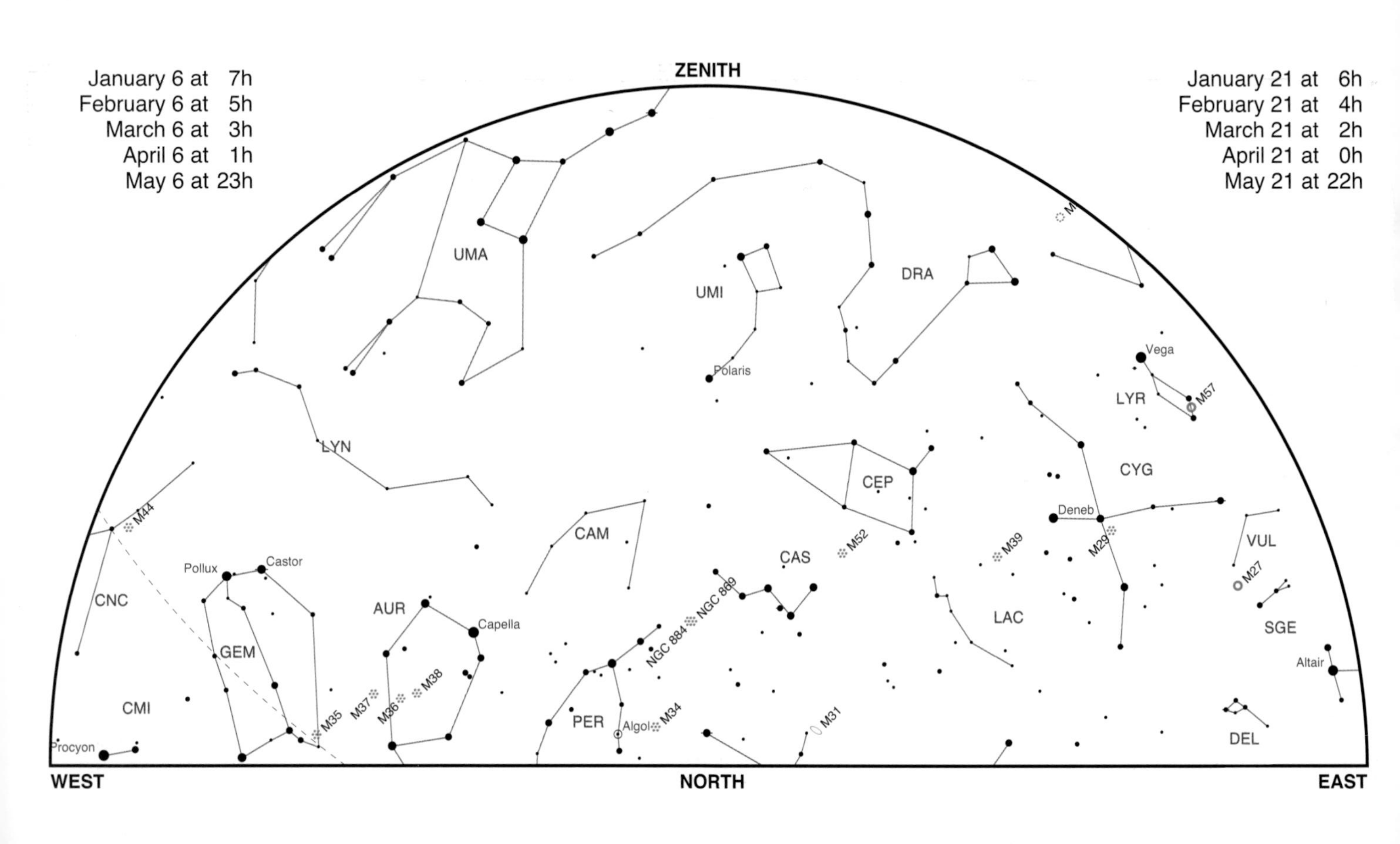

5S

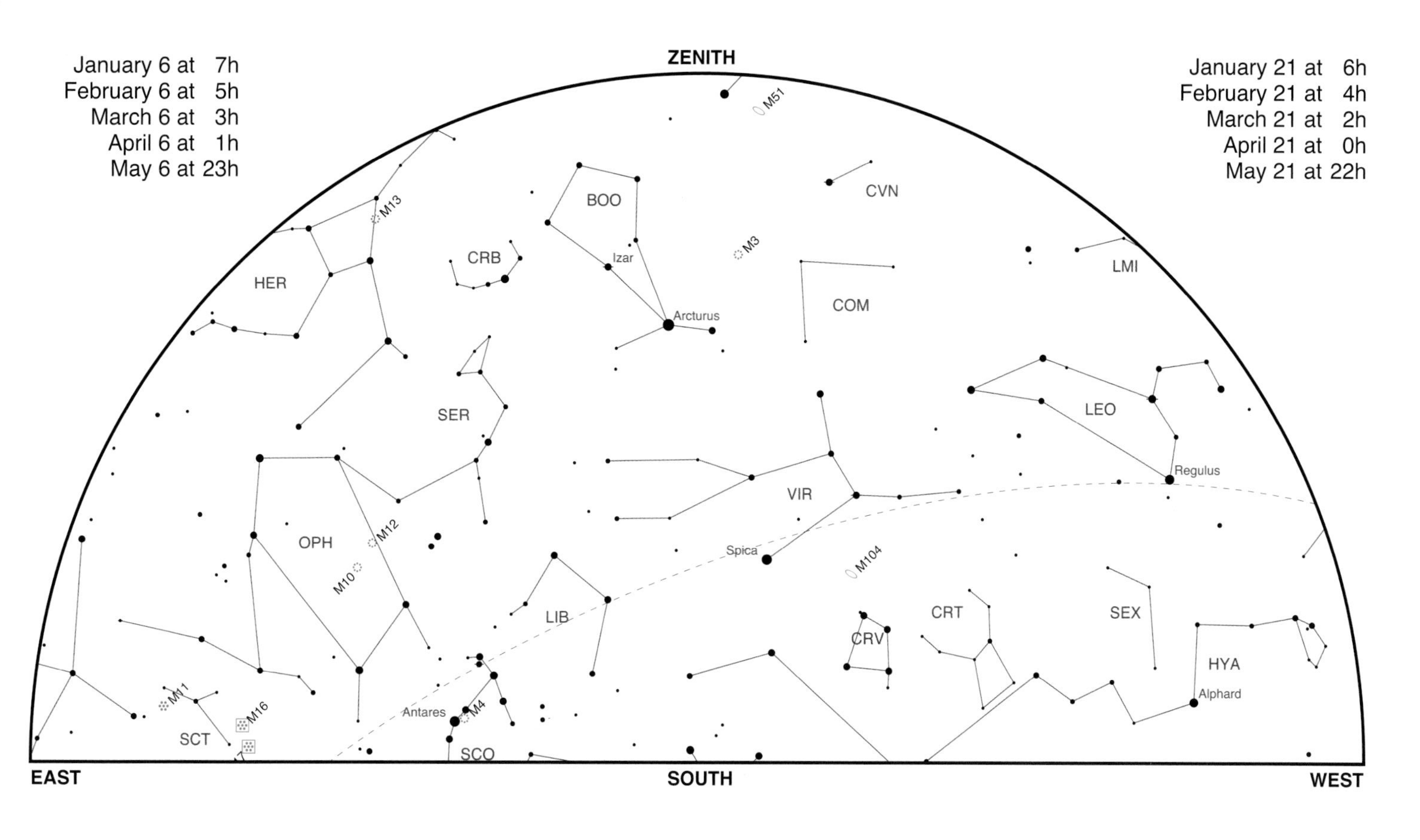
January 6 at 7h
February 6 at 5h
March 6 at 3h
April 6 at 1h
May 6 at 23h
January 21 at 6h
February 21 at 4h
March 21 at 2h
April 21 at 0h
May 21 at 22h
ZENITH
EAST
SOUTH
WEST
M51
BOO
CVN
M13
CRB
Izar
M3
LMI
HER
COM
Arcturus
LEO
SER
Regulus
VIR
OPH
M12
Spica
M104
M10
LIB
CRT
SEX
CRV
HYA
Alphard
M11
M16
M4
Antares
SCT
SCO

6N

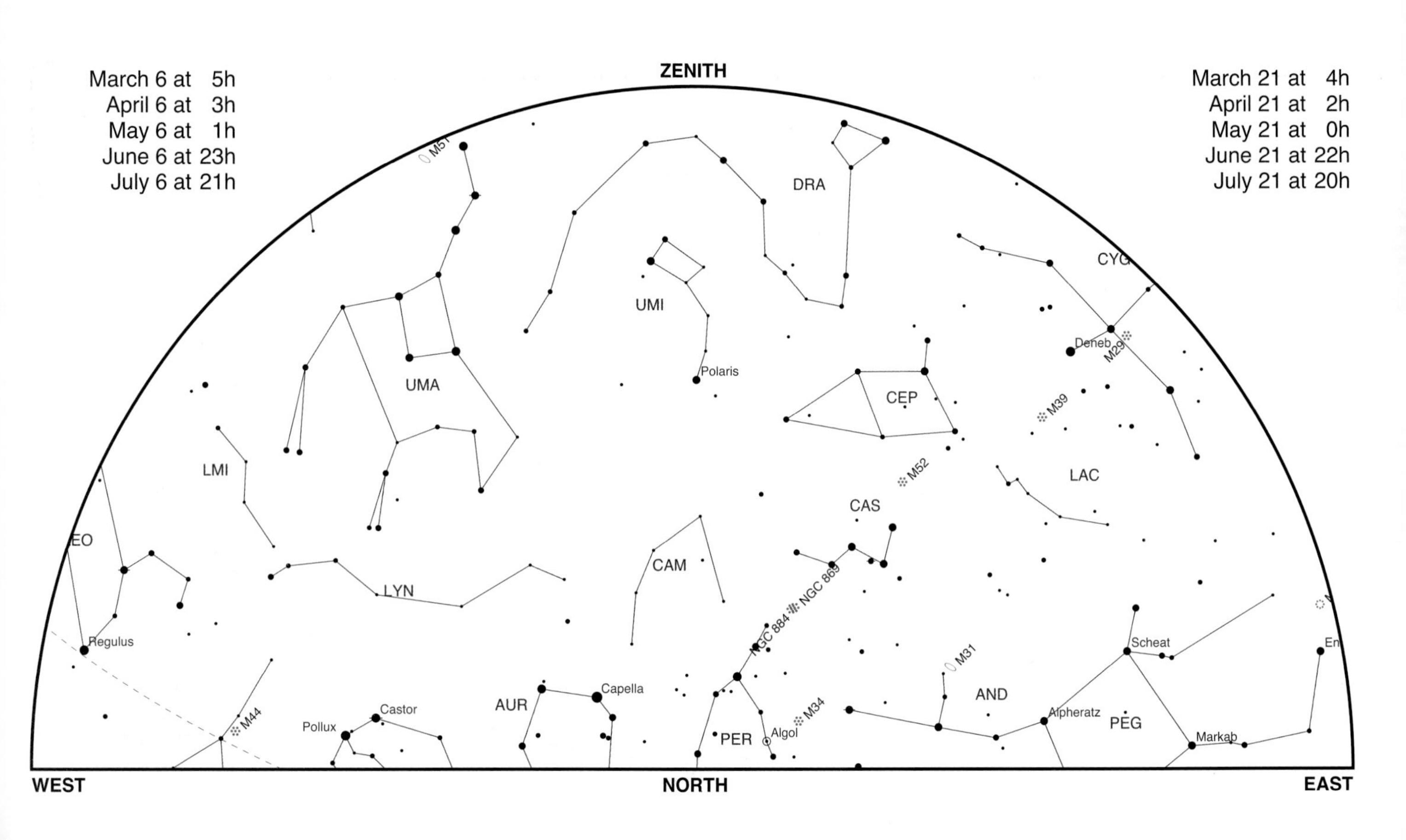
March 6 at 5h
April 6 at 3h
May 6 at 1h
June 6 at 23h
July 6 at 21h
March 21 at 4h
April 21 at 2h
May 21 at 0h
June 21 at 22h
July 21 at 20h
ZENITH
WEST
NORTH
EAST
M51
DRA
UMI
CYG
Deneb
M29
Polaris
UMA
CEP
M39
LMI
M52
LAC
CAS
EO
CAM
LYN
NGC 884
NGC 869
Regulus
Scheat
M31
AND
Capella
Castor
AUR
Alpheratz
PEG
M44
Pollux
M34
PER
Algol
Markab

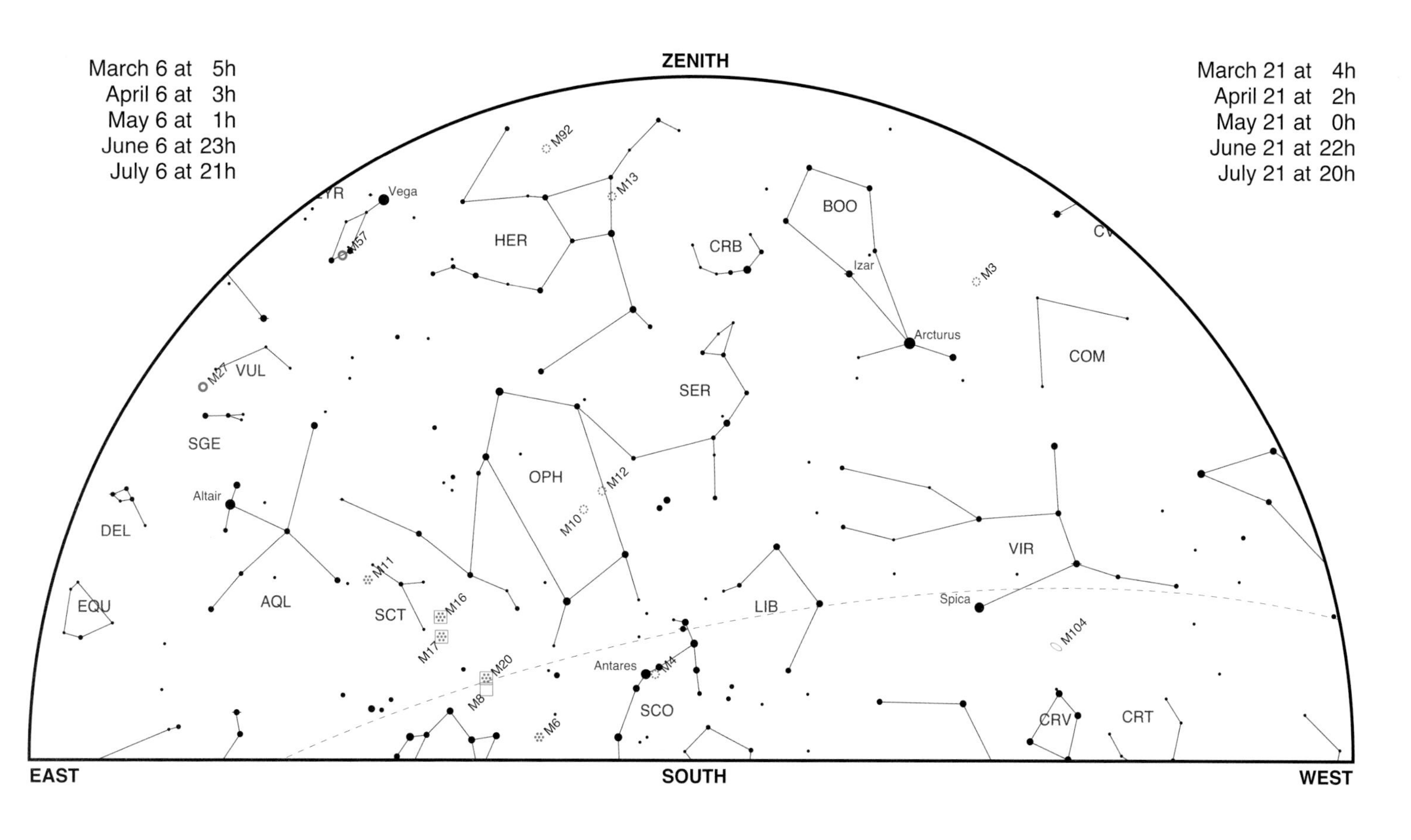
March 6 at 5h
April 6 at 3h
May 6 at 1h
June 6 at 23h
July 6 at 21h
ZENITH
March 21 at 4h
April 21 at 2h
May 21 at 0h
June 21 at 22h
July 21 at 20h
Vega
M92
M13
HER
BOO
CRB
Izar
M3
M57
Arcturus
COM
M27
VUL
SER
SGE
OPH
M12
Altair
M10
DEL
VIR
M11
EQU
AQL
SCT
M16
LIB
Spica
M17
M104
M20
Antares
M4
M8
SCO
M6
CRV
CRT
EAST
SOUTH
WEST

6S

7N

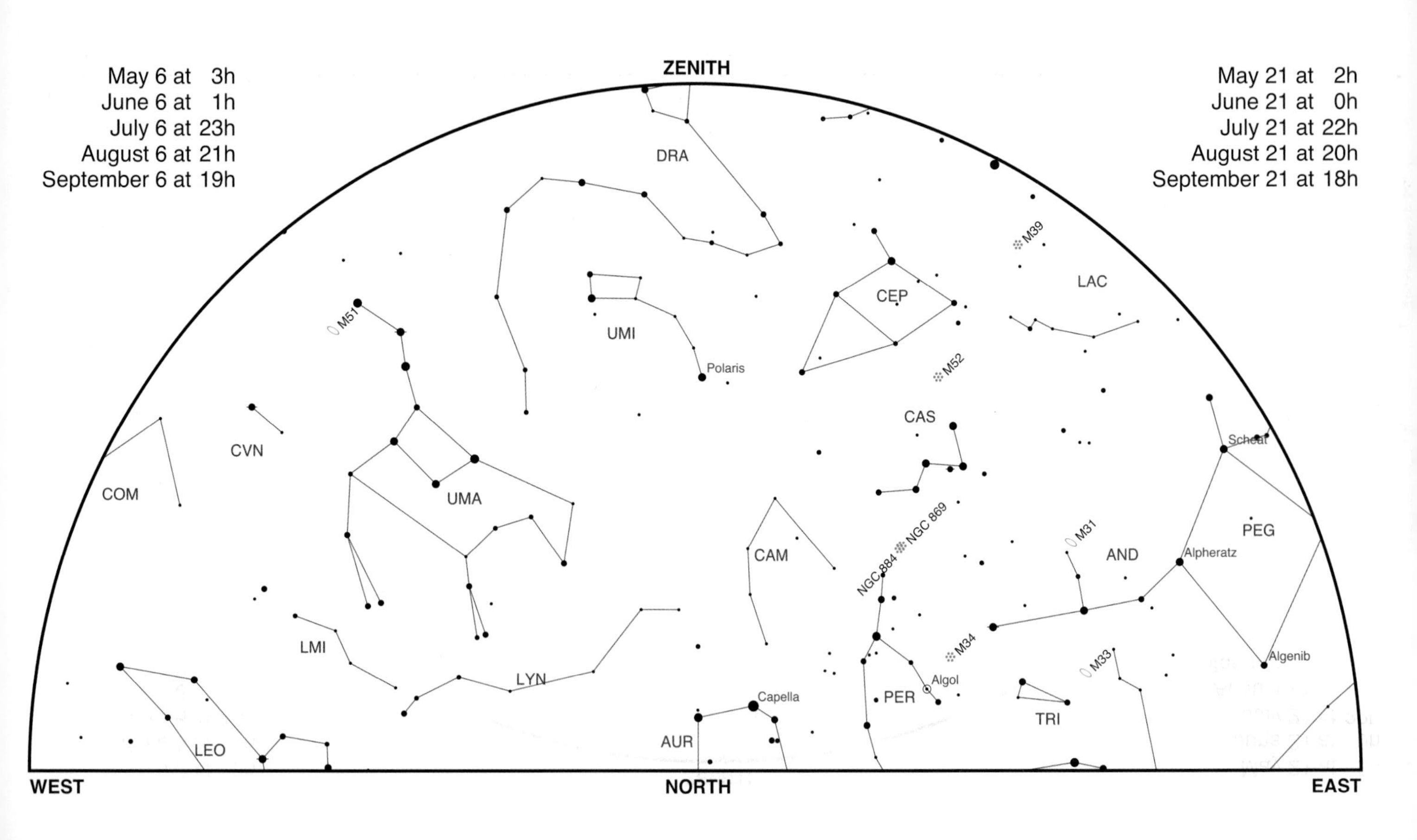
May 6 at 3h
June 6 at 1h
July 6 at 23h
August 6 at 21h
September 6 at 19h
May 21 at 2h
June 21 at 0h
July 21 at 22h
August 21 at 20h
September 21 at 18h
ZENITH
WEST
NORTH
EAST
DRA
UMI
Polaris
CEP
M39
LAC
M52
CAS
M51
CVN
COM
UMA
CAM
NGC 884
NGC 869
M31
AND
PEG
Scheat
Alpheratz
Algenib
M34
M33
Algol
PER
TRI
LMI
LYN
Capella
AUR
LEO

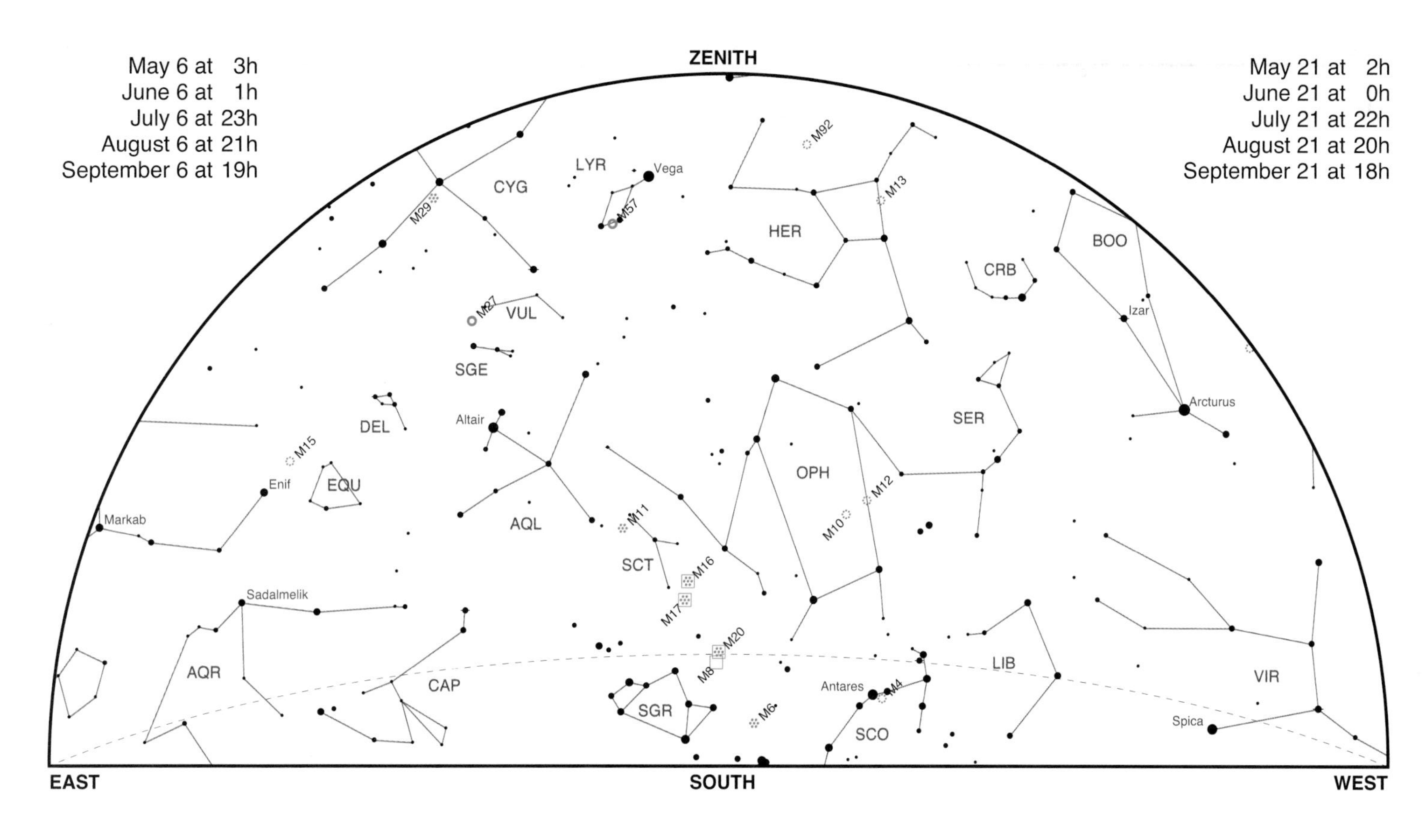
May 6 at 3h
June 6 at 1h
July 6 at 23h
August 6 at 21h
September 6 at 19h
ZENITH
May 21 at 2h
June 21 at 0h
July 21 at 22h
August 21 at 20h
September 21 at 18h
CYG
M29
LYR
Vega
M57
M92
M13
HER
BOO
CRB
Izar
M27
VUL
SGE
Altair
DEL
M15
EQU
Enif
Markab
AQL
M11
SCT
M16
M17
OPH
M10
M12
SER
Arcturus
Sadalmelik
AQR
CAP
M20
M8
SGR
M6
Antares
M4
SCO
LIB
VIR
Spica
EAST
SOUTH
WEST

7S

8N

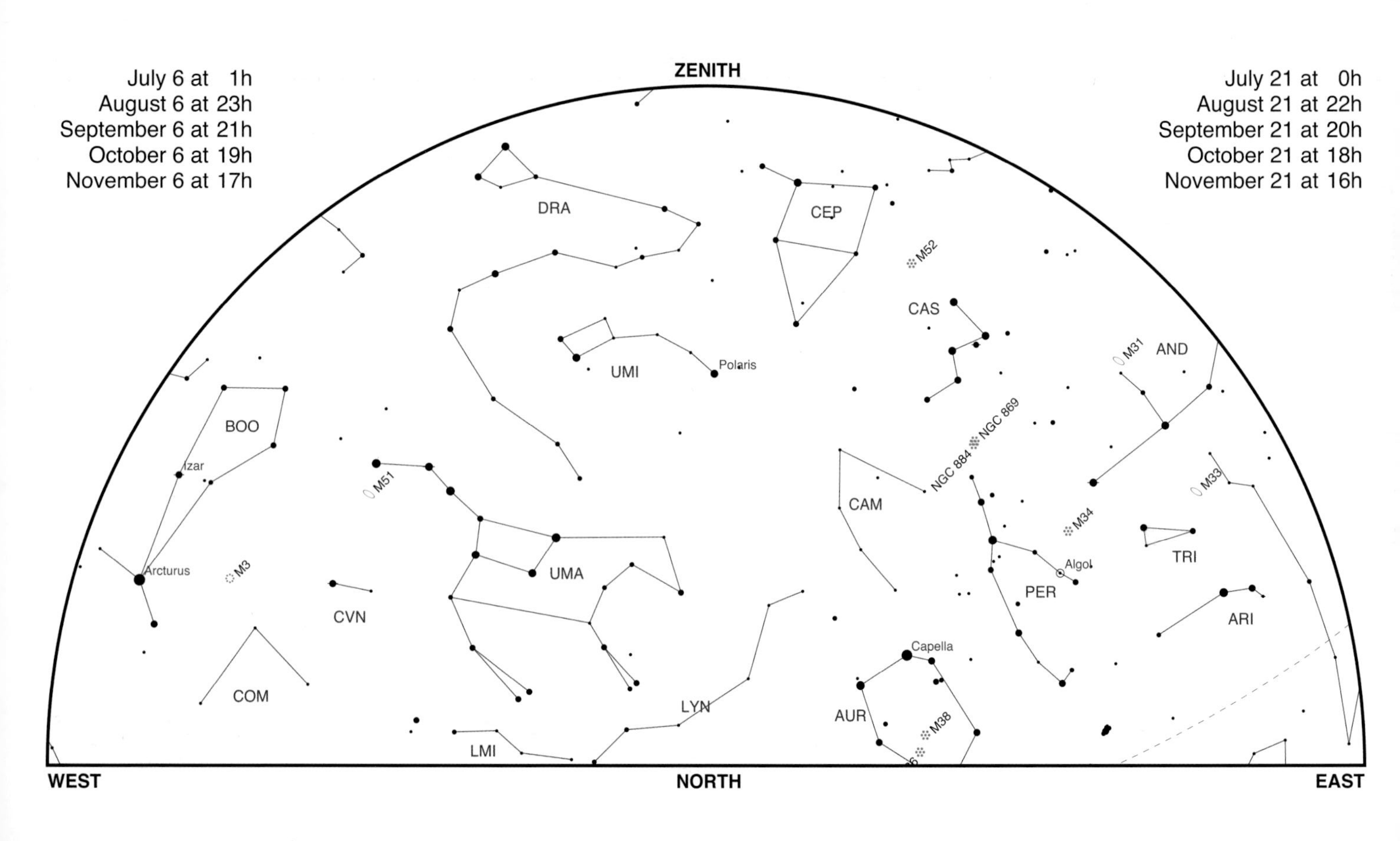
July 6 at 1h
August 6 at 23h
September 6 at 21h
October 6 at 19h
November 6 at 17h
July 21 at 0h
August 21 at 22h
September 21 at 20h
October 21 at 18h
November 21 at 16h
ZENITH
WEST
NORTH
EAST
DRA
CEP
M52
CAS
UMI
Polaris
M31
AND
BOO
Izar
NGC 884
NGC 869
M51
CAM
M33
M34
TRI
Algol
Arcturus
M3
UMA
PER
ARI
CVN
Capella
COM
LYN
AUR
M38
LMI

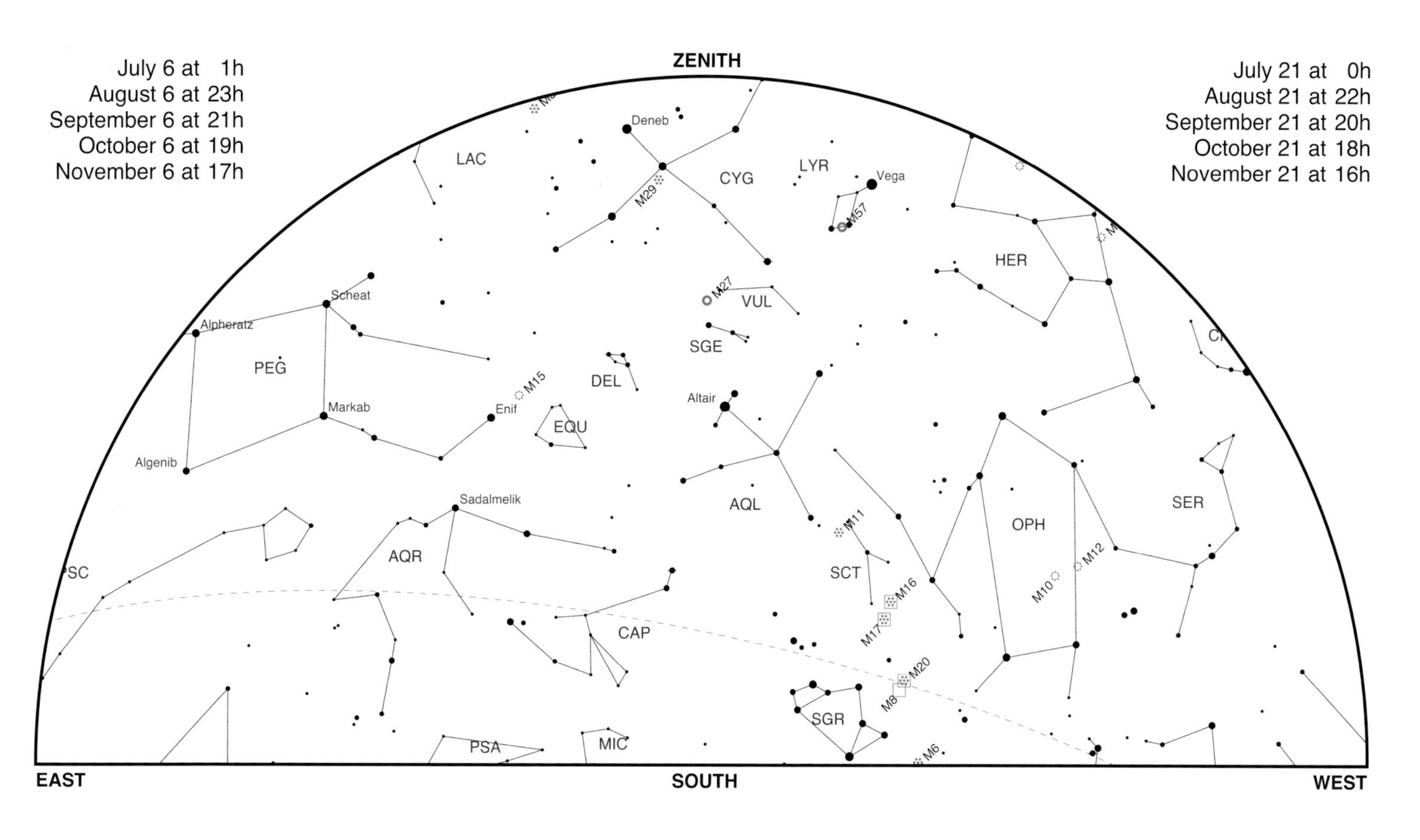
8S
July 6 at 1h
August 6 at 23h
September 6 at 21h
October 6 at 19h
November 6 at 17h
July 21 at 0h
August 21 at 22h
September 21 at 20h
October 21 at 18h
November 21 at 16h
ZENITH
EAST
SOUTH
WEST
LAC
Deneb
CYG
M29
LYR
Vega
M57
HER
M27
VUL
SGE
Scheat
Alpheratz
PEG
Markab
Algenib
M15
Enif
DEL
EQU
Altair
AQL
OPH
SER
M10
M12
M11
SCT
M16
M17
M20
M8
SGR
M6
Sadalmelik
AQR
CAP
PSA
MIC

9N

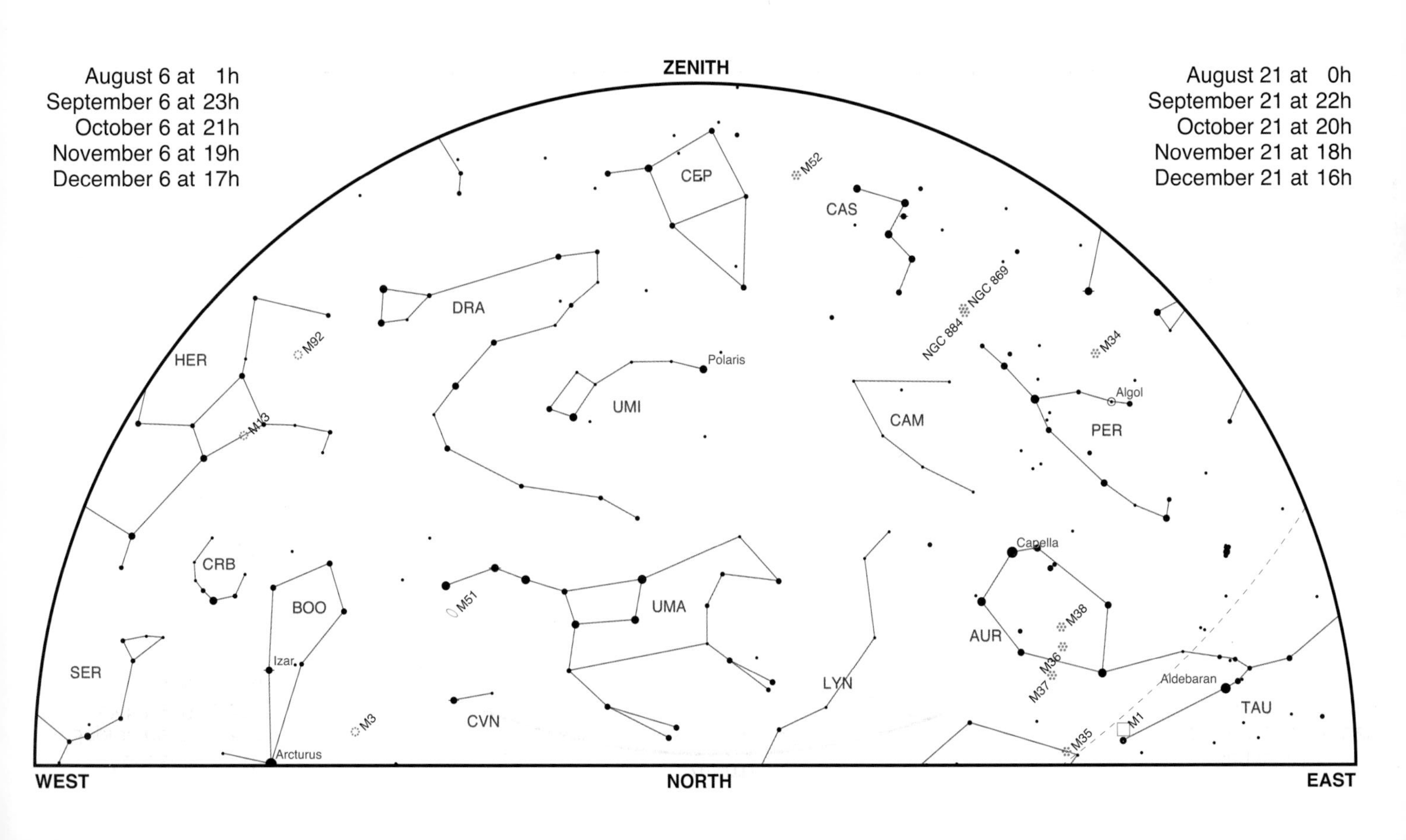

August 6 at 1h
September 6 at 23h
October 6 at 21h
November 6 at 19h
December 6 at 17h
August 21 at 0h
September 21 at 22h
October 21 at 20h
November 21 at 18h
December 21 at 16h
ZENITH
WEST
NORTH
EAST
CEP
M52
CAS
DRA
NGC 884
NGC 869
HER
M92
Polaris
UMI
M34
CAM
Algol
PER
M13
CRB
Capella
BOO
M51
UMA
M38
AUR
Izar
M36
M37
LYN
Aldebaran
SER
TAU
M3
CVN
M1
M35
Arcturus

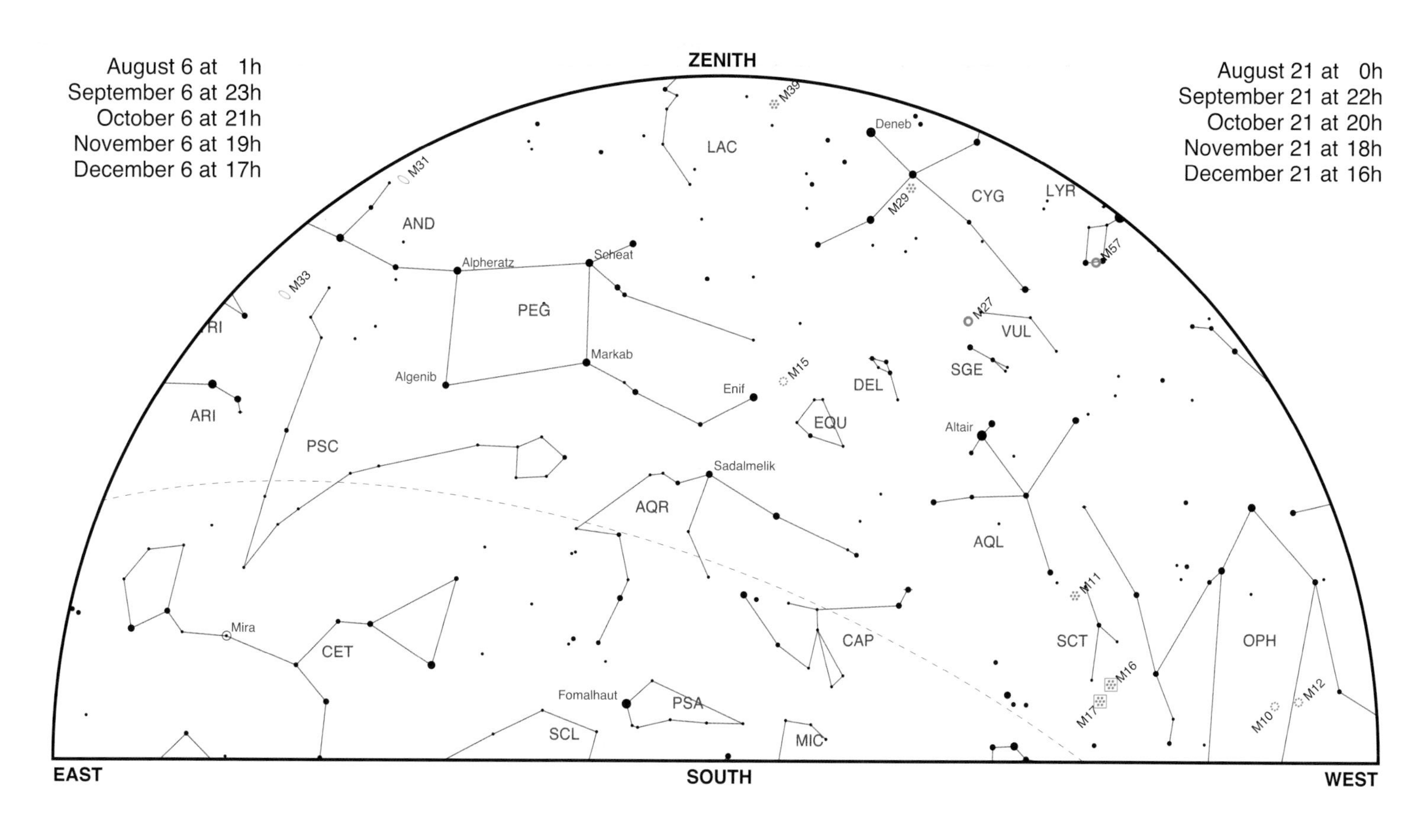
August 6 at 1h
September 6 at 23h
October 6 at 21h
November 6 at 19h
December 6 at 17h
ZENITH
August 21 at 0h
September 21 at 22h
October 21 at 20h
November 21 at 18h
December 21 at 16h
M39
LAC
Deneb
M31
AND
M29
CYG
LYR
M57
Alpheratz
Scheat
M33
PEG
M27
VUL
RI
Markab
SGE
M15
DEL
Algenib
Enif
ARI
EQU
Altair
PSC
Sadalmelik
AQR
AQL
M11
Mira
CET
CAP
SCT
OPH
M16
Fomalhaut
PSA
M17
M10
M12
SCL
MIC
EAST
SOUTH
WEST

S6

10N

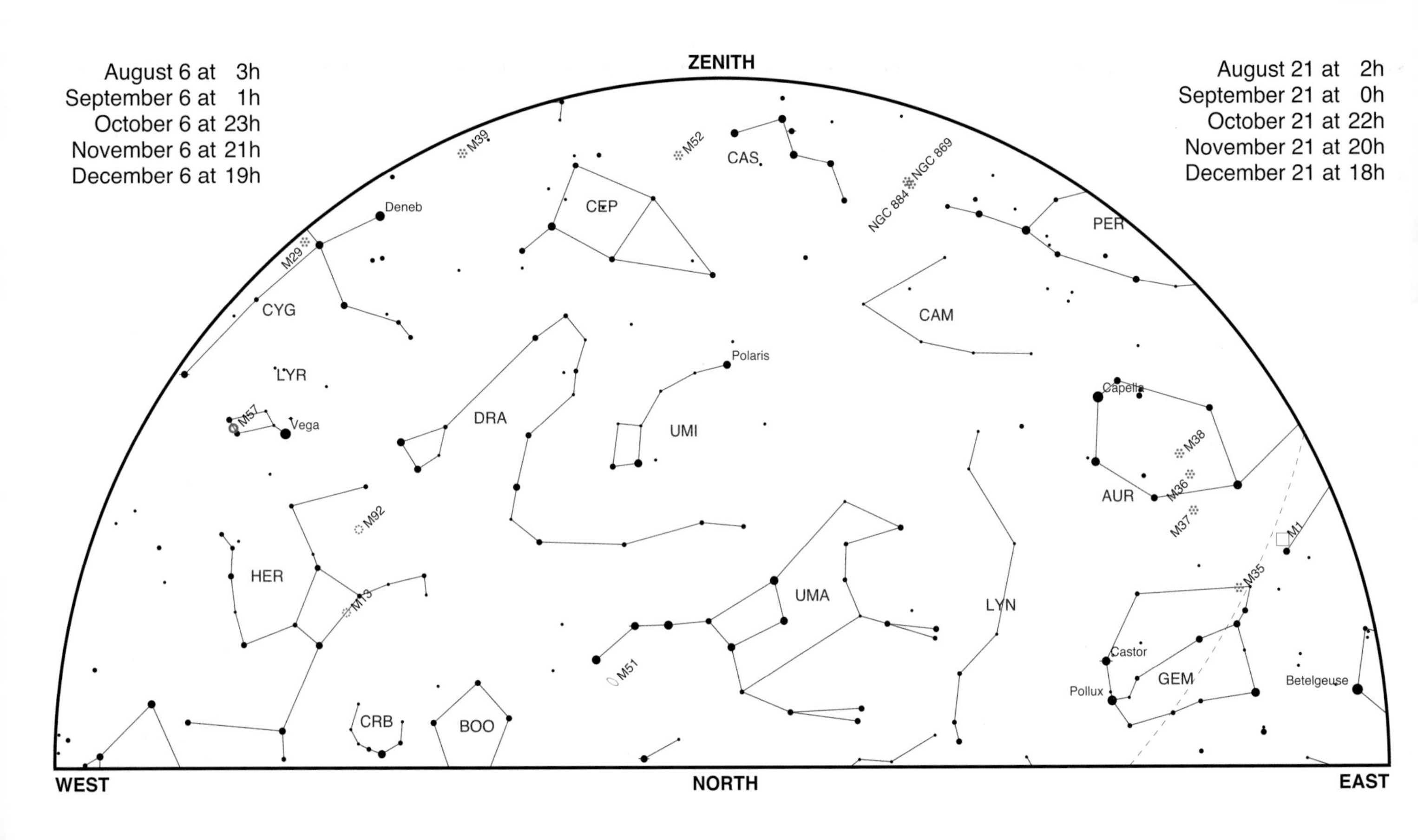

August 6 at 3h
September 6 at 1h
October 6 at 23h
November 6 at 21h
December 6 at 19h
ZENITH
August 21 at 2h
September 21 at 0h
October 21 at 22h
November 21 at 20h
December 21 at 18h
M39
M52
CAS
NGC 884
NGC 869
CEP
PER
Deneb
M29
CYG
CAM
Polaris
LYR
Capella
M57
Vega
DRA
UMI
M38
AUR
M36
M92
M37
M1
M35
HER
UMA
LYN
M13
Castor
M51
GEM
Pollux
Betelgeuse
CRB
BOO
WEST
NORTH
EAST

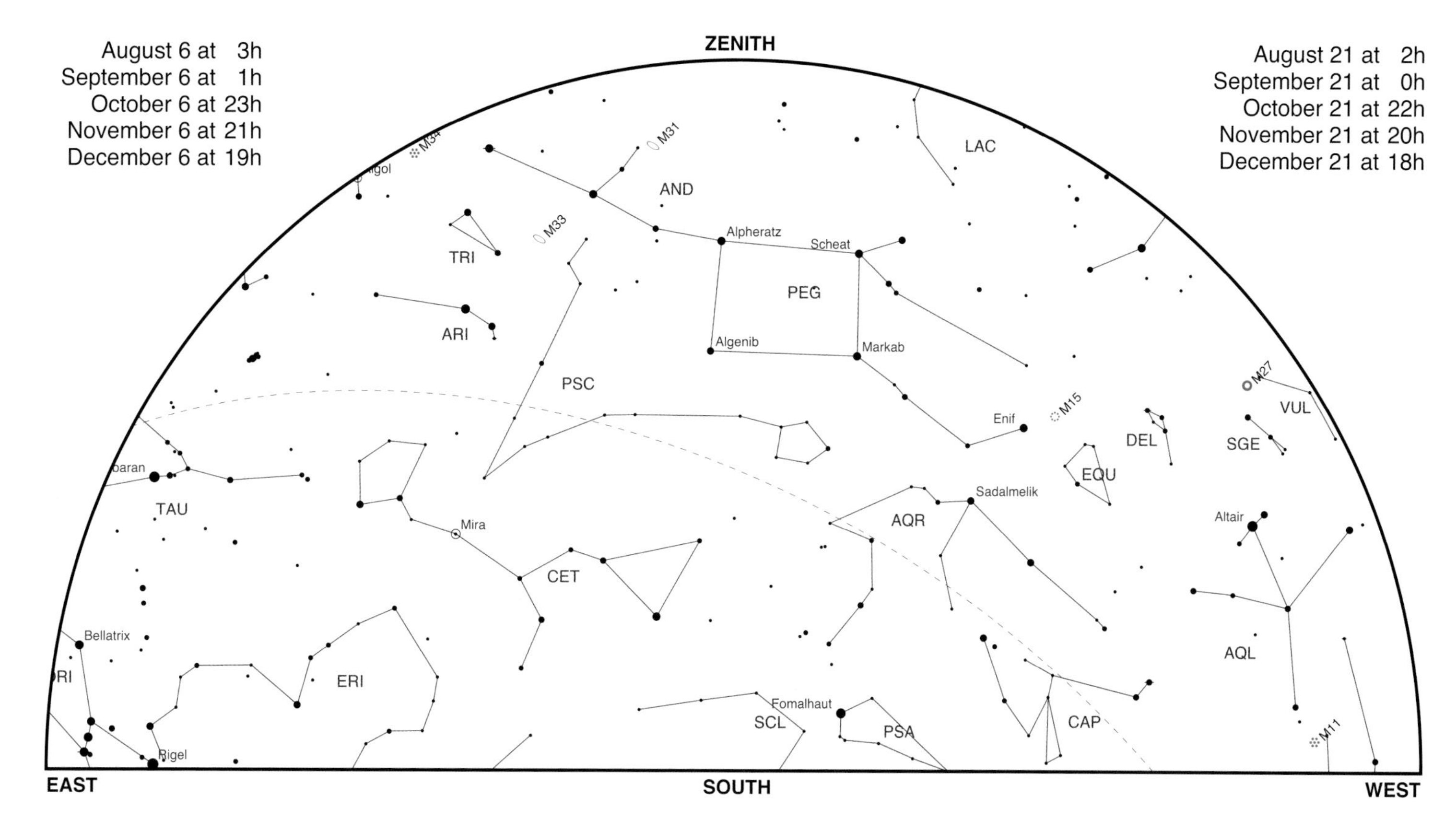
10S
ZENITH
August 6 at 3h
September 6 at 1h
October 6 at 23h
November 6 at 21h
December 6 at 19h
August 21 at 2h
September 21 at 0h
October 21 at 22h
November 21 at 20h
December 21 at 18h
EAST
SOUTH
WEST
AND
M31
M33
M34
LAC
TRI
ARI
PEG
Alpheratz
Scheat
Algenib
Markab
Enif
M15
M27
VUL
DEL
SGE
EQU
PSC
TAU
Bellatrix
Rigel
Mira
CET
ERI
AQR
Sadalmelik
Altair
AQL
M11
SCL
Fomalhaut
PSA
CAP

11N

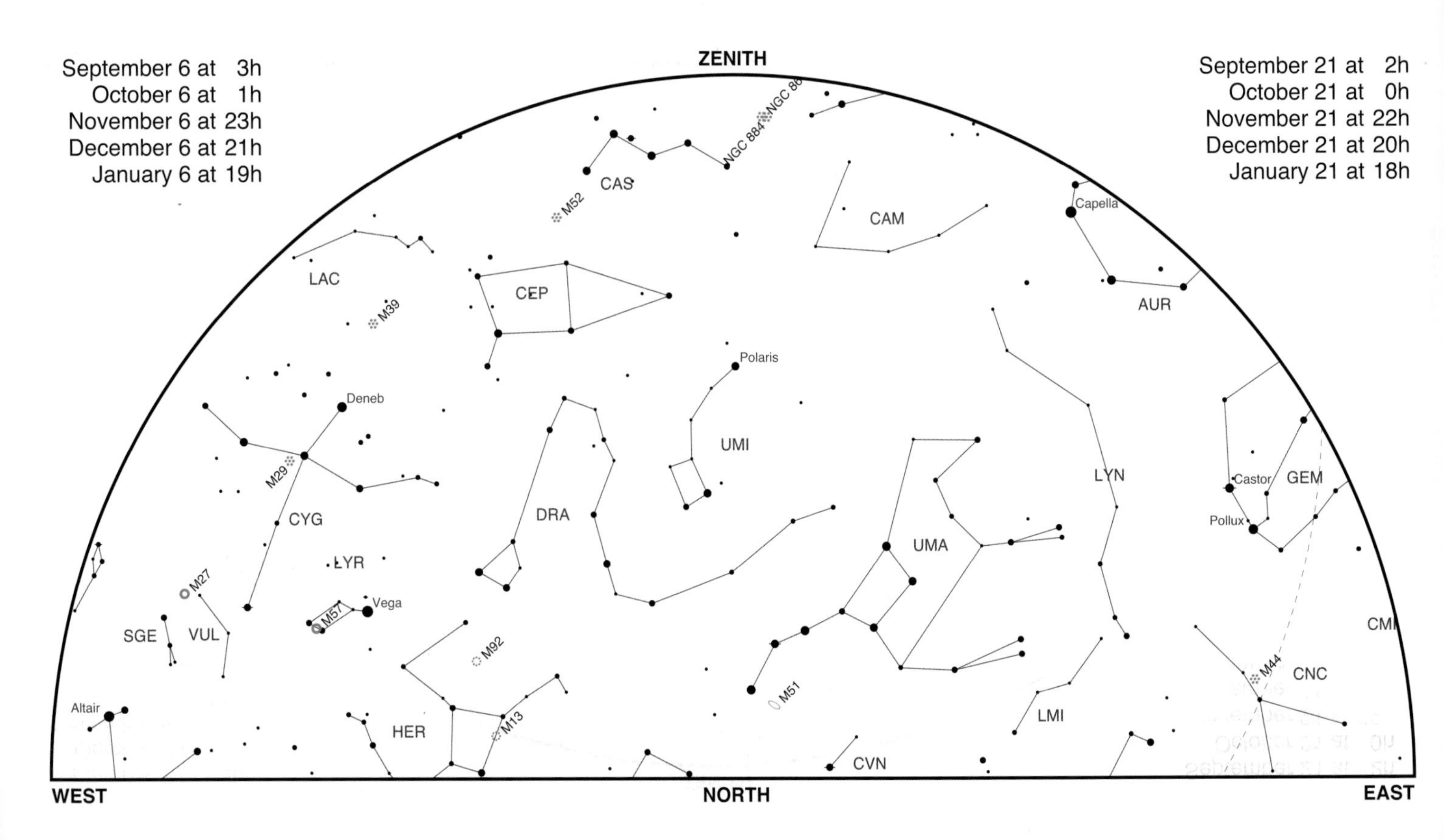
September 6 at 3h
October 6 at 1h
November 6 at 23h
December 6 at 21h
January 6 at 19h
ZENITH
September 21 at 2h
October 21 at 0h
November 21 at 22h
December 21 at 20h
January 21 at 18h
NGC 884
NGC 86
CAS
M52
CAM
Capella
AUR
LAC
CEP
M39
Polaris
Deneb
UMI
M29
LYN
Castor
GEM
Pollux
CYG
DRA
UMA
LYR
M27
Vega
M57
CMI
SGE
VUL
M92
M44
CNC
M51
Altair
HER
M13
LMI
CVN
WEST
NORTH
EAST

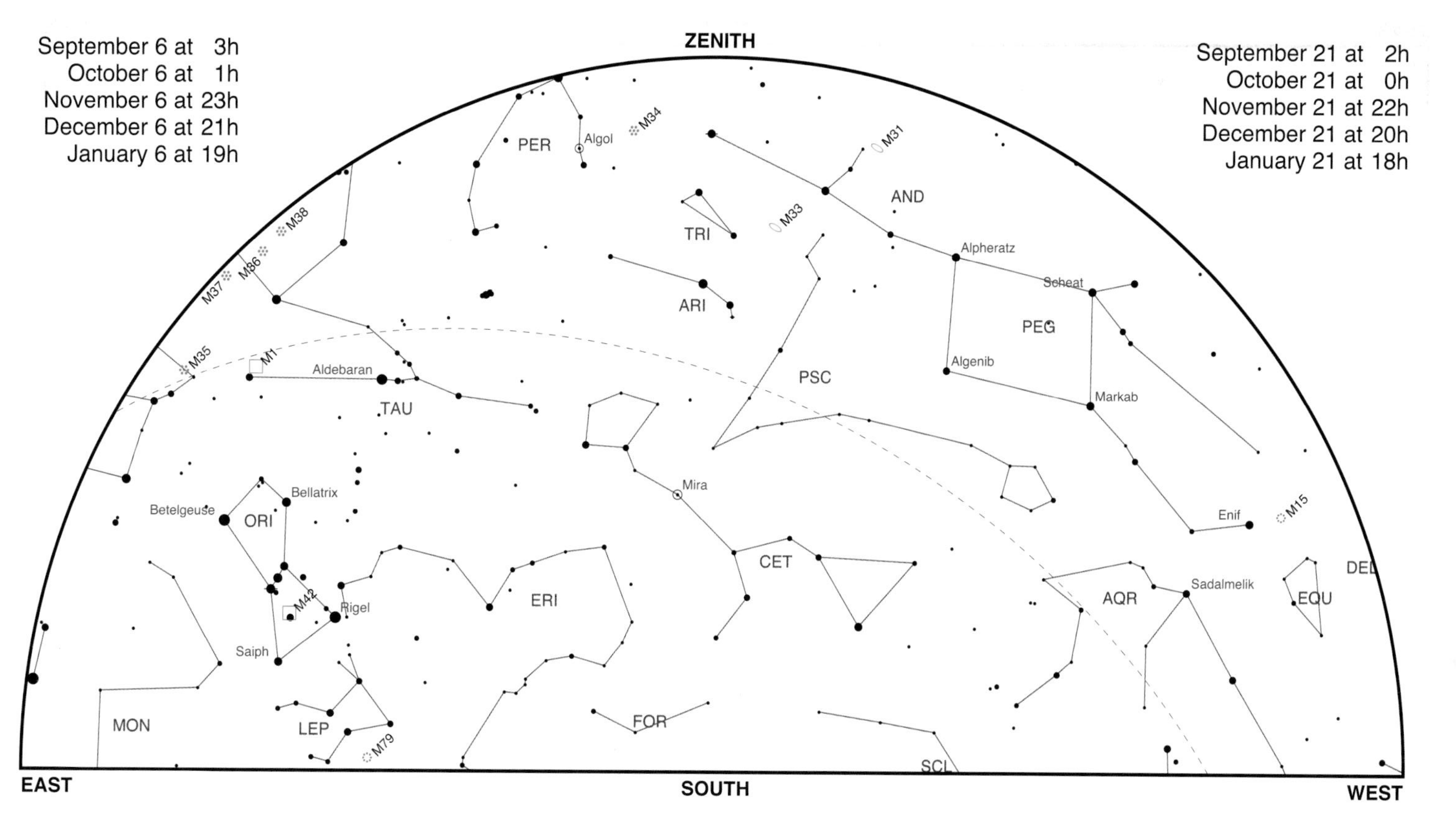
September 6 at 3h
October 6 at 1h
November 6 at 23h
December 6 at 21h
January 6 at 19h
ZENITH
September 21 at 2h
October 21 at 0h
November 21 at 22h
December 21 at 20h
January 21 at 18h
PER
Algol
M34
M31
AND
TRI
M33
M38
M36
M37
Alpheratz
Scheat
ARI
PEG
M35
M1
Aldebaran
Algenib
PSC
TAU
Markab
Mira
Bellatrix
Betelgeuse
ORI
Enif
M15
CET
DEL
Sadalmelik
M42
Rigel
ERI
AQR
EQU
Saiph
MON
LEP
M79
FOR
SCL
EAST
SOUTH
WEST

11S

12N

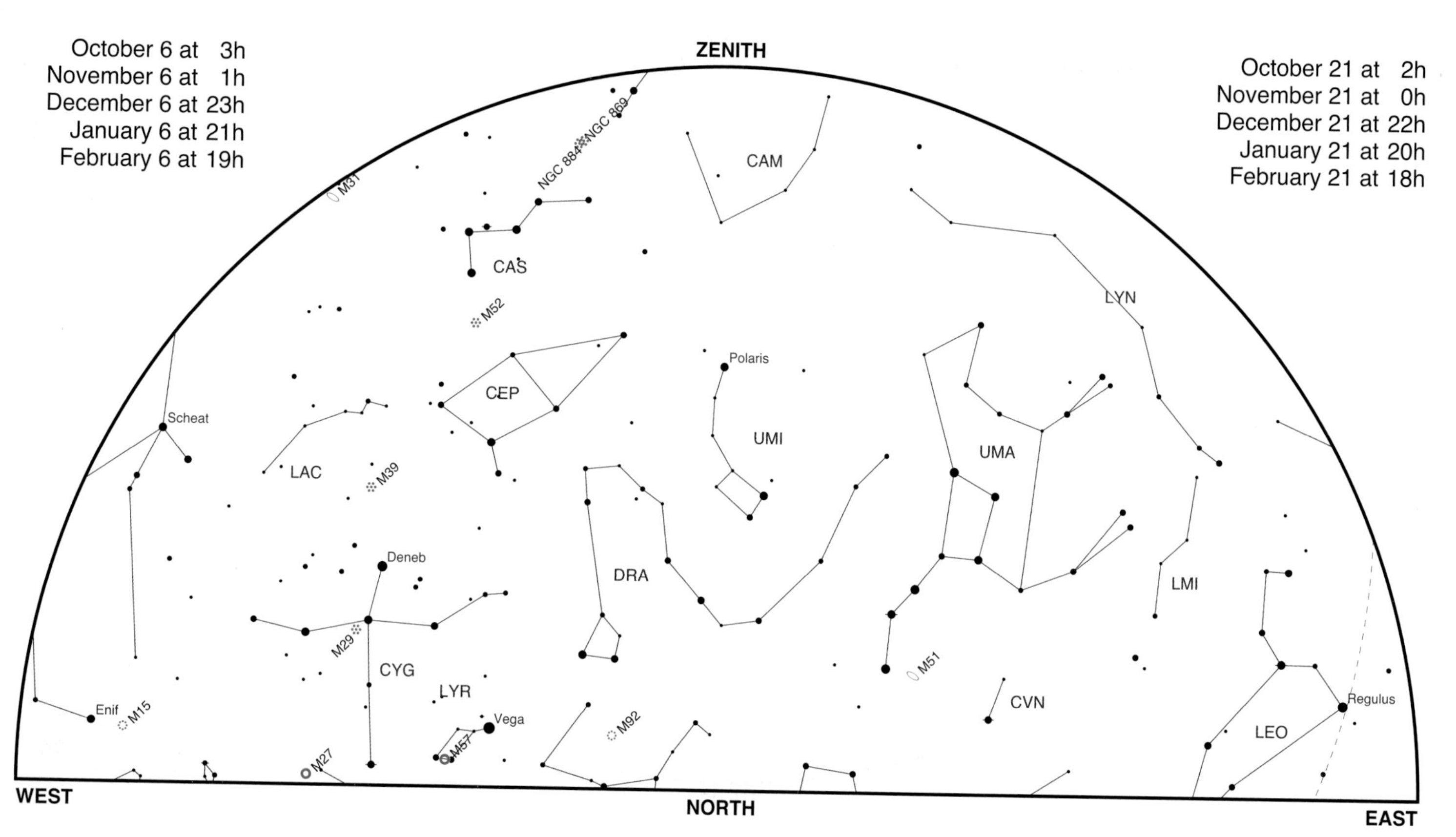
October 6 at 3h
November 6 at 1h
December 6 at 23h
January 6 at 21h
February 6 at 19h
October 21 at 2h
November 21 at 0h
December 21 at 22h
January 21 at 20h
February 21 at 18h
ZENITH
WEST
NORTH
EAST
CAM
CAS
NGC 884
NGC 869
M31
M52
LYN
CEP
Polaris
UMI
UMA
Scheat
LAC
M39
DRA
LMI
Deneb
M29
CYG
LYR
Vega
M57
M27
Enif
M15
M92
M51
CVN
LEO
Regulus

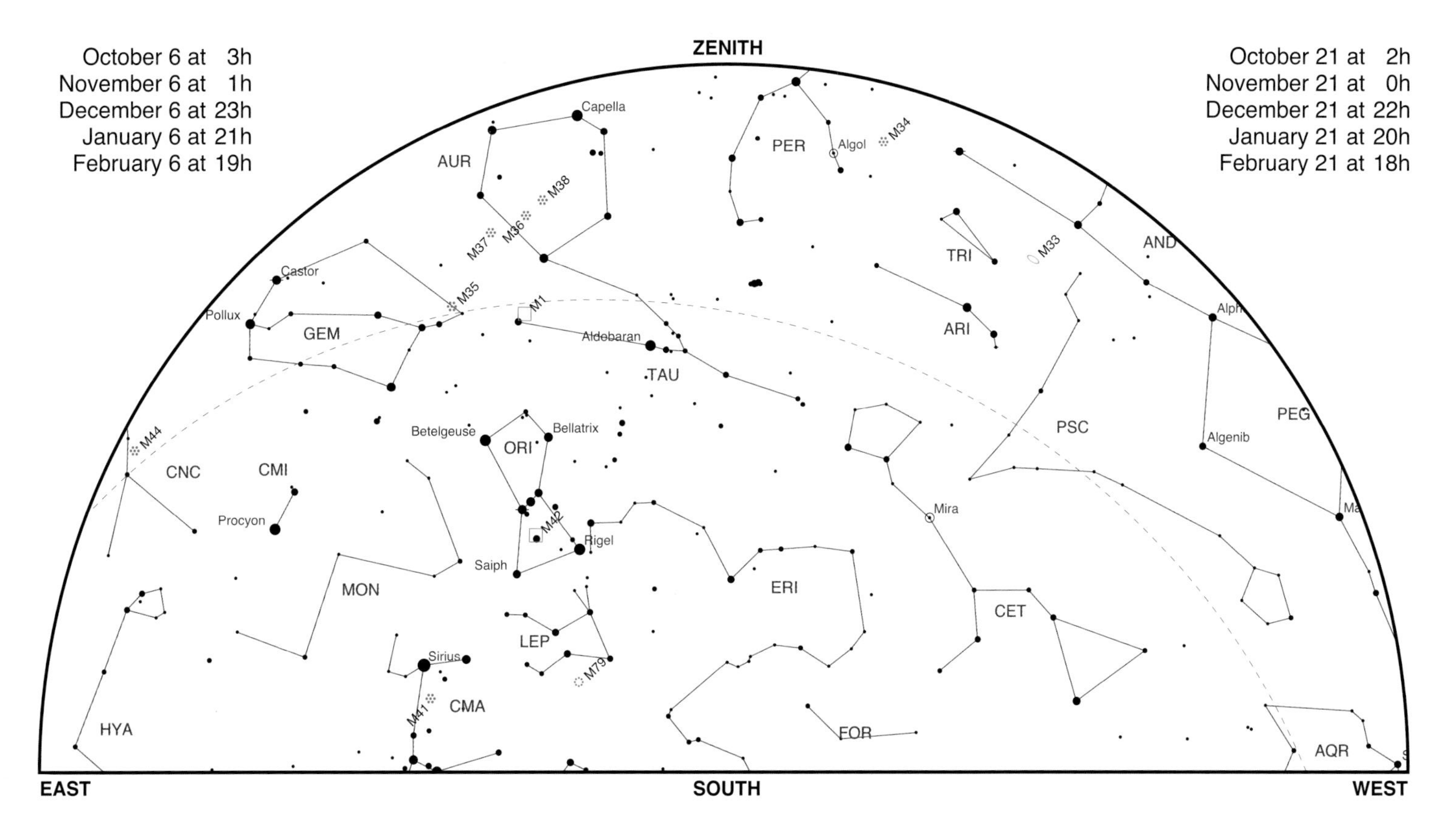
October 6 at 3h
November 6 at 1h
December 6 at 23h
January 6 at 21h
February 6 at 19h
ZENITH
October 21 at 2h
November 21 at 0h
December 21 at 22h
January 21 at 20h
February 21 at 18h
Capella
AUR
PER
Algol
M34
M38
M36
M37
TRI
M33
AND
Castor
M35
M1
Pollux
GEM
Aldebaran
ARI
Alph
TAU
PSC
PEG
Algenib
Betelgeuse
Bellatrix
M44
ORI
CNC
CMI
Mira
Procyon
M42
Rigel
Saiph
MON
ERI
CET
LEP
Sirius
M79
CMA
M41
HYA
FOR
AQR
EAST
SOUTH
WEST

12S

Southern Hemisphere Star Charts

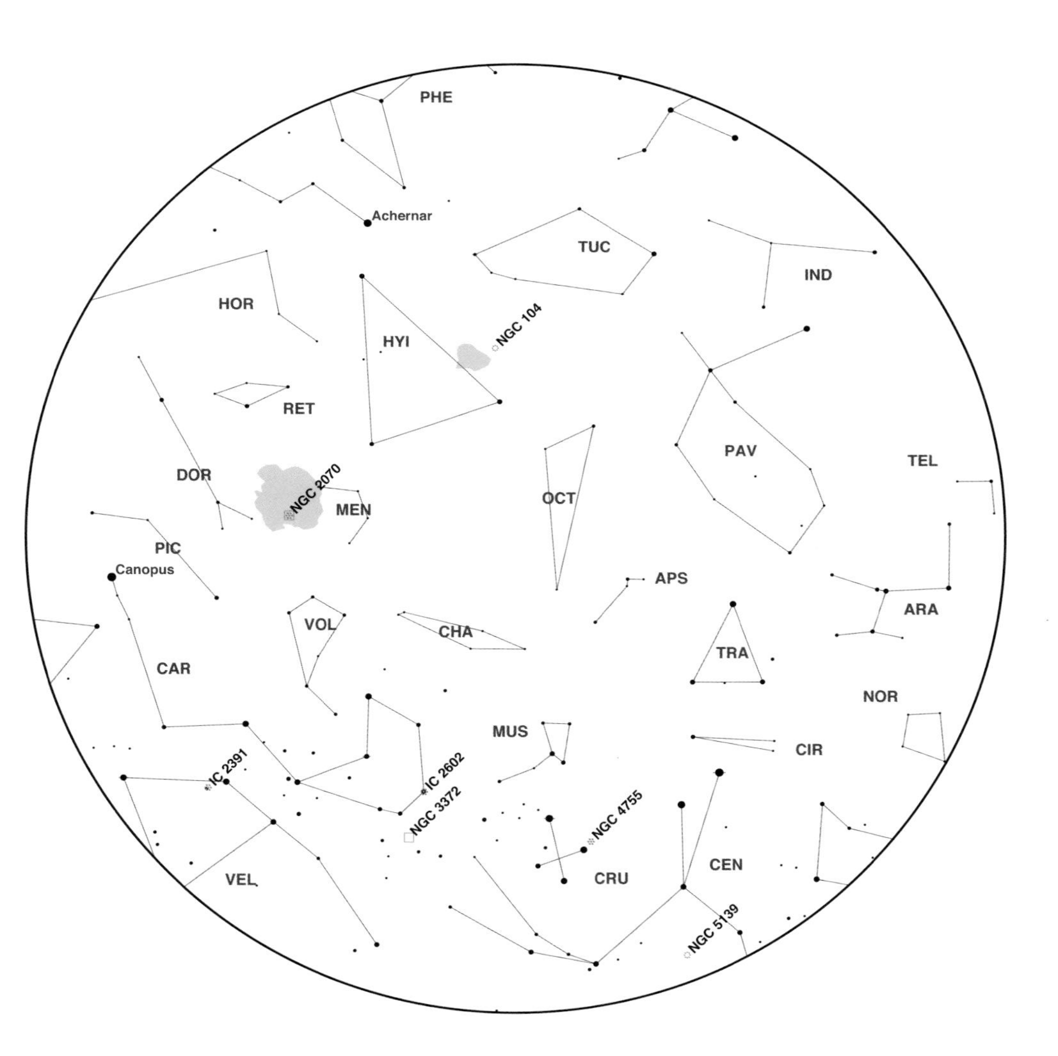

This chart shows stars lying at declinations between −45 and −90 degrees. These constellations are circumpolar for observers in Australia and New Zealand.

1N

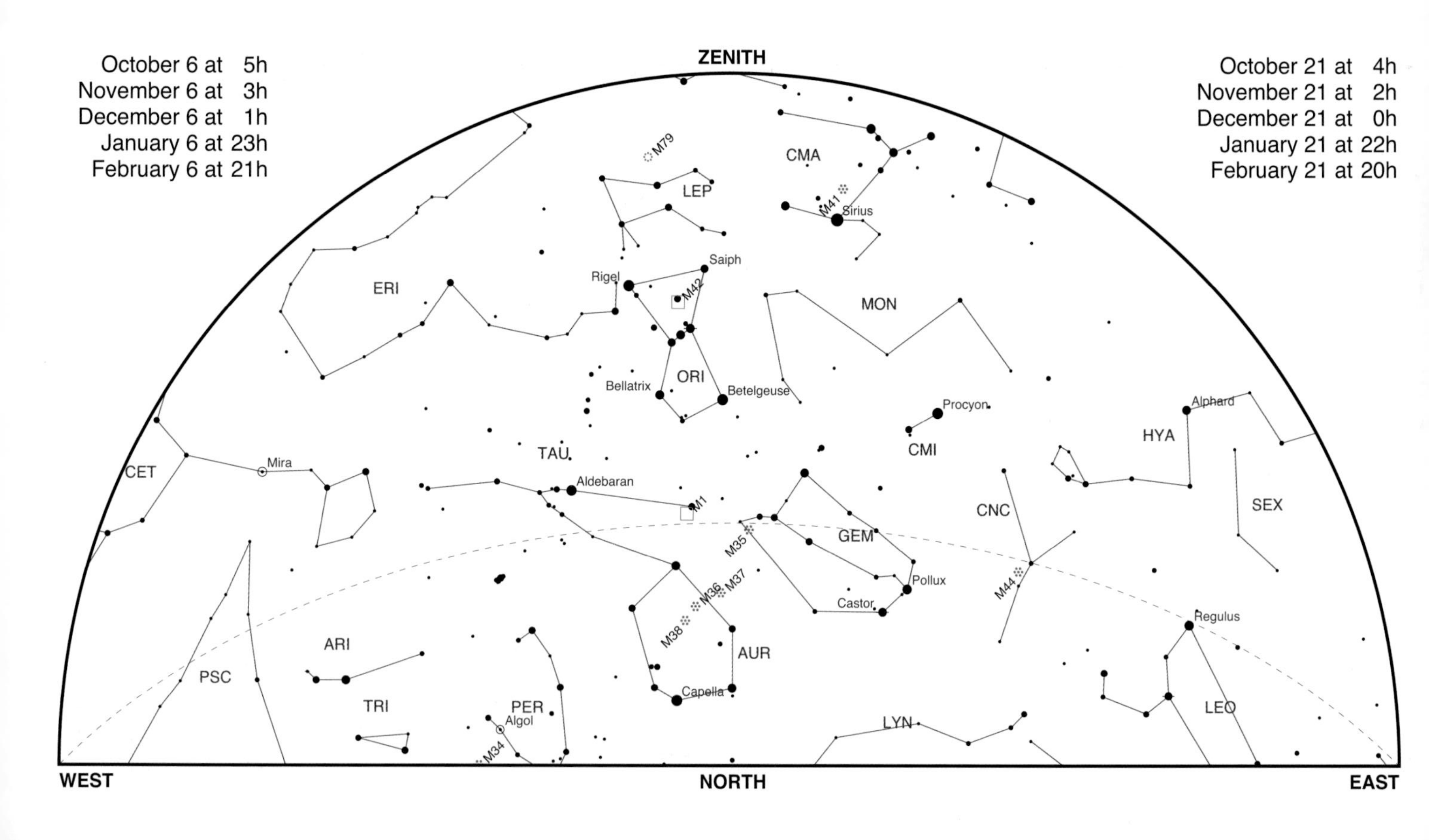

October 6 at 5h
November 6 at 3h
December 6 at 1h
January 6 at 23h
February 6 at 21h
October 21 at 4h
November 21 at 2h
December 21 at 0h
January 21 at 22h
February 21 at 20h
ZENITH
WEST
NORTH
EAST
M79
CMA
LEP
M41
Sirius
ERI
Saiph
Rigel
M42
MON
ORI
Bellatrix
Betelgeuse
Procyon
Alphard
HYA
CMI
TAU
CET
Mira
Aldebaran
M1
SEX
CNC
GEM
M35
Pollux
M44
M36
M37
Castor
Regulus
M38
ARI
AUR
PSC
Capella
TRI
PER
LEO
Algol
LYN
M34

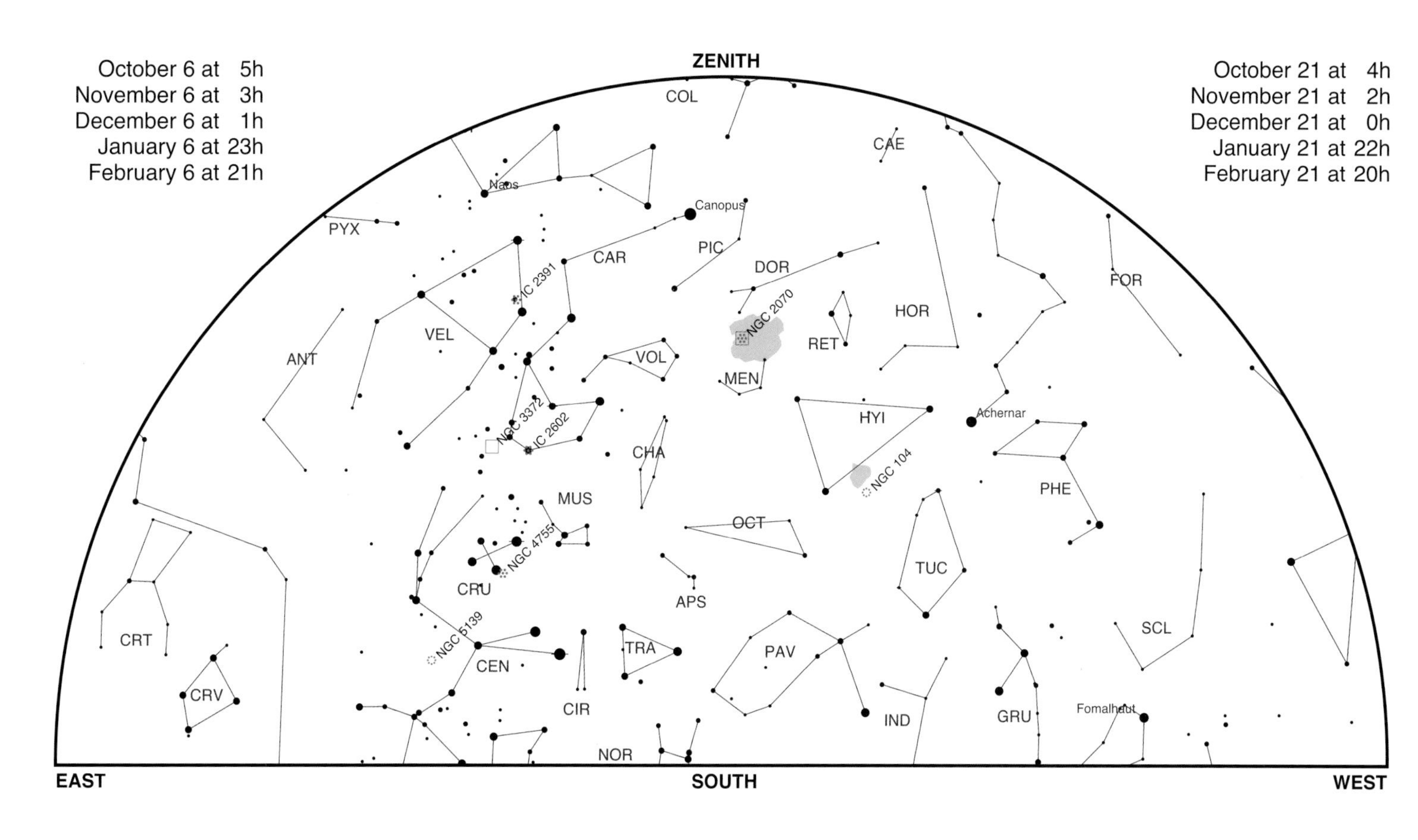
October 6 at 5h
November 6 at 3h
December 6 at 1h
January 6 at 23h
February 6 at 21h
ZENITH
October 21 at 4h
November 21 at 2h
December 21 at 0h
January 21 at 22h
February 21 at 20h
COL
CAE
Naos
Canopus
PYX
CAR
PIC
DOR
FOR
IC 2391
NGC 2070
HOR
VEL
RET
ANT
VOL
MEN
Achernar
HYI
NGC 3372
IC 2602
CHA
NGC 104
PHE
MUS
OCT
NGC 4755
TUC
CRU
APS
SCL
NGC 5139
CRT
TRA
PAV
CEN
CRV
CIR
IND
GRU
Fomalhaut
NOR
EAST
SOUTH
WEST

1S

2N

November 6 at 5h
December 6 at 3h
January 6 at 1h
February 6 at 23h
March 6 at 21h

November 21 at 4h
December 21 at 2h
January 21 at 0h
February 21 at 22h
March 21 at 20h

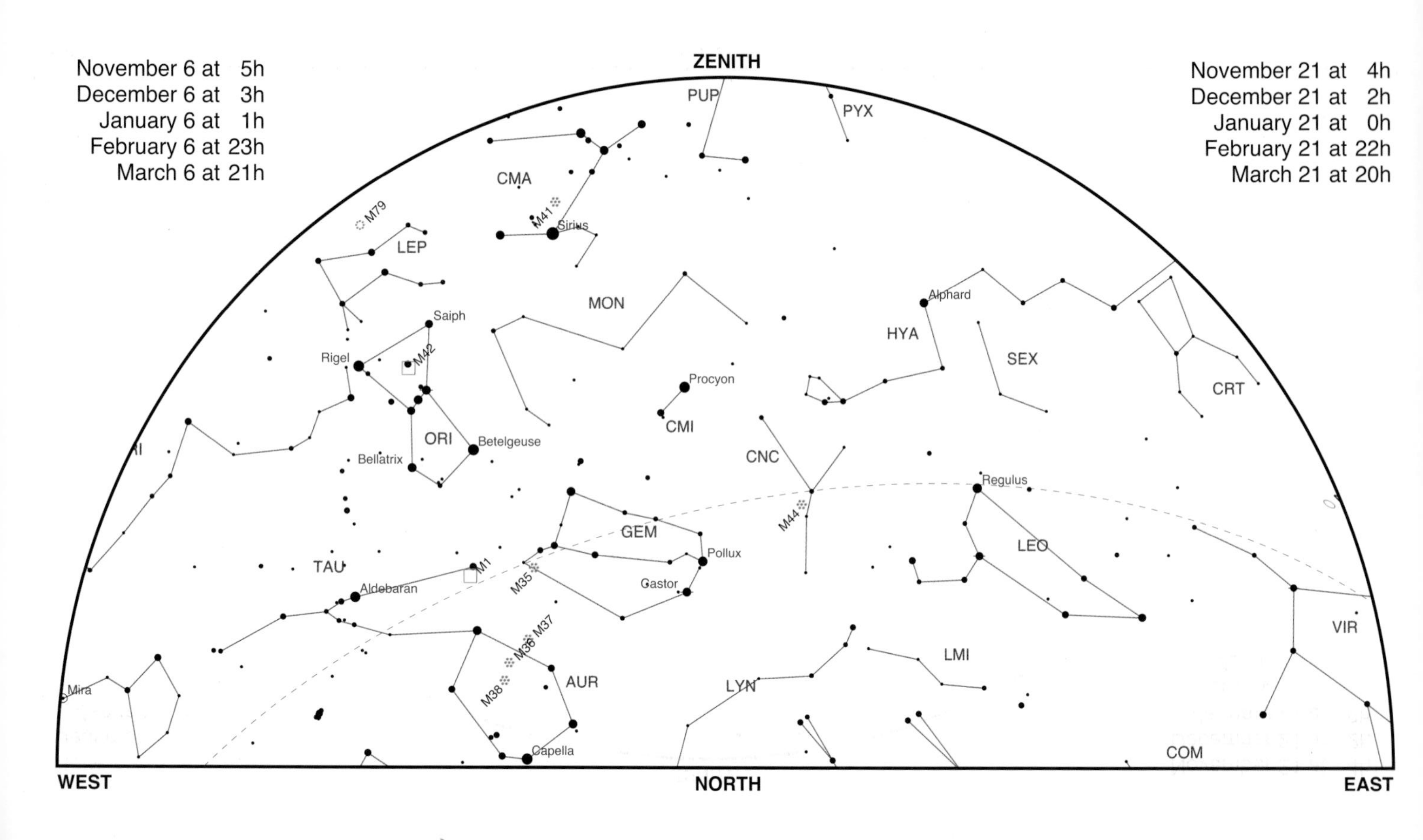

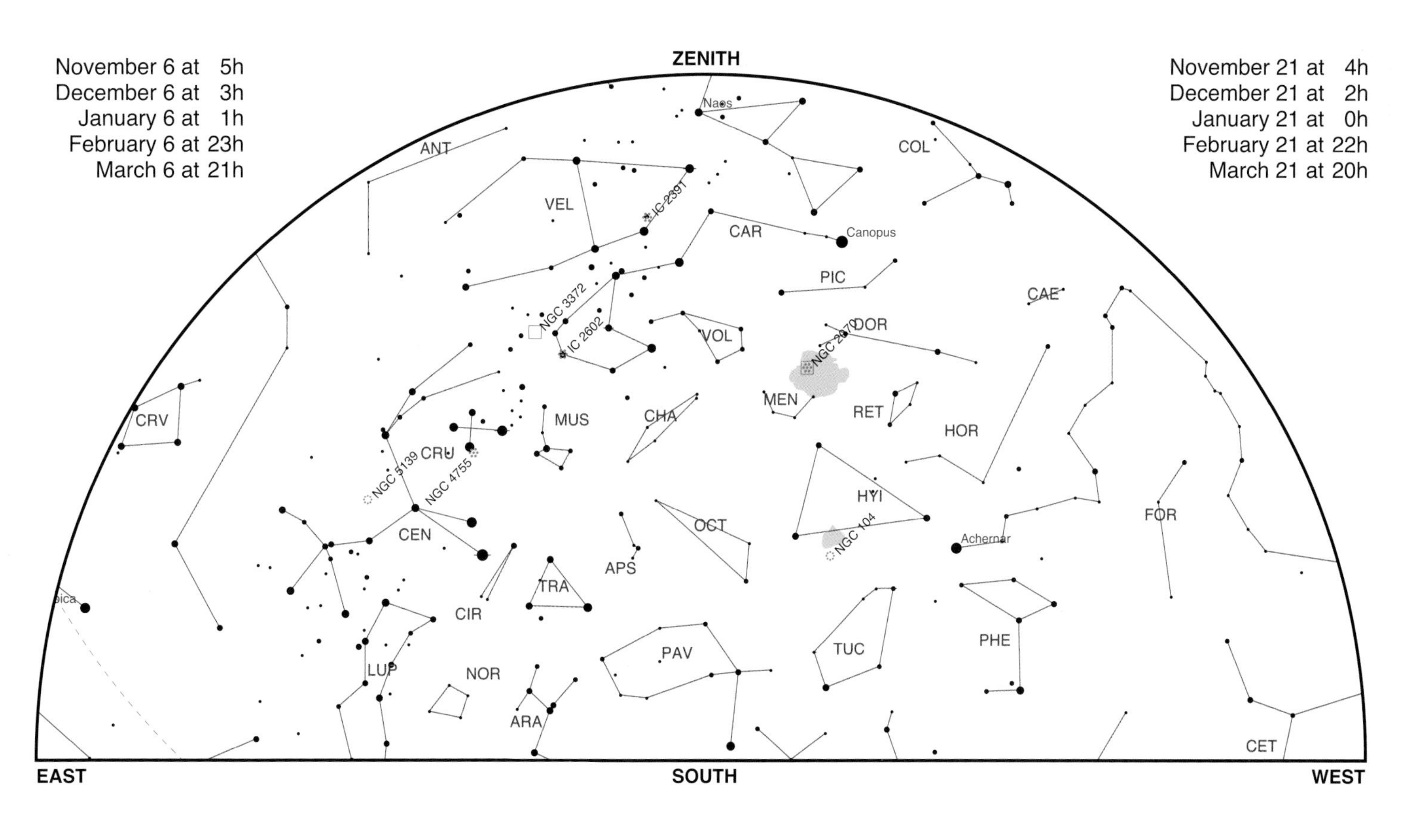

November 6 at 5h
December 6 at 3h
January 6 at 1h
February 6 at 23h
March 6 at 21h
November 21 at 4h
December 21 at 2h
January 21 at 0h
February 21 at 22h
March 21 at 20h
ZENITH
EAST
SOUTH
WEST
2S
Naos
Canopus
Achernar
ANT
VEL
CAR
COL
PIC
CAE
DOR
VOL
MEN
RET
HOR
CRV
MUS
CHA
CRU
CEN
HYI
OCT
FOR
APS
TRA
CIR
PAV
TUC
PHE
LUP
NOR
ARA
CET
IC 2391
NGC 3372
IC 2602
NGC 2070
NGC 5139
NGC 4755
NGC 104

3N

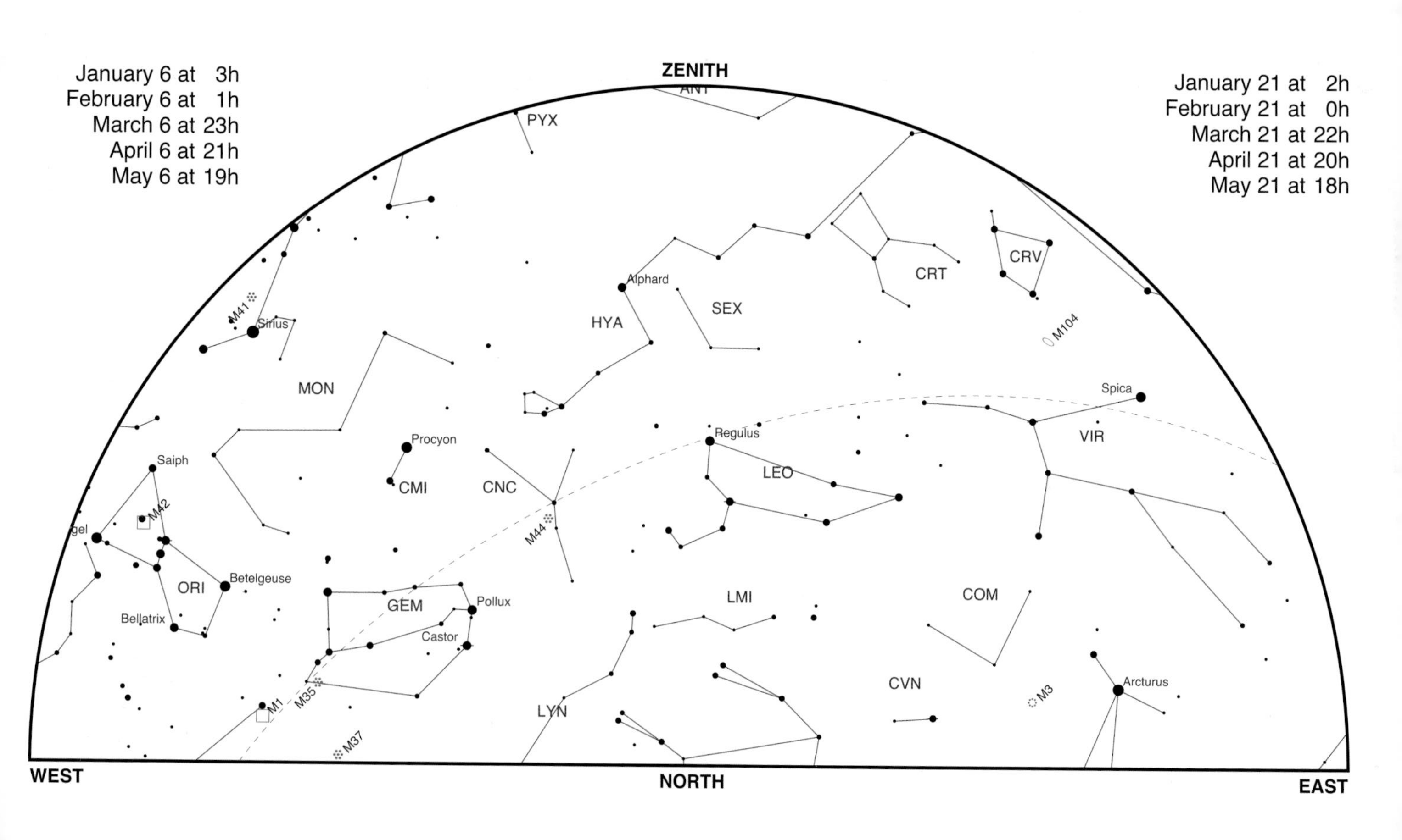
January 6 at 3h
February 6 at 1h
March 6 at 23h
April 6 at 21h
May 6 at 19h
January 21 at 2h
February 21 at 0h
March 21 at 22h
April 21 at 20h
May 21 at 18h
ZENITH
WEST
NORTH
EAST
PYX
HYA
Alphard
SEX
CRT
CRV
M104
M41
Sirius
MON
Spica
VIR
Procyon
CMI
CNC
Regulus
LEO
Saiph
M42
ORI
Betelgeuse
Bellatrix
M44
GEM
Pollux
Castor
LMI
COM
CVN
M3
Arcturus
LYN
M1
M35
M37

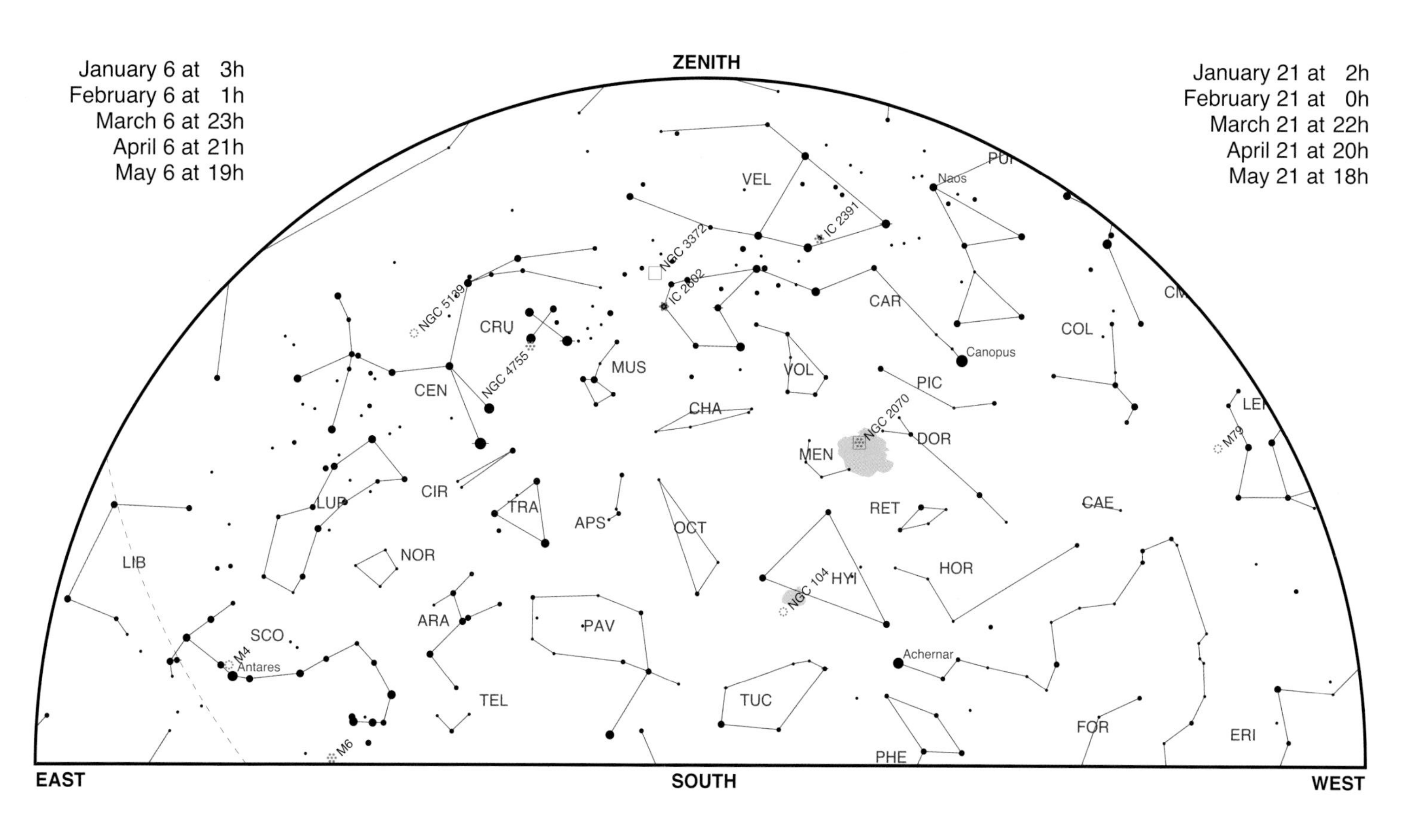
January 6 at 3h
February 6 at 1h
March 6 at 23h
April 6 at 21h
May 6 at 19h
ZENITH
January 21 at 2h
February 21 at 0h
March 21 at 22h
April 21 at 20h
May 21 at 18h
VEL
PUP
Naos
NGC 3372
IC 2391
CAR
NGC 2602
NGC 5139
CRU
COL
Canopus
MUS
NGC 4755
VOL
CEN
PIC
LEP
CHA
NGC 2070
DOR
M79
MEN
CIR
LUP
TRA
RET
CAE
APS
OCT
LIB
NOR
HOR
HYI
NGC 104
ARA
PAV
SCO
M4
Antares
Achernar
TEL
TUC
FOR
ERI
M6
PHE
EAST
SOUTH
WEST

3S

4N

February 21 at 2h
March 21 at 0h
April 21 at 22h
May 21 at 20h
June 21 at 18h

February 6 at 3h
March 6 at 1h
April 6 at 23h
May 6 at 21h
June 6 at 19h

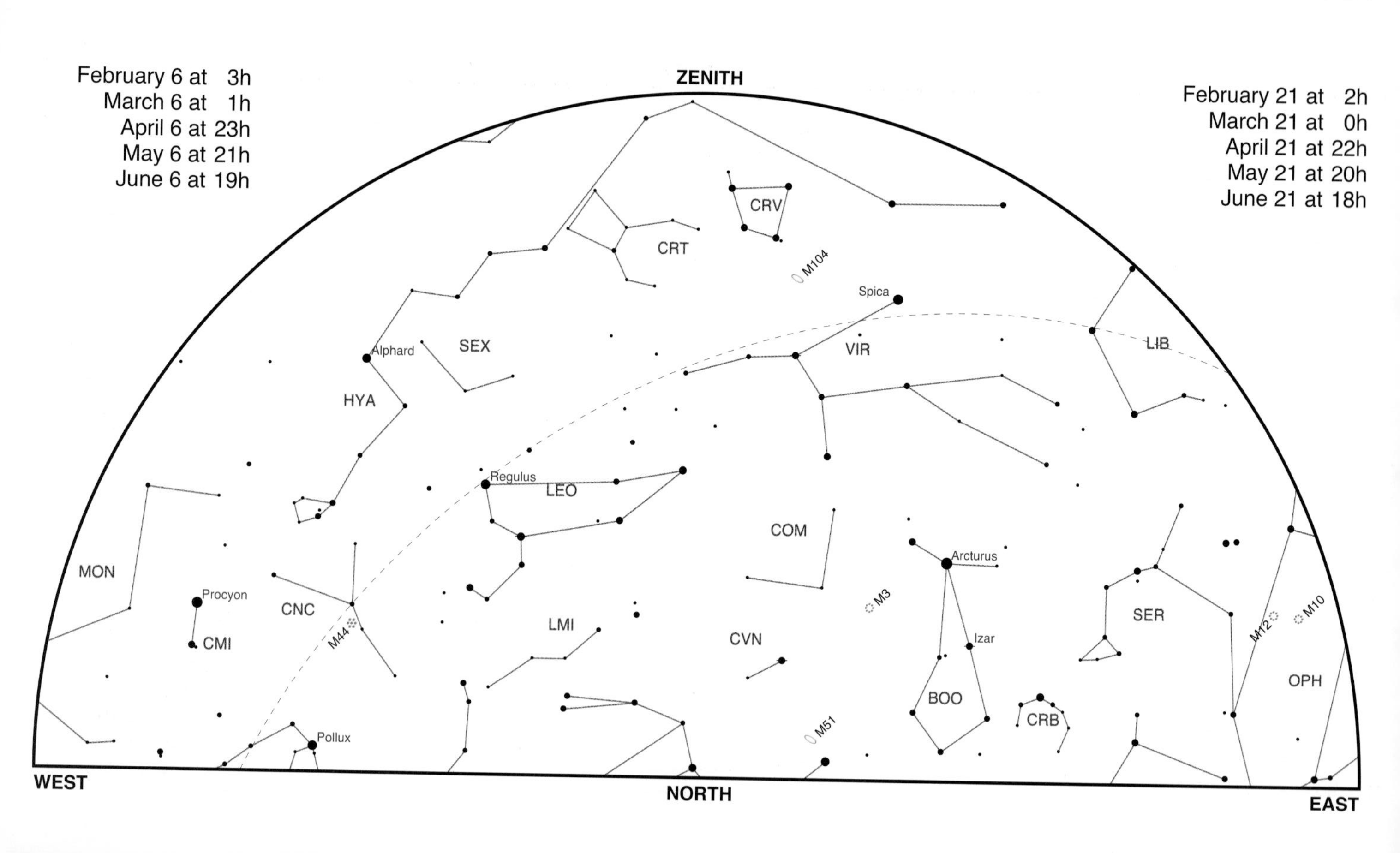

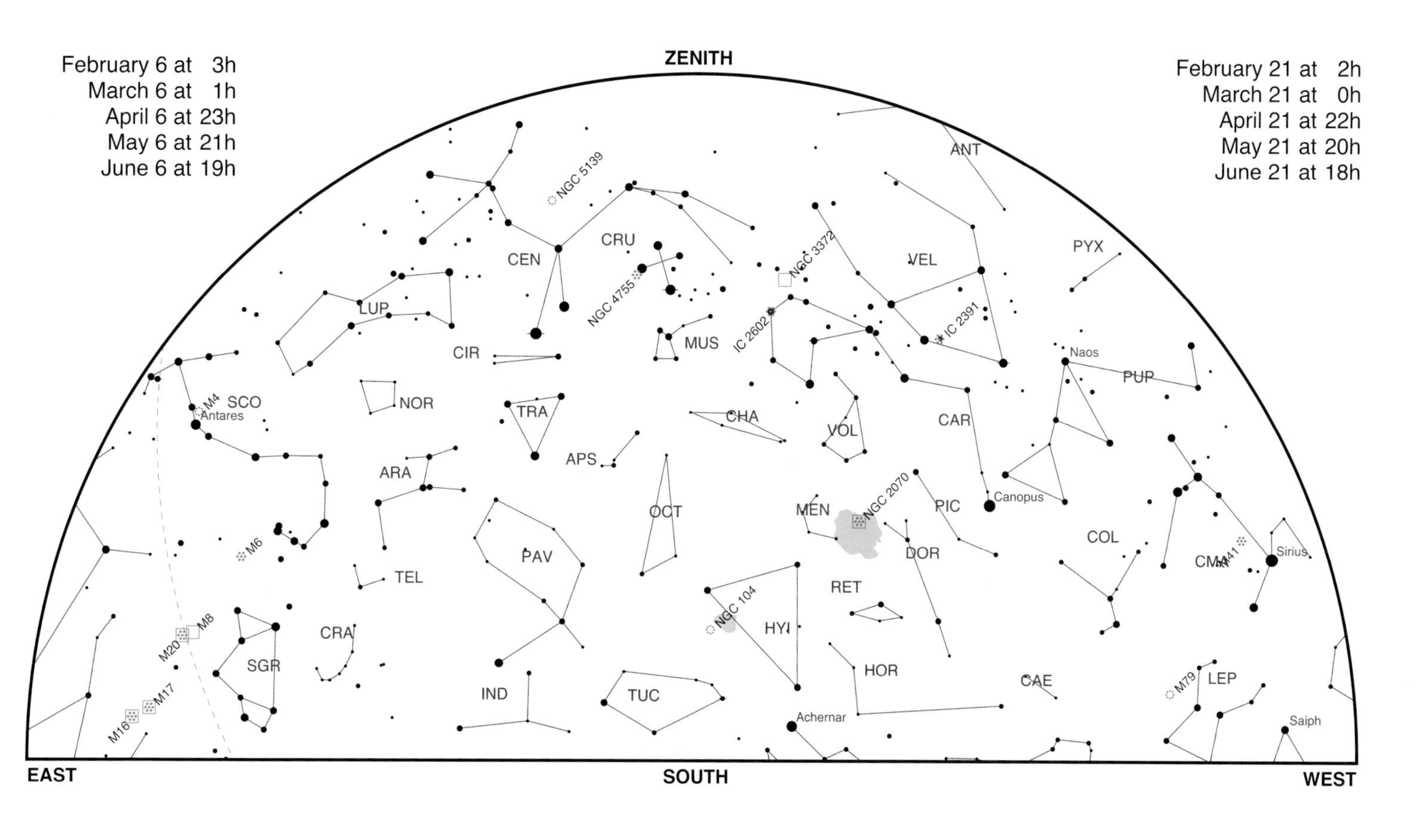
February 6 at 3h
March 6 at 1h
April 6 at 23h
May 6 at 21h
June 6 at 19h
ZENITH
February 21 at 2h
March 21 at 0h
April 21 at 22h
May 21 at 20h
June 21 at 18h
ANT
PYX
VEL
CEN
CRU
NGC 5139
NGC 4755
NGC 3372
IC 2602
IC 2391
LUP
CIR
MUS
Naos
PUP
SCO
M4
Antares
NOR
TRA
CHA
VOL
CAR
APS
ARA
OCT
MEN
NGC 2070
PIC
Canopus
COL
M6
PAV
DOR
CMA
M41
Sirius
TEL
RET
NGC 104
HYI
CRA
M8
M20
SGR
M17
M16
IND
TUC
HOR
CAE
M79
LEP
Achernar
Saiph
EAST
SOUTH
WEST
4S

5N

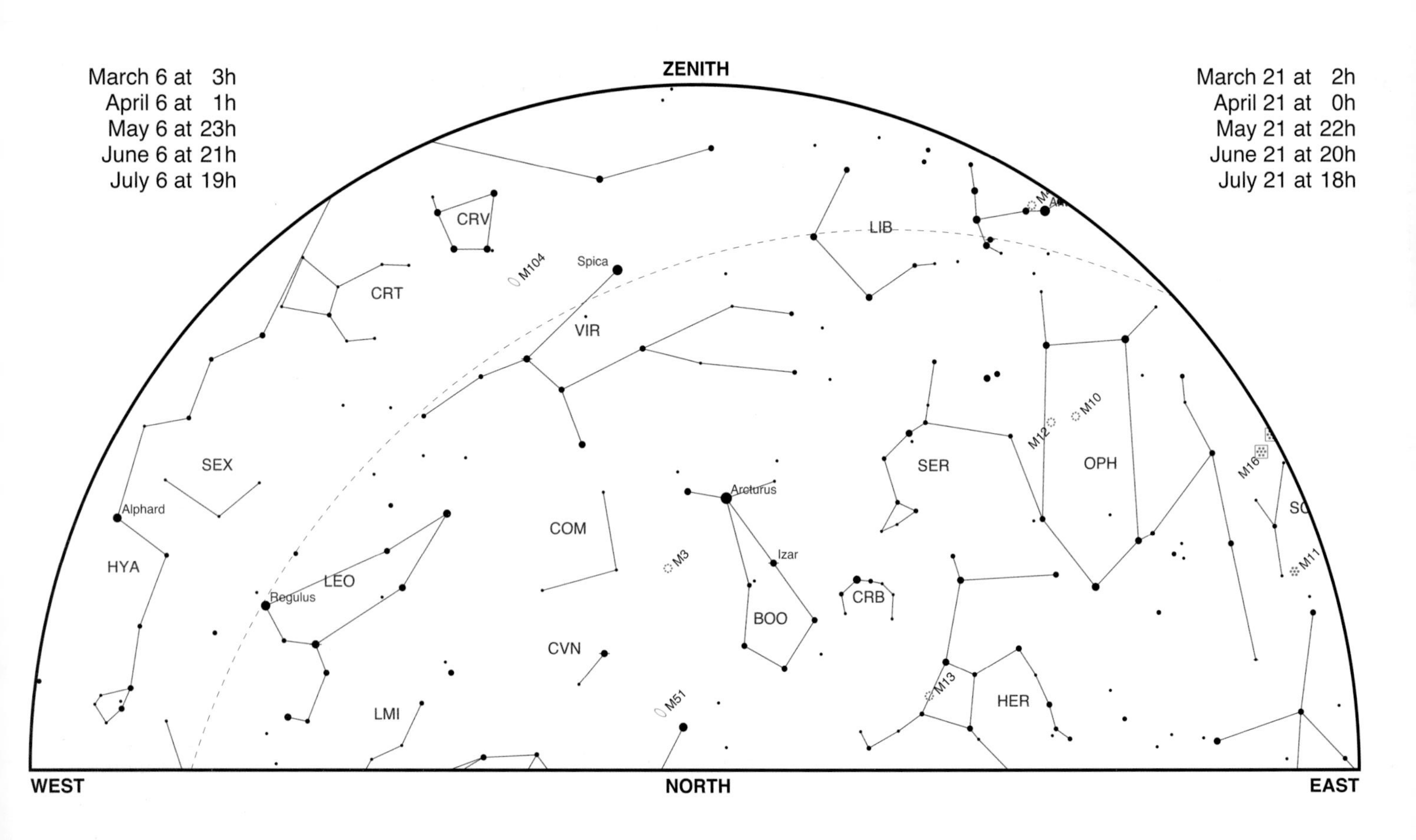
March 6 at 3h
April 6 at 1h
May 6 at 23h
June 6 at 21h
July 6 at 19h
ZENITH
March 21 at 2h
April 21 at 0h
May 21 at 22h
June 21 at 20h
July 21 at 18h
CRV
CRT
M104
Spica
VIR
LIB
SEX
Alphard
HYA
LEO
Regulus
COM
CVN
LMI
M3
M51
Arcturus
Izar
BOO
CRB
SER
M12
M10
OPH
M16
M11
M13
HER
WEST
NORTH
EAST

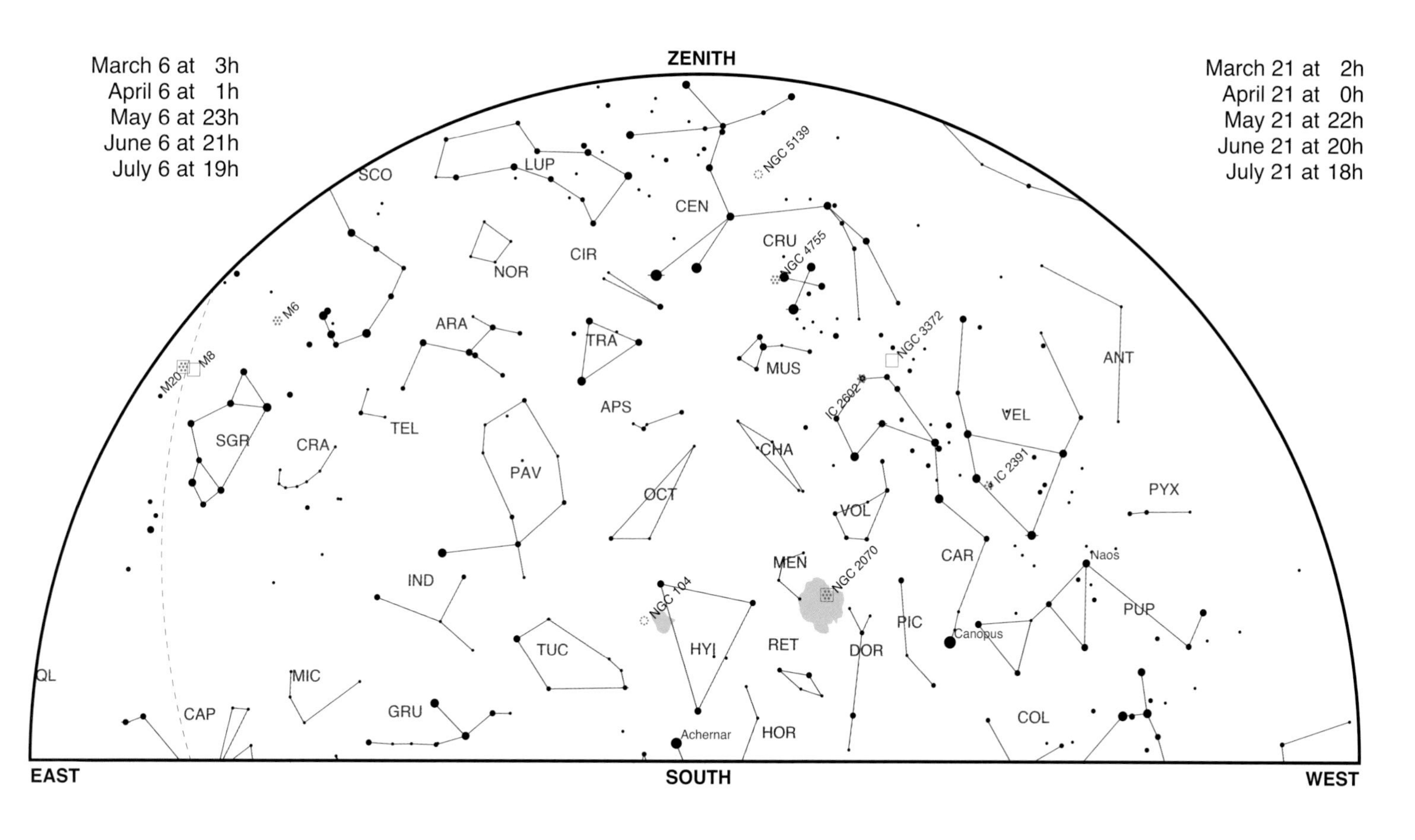
March 6 at 3h
April 6 at 1h
May 6 at 23h
June 6 at 21h
July 6 at 19h
ZENITH
March 21 at 2h
April 21 at 0h
May 21 at 22h
June 21 at 20h
July 21 at 18h
SCO
LUP
CEN
NGC 5139
CRU
NGC 4755
NOR
CIR
ANT
M6
ARA
TRA
MUS
NGC 3372
M20
M8
IC 2602
VEL
APS
TEL
SGR
CRA
CHA
PAV
OCT
IC 2391
PYX
VOL
CAR
Naos
MEN
NGC 2070
IND
NGC 104
PUP
PIC
TUC
HYI
RET
DOR
Canopus
QL
MIC
CAP
GRU
COL
Achernar
HOR
EAST
SOUTH
WEST

5S

6N

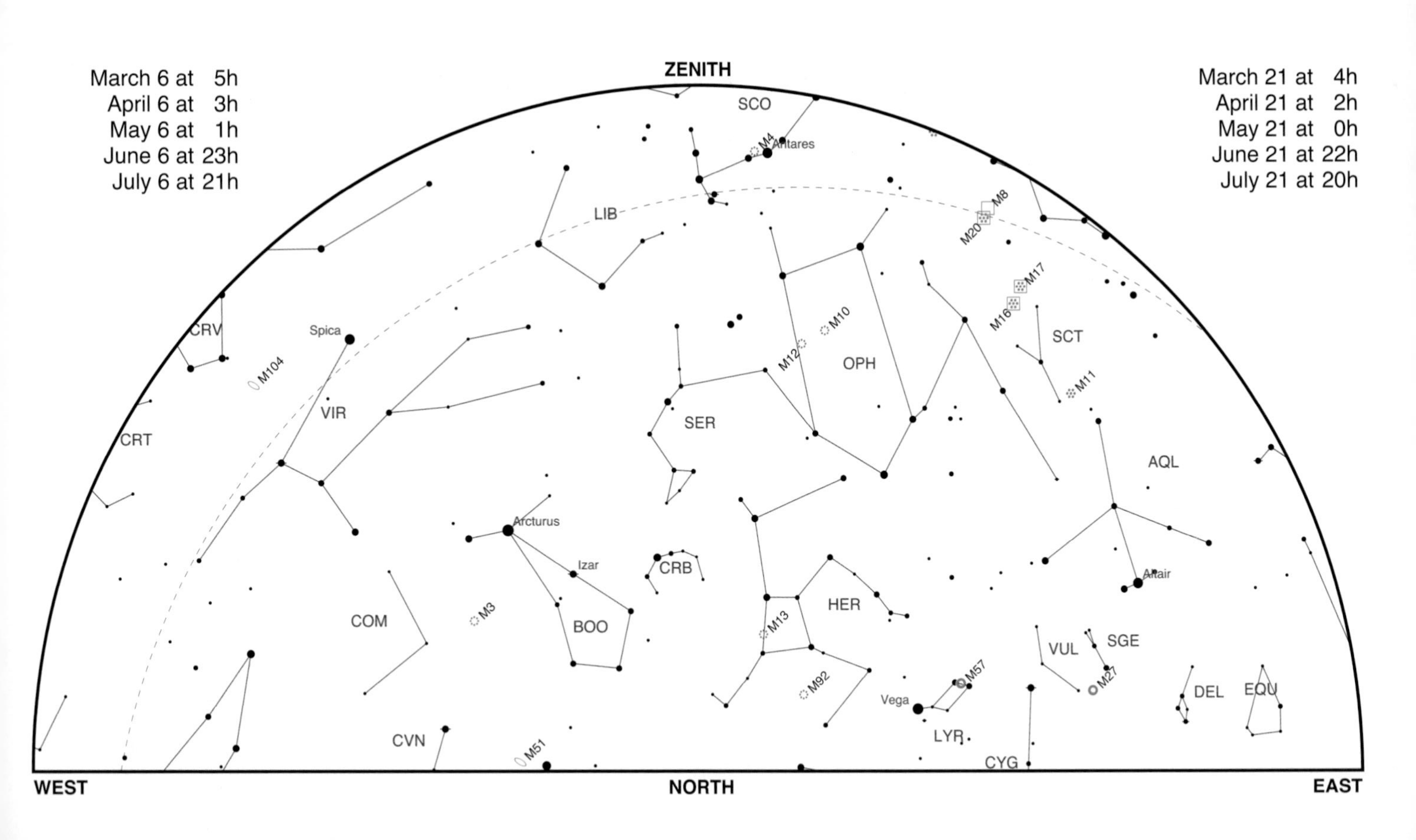
March 6 at 5h
April 6 at 3h
May 6 at 1h
June 6 at 23h
July 6 at 21h
March 21 at 4h
April 21 at 2h
May 21 at 0h
June 21 at 22h
July 21 at 20h
ZENITH
WEST
NORTH
EAST
SCO
Antares
M4
LIB
M8
M20
M17
M16
SCT
M11
M10
M12
OPH
CRV
Spica
M104
VIR
CRT
SER
AQL
Arcturus
Izar
CRB
Altair
HER
M13
COM
M3
BOO
VUL
SGE
M57
M27
M92
DEL
EQU
Vega
LYR
CVN
M51
CYG

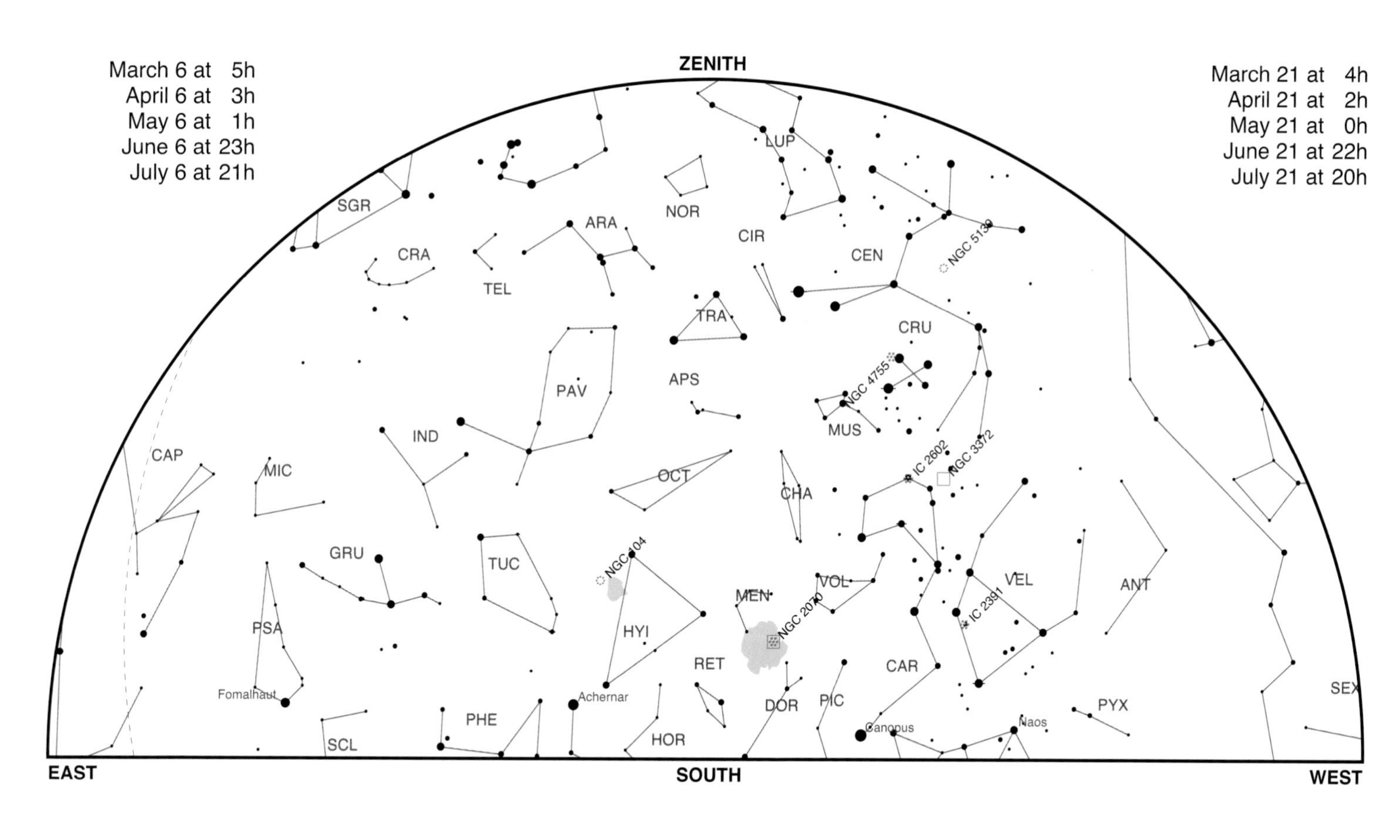

March 6 at 5h
April 6 at 3h
May 6 at 1h
June 6 at 23h
July 6 at 21h
ZENITH
March 21 at 4h
April 21 at 2h
May 21 at 0h
June 21 at 22h
July 21 at 20h
SGR
CRA
TEL
ARA
NOR
LUP
CIR
CEN
NGC 5139
TRA
CRU
NGC 4755
PAV
APS
MUS
IND
CAP
MIC
OCT
CHA
IC 2602
NGC 3372
GRU
TUC
NGC104
HYI
MEN
NGC 2070
VOL
VEL
IC 2391
ANT
PSA
RET
CAR
Fomalhaut
Achernar
DOR
PIC
PYX
SEX
PHE
SCL
HOR
Canopus
Naos
EAST
SOUTH
WEST
6S

7N

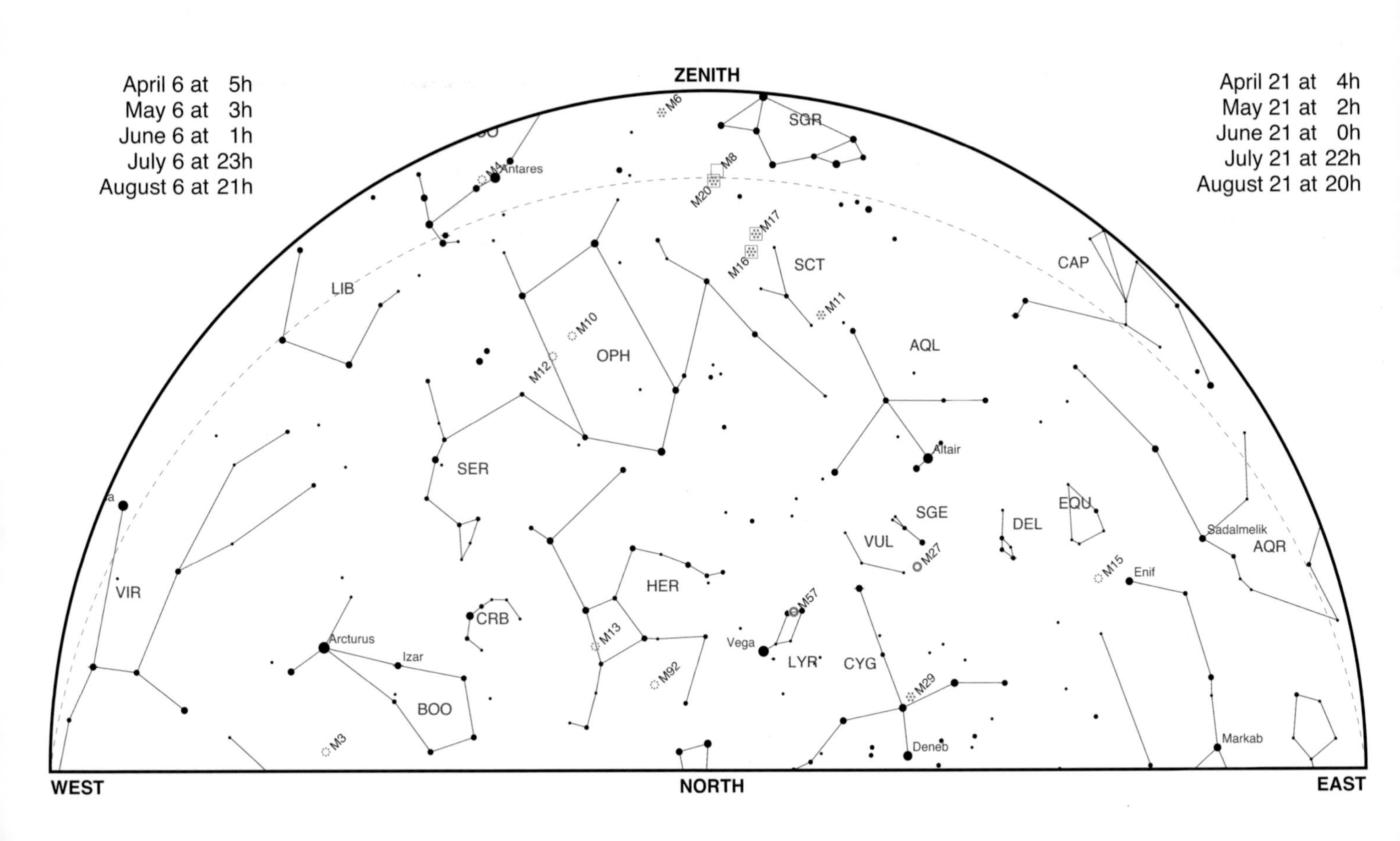
April 6 at 5h
May 6 at 3h
June 6 at 1h
July 6 at 23h
August 6 at 21h
April 21 at 4h
May 21 at 2h
June 21 at 0h
July 21 at 22h
August 21 at 20h
ZENITH
WEST
NORTH
EAST
SGR
M6
M8
M20
M17
M16
SCT
M11
CAP
Antares
M4
LIB
M10
M12
OPH
AQL
Altair
SER
SGE
DEL
EQU
VUL
M27
Sadalmelik
AQR
M15
Enif
HER
M13
M92
M57
Vega
LYR
CYG
M29
Deneb
Markab
VIR
CRB
Arcturus
Izar
BOO
M3

7S

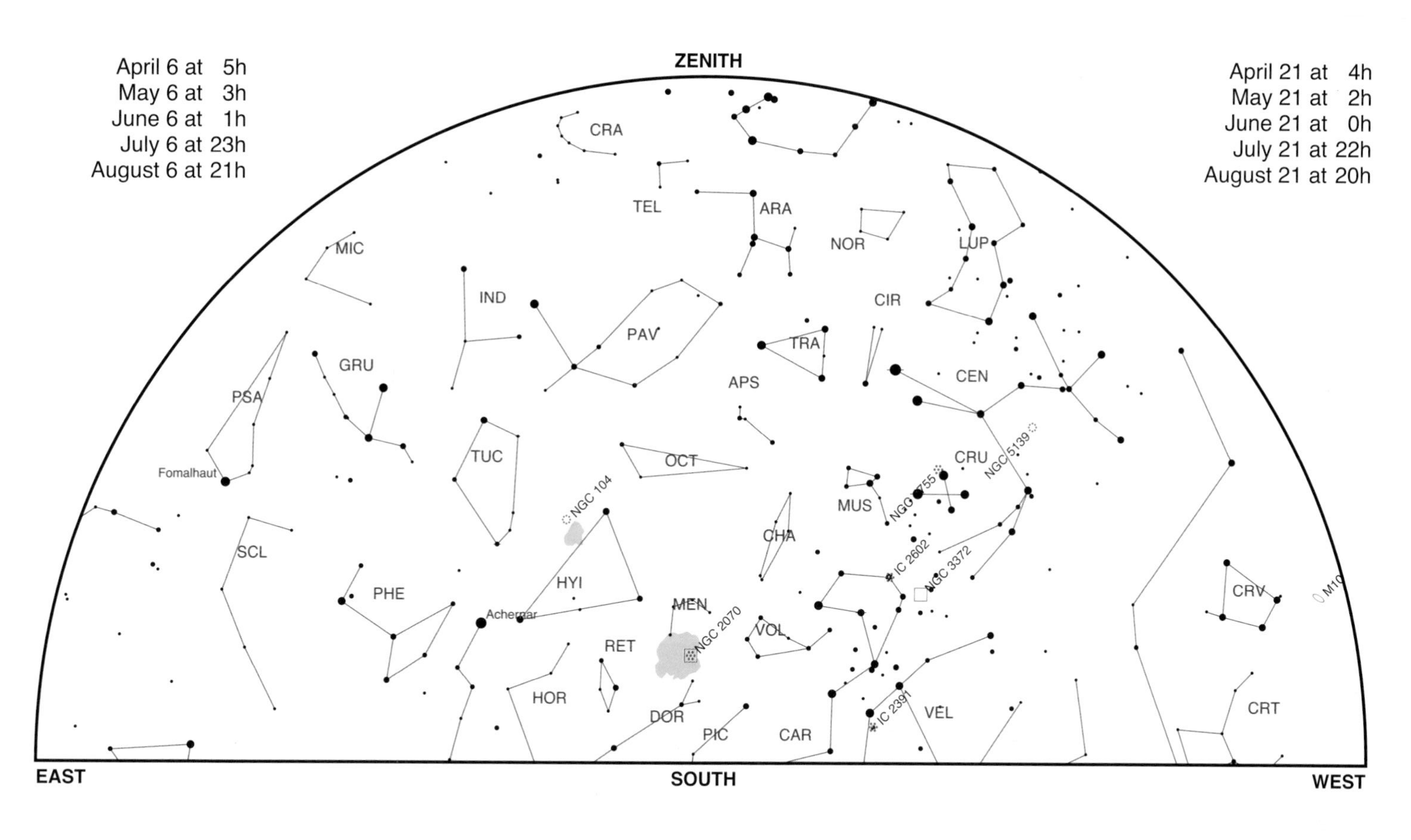
April 6 at 5h
May 6 at 3h
June 6 at 1h
July 6 at 23h
August 6 at 21h
ZENITH
April 21 at 4h
May 21 at 2h
June 21 at 0h
July 21 at 22h
August 21 at 20h
CRA
TEL
ARA
NOR
LUP
MIC
IND
PAV
CIR
GRU
TRA
CEN
PSA
APS
TUC
OCT
CRU
NGC 5139
Fomalhaut
MUS
NGC 4755
NGC 104
CHA
SCL
HYI
IC 2602
NGC 3372
PHE
CRV
M104
Achernar
MEN
NGC 2070
VOL
RET
HOR
DOR
PIC
CAR
IC 2391
VEL
CRT
EAST
SOUTH
WEST

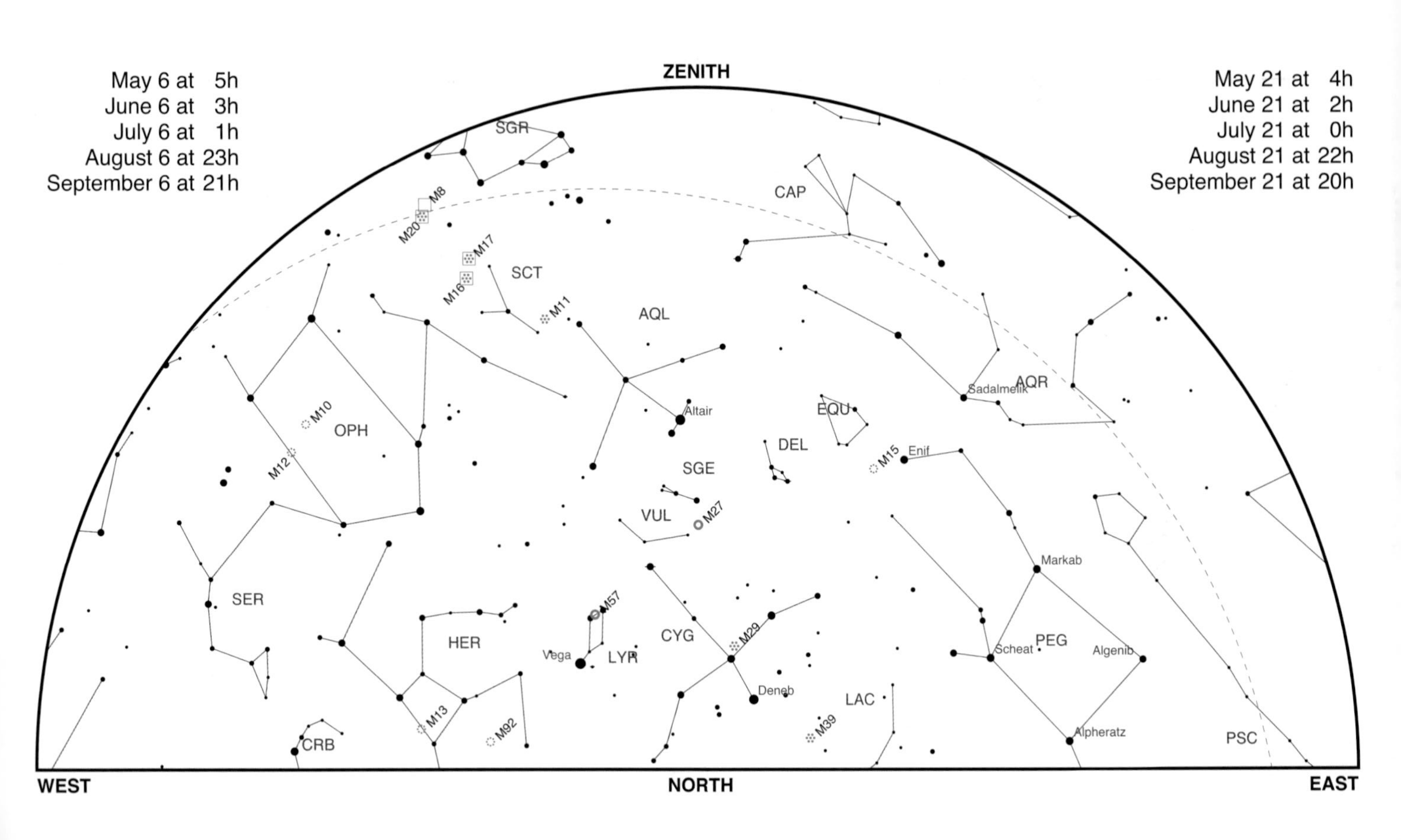
8N
May 6 at 5h
June 6 at 3h
July 6 at 1h
August 6 at 23h
September 6 at 21h
May 21 at 4h
June 21 at 2h
July 21 at 0h
August 21 at 22h
September 21 at 20h
ZENITH
WEST
NORTH
EAST
SGR
M8
M20
M17
M16
SCT
M11
AQL
CAP
AQR
Sadalmelik
OPH
M10
M12
Altair
EQU
DEL
SGE
M15
Enif
VUL
M27
SER
HER
M57
Vega
LYR
CYG
M29
Deneb
LAC
M39
M13
M92
CRB
Markab
Scheat
PEG
Algenib
Alpheratz
PSC

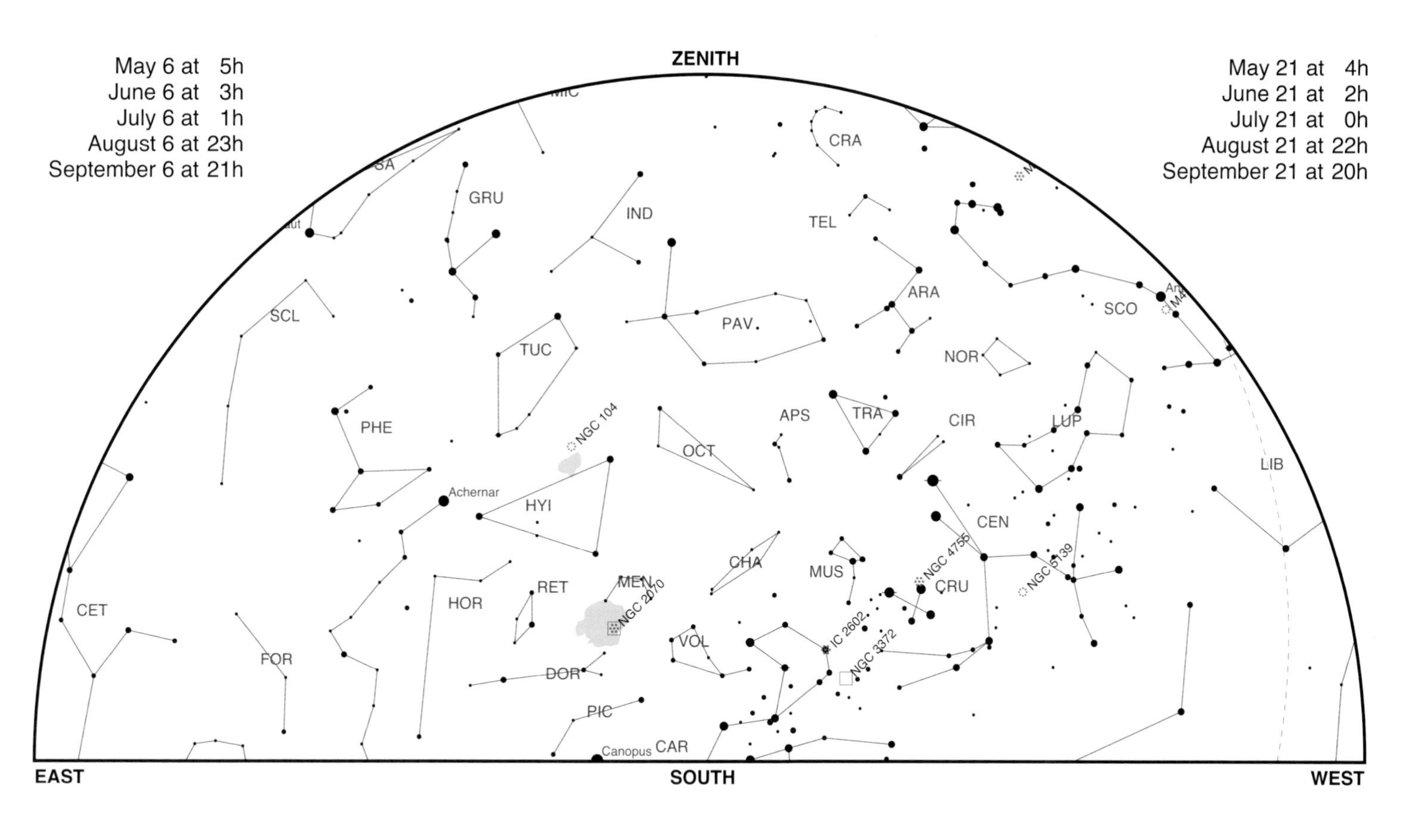

May 6 at 5h
June 6 at 3h
July 6 at 1h
August 6 at 23h
September 6 at 21h
ZENITH
May 21 at 4h
June 21 at 2h
July 21 at 0h
August 21 at 22h
September 21 at 20h
GRU
IND
CRA
TEL
ARA
SCO
SCL
PAV
TUC
NOR
PHE
NGC 104
APS
TRA
CIR
LUP
OCT
LIB
Achernar
HYI
CEN
NGC 4755
NGC 5139
CHA
MUS
CRU
CET
RET
MEN
HOR
NGC 2070
IC 2602
VOL
NGC 3372
FOR
DOR
PIC
Canopus
CAR
EAST
SOUTH
WEST
8S

9N

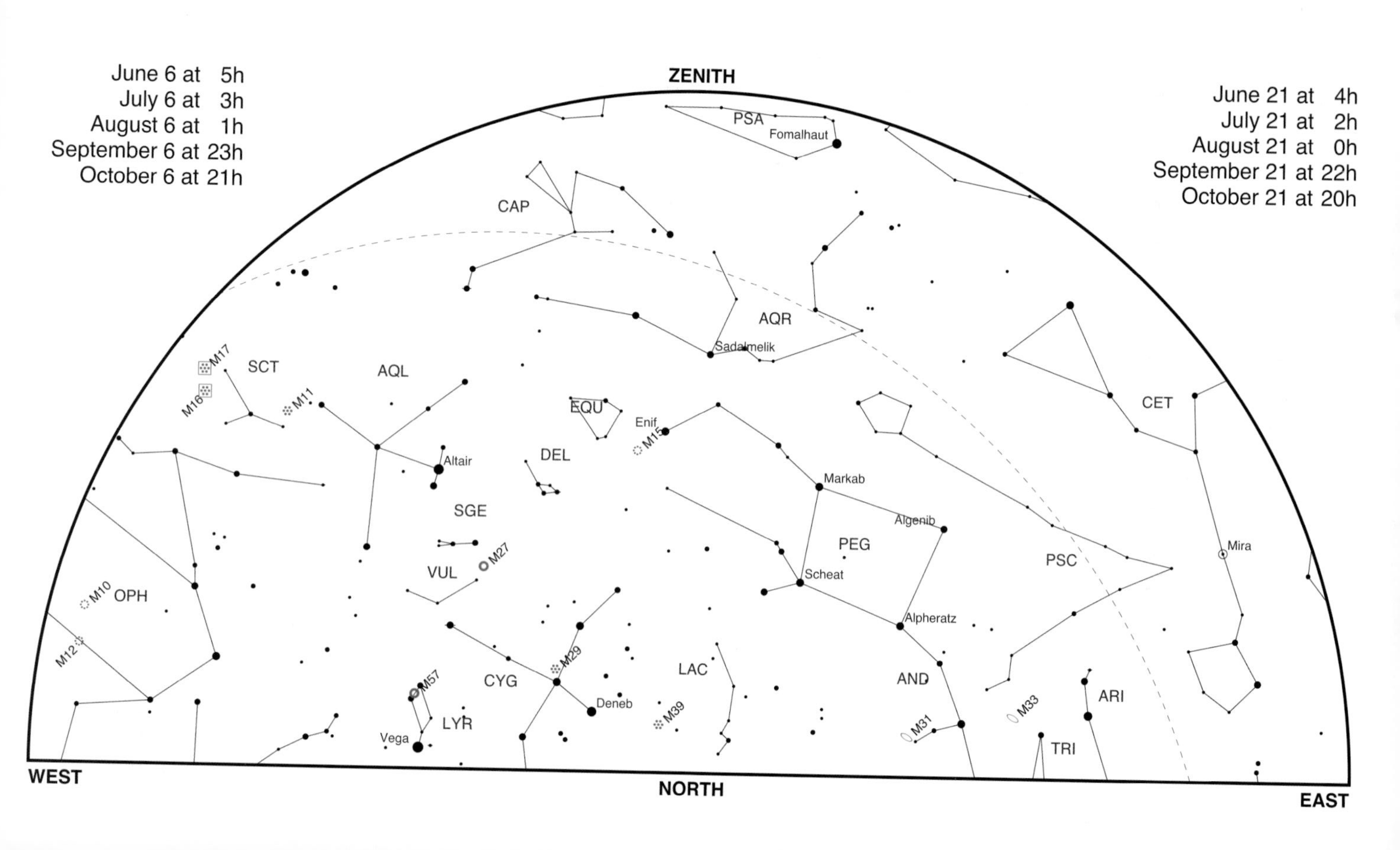
June 6 at 5h
July 6 at 3h
August 6 at 1h
September 6 at 23h
October 6 at 21h
June 21 at 4h
July 21 at 2h
August 21 at 0h
September 21 at 22h
October 21 at 20h
ZENITH
WEST
NORTH
EAST
PSA
Fomalhaut
CAP
AQR
Sadalmelik
CET
SCT
M17
M16
M11
AQL
EQU
Enif
M15
DEL
Altair
SGE
VUL
M27
Markab
Algenib
PEG
Scheat
Alpheratz
PSC
Mira
OPH
M10
M12
CYG
M29
M57
LYR
Vega
Deneb
LAC
M39
AND
M31
M33
ARI
TRI

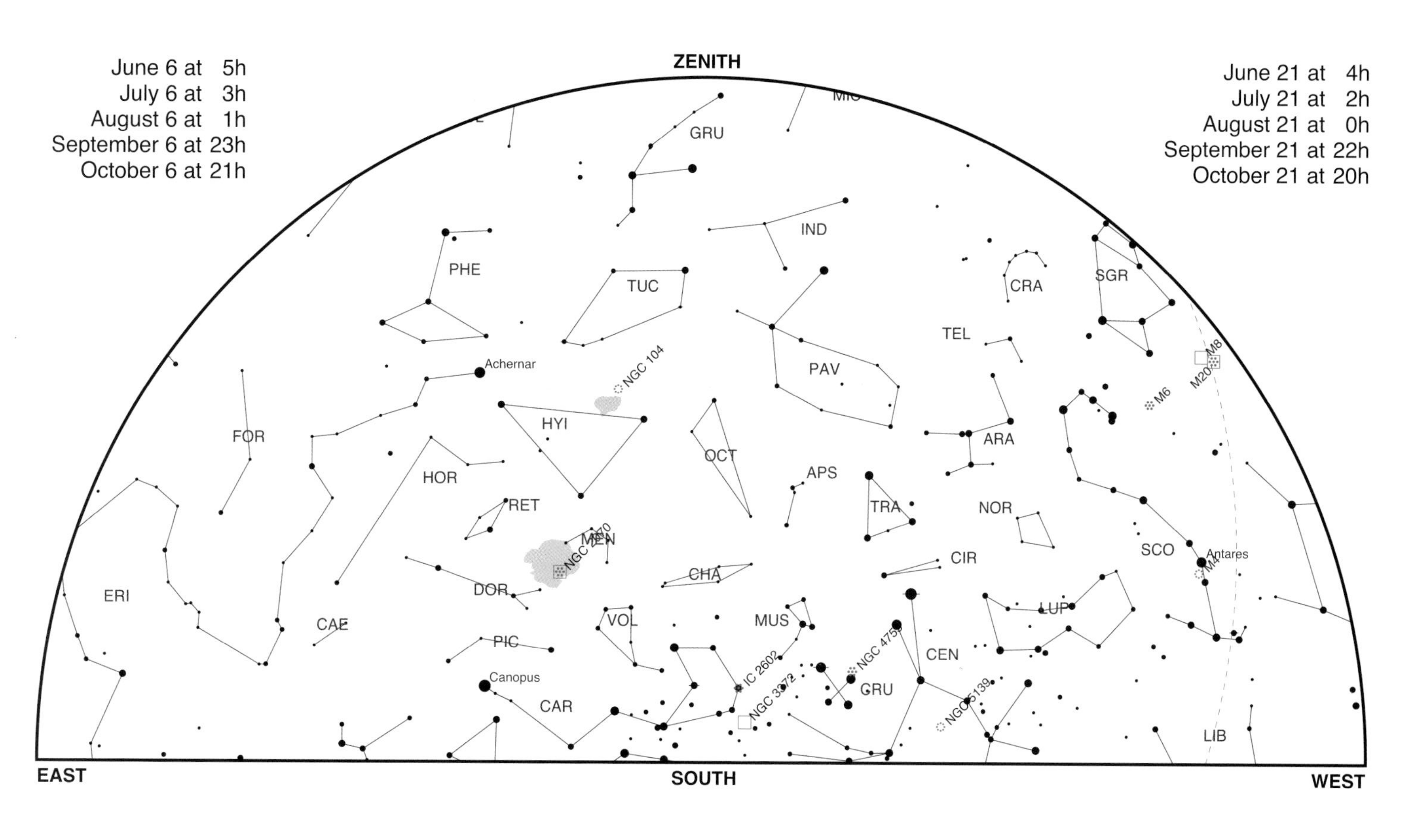
June 6 at 5h
July 6 at 3h
August 6 at 1h
September 6 at 23h
October 6 at 21h
ZENITH
June 21 at 4h
July 21 at 2h
August 21 at 0h
September 21 at 22h
October 21 at 20h
GRU
IND
PHE
TUC
CRA
SGR
TEL
PAV
M8
M20
M6
Achernar
NGC 104
HYI
FOR
ARA
OCT
HOR
APS
TRA
NOR
RET
MEN
NGC 2070
SCO
Antares
M4
CIR
CHA
DOR
ERI
LUP
CAE
VOL
MUS
PIC
NGC 4755
CEN
IC 2602
Canopus
NGC 3372
GRU
NGC 5139
CAR
LIB
EAST
SOUTH
WEST

S6

10N

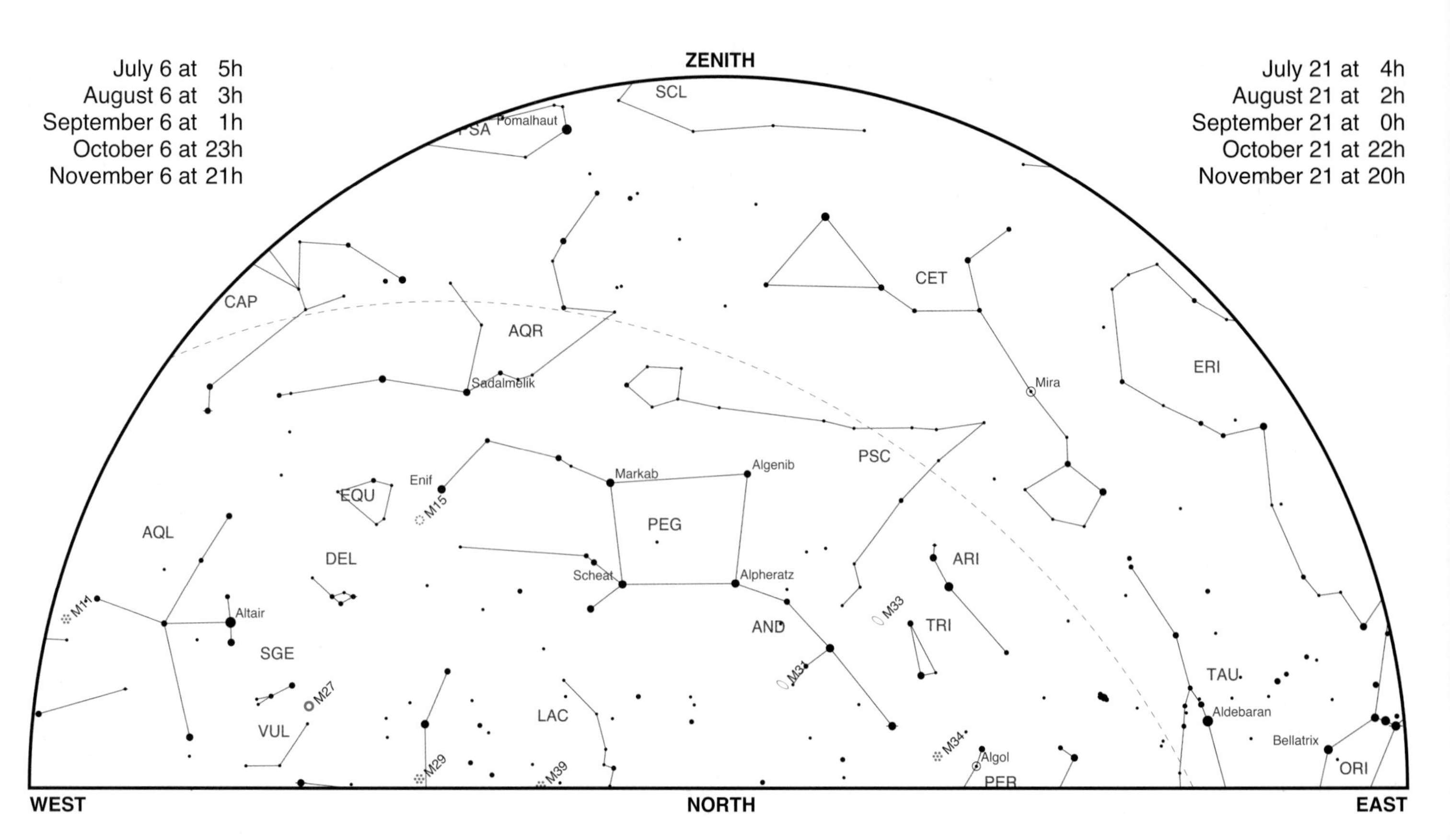
July 6 at 5h
August 6 at 3h
September 6 at 1h
October 6 at 23h
November 6 at 21h
July 21 at 4h
August 21 at 2h
September 21 at 0h
October 21 at 22h
November 21 at 20h
ZENITH
WEST
NORTH
EAST
SCL
PSA
Fomalhaut
CAP
AQR
Sadalmelik
CET
ERI
Mira
PSC
EQU
Enif
M15
Markab
Algenib
PEG
AQL
DEL
Scheat
Alpheratz
ARI
M14
Altair
AND
M33
TRI
SGE
M31
TAU
M27
LAC
Aldebaran
VUL
M34
Bellatrix
M29
M39
Algol
PER
ORI

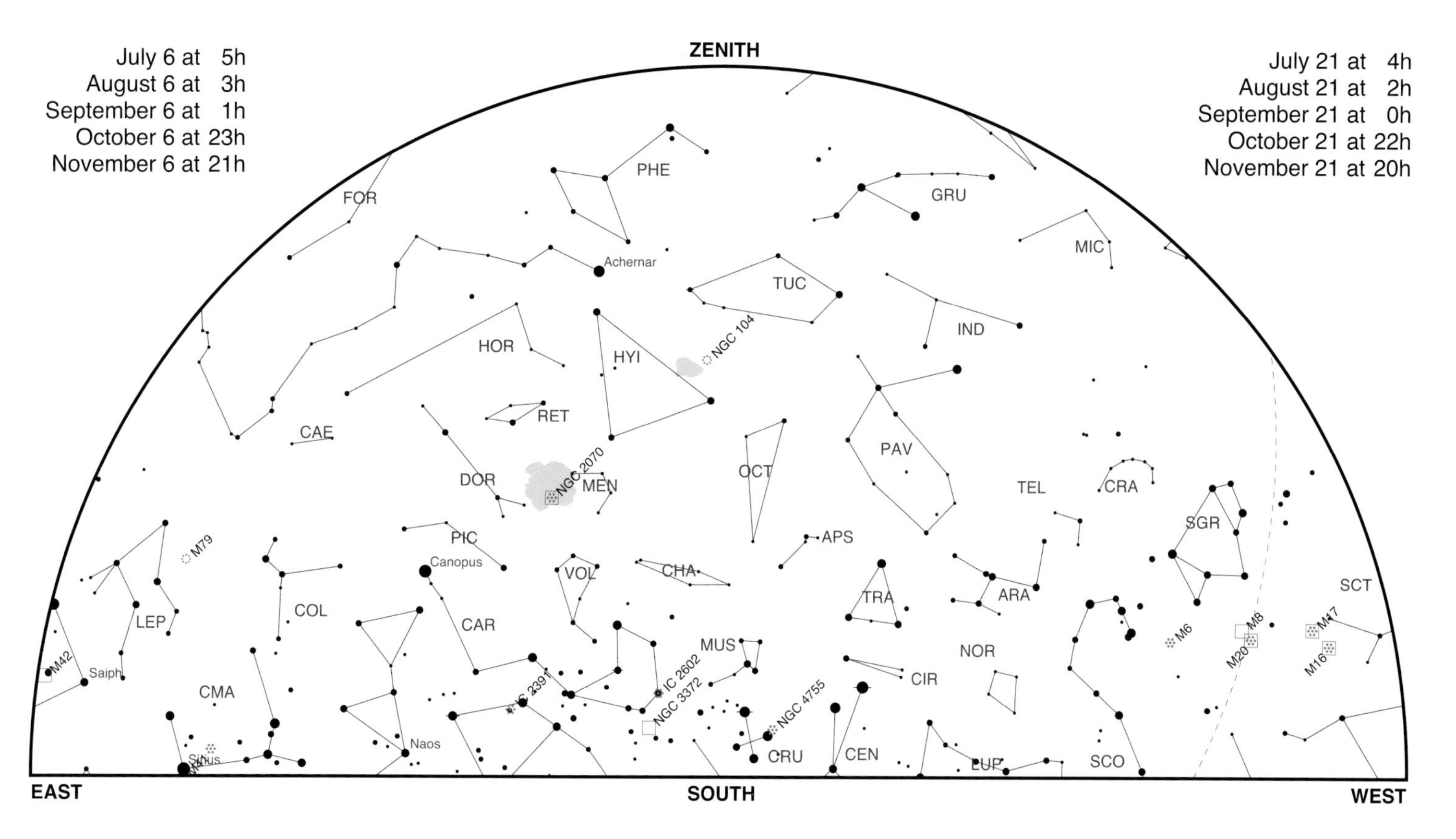
July 6 at 5h
August 6 at 3h
September 6 at 1h
October 6 at 23h
November 6 at 21h
ZENITH
July 21 at 4h
August 21 at 2h
September 21 at 0h
October 21 at 22h
November 21 at 20h
PHE
GRU
FOR
MIC
Achernar
TUC
IND
HOR
HYI
NGC 104
RET
CAE
PAV
DOR
NGC 2070
MEN
OCT
TEL
CRA
SGR
PIC
M79
Canopus
APS
VOL
CHA
TRA
ARA
SCT
LEP
COL
CAR
MUS
NOR
M6
M8
M17
M20
M16
M42
Saiph
CMA
IC 2602
NGC 3372
CIR
NGC 4755
Naos
CRU
CEN
LUP
SCO
EAST
SOUTH
WEST

10S

11N

August 6 at 5h
September 6 at 3h
October 6 at 1h
November 6 at 23h
December 6 at 21h

August 21 at 4h
September 21 at 2h
October 21 at 0h
November 21 at 22h
December 21 at 20h

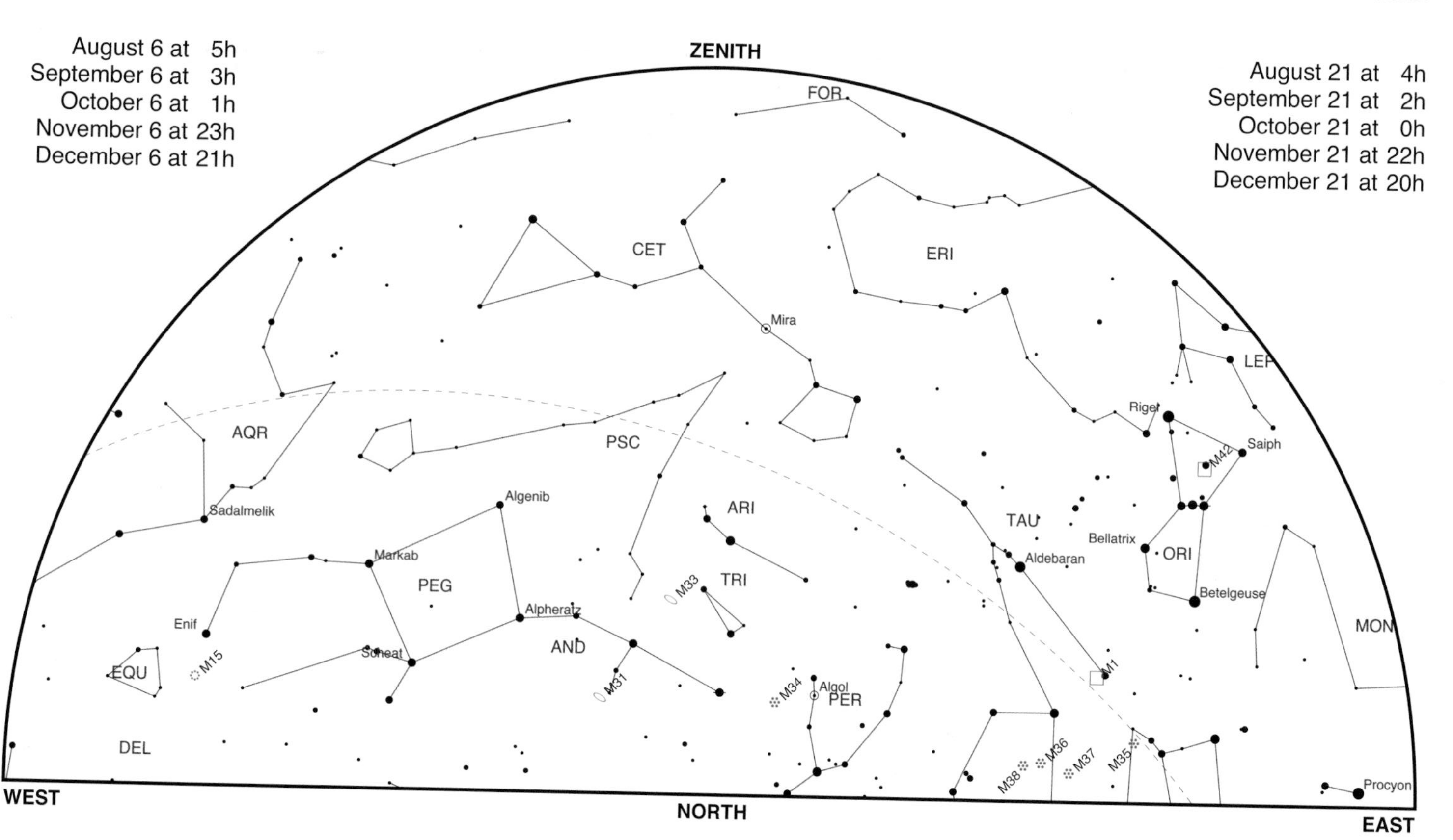

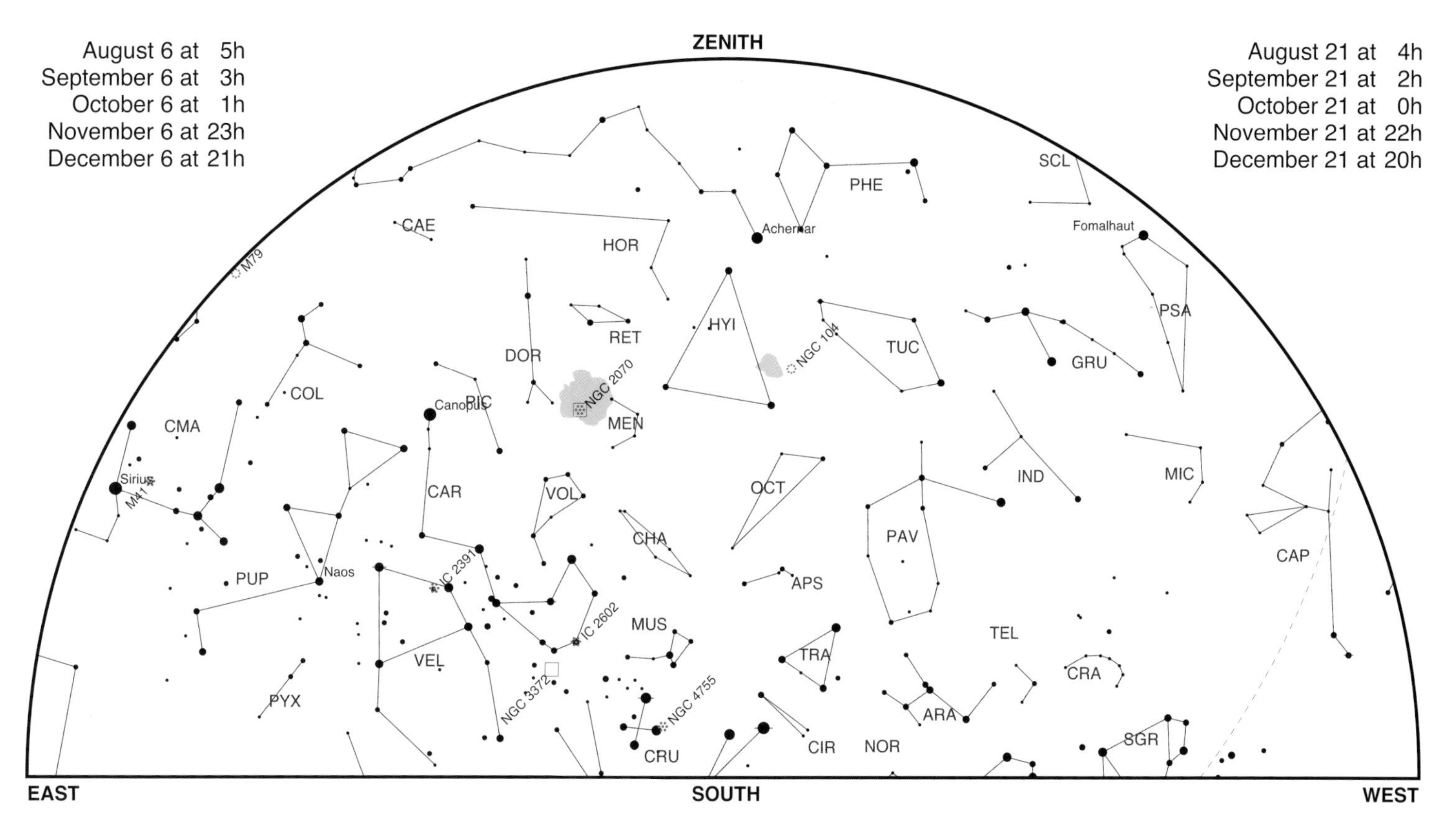
11S
August 6 at 5h
September 6 at 3h
October 6 at 1h
November 6 at 23h
December 6 at 21h
ZENITH
August 21 at 4h
September 21 at 2h
October 21 at 0h
November 21 at 22h
December 21 at 20h
EAST
SOUTH
WEST
CAE
HOR
PHE
SCL
Achernar
Fomalhaut
M79
HYI
RET
DOR
NGC 2070
NGC 104
TUC
GRU
PSA
COL
CMA
Canopus
PIC
MEN
Sirius
M41
CAR
VOL
OCT
IND
MIC
PAV
CAP
CHA
PUP
Naos
IC 2391
APS
MUS
IC 2602
TEL
VEL
TRA
CRA
PYX
NGC 3372
NGC 4755
ARA
SGR
CRU
CIR
NOR

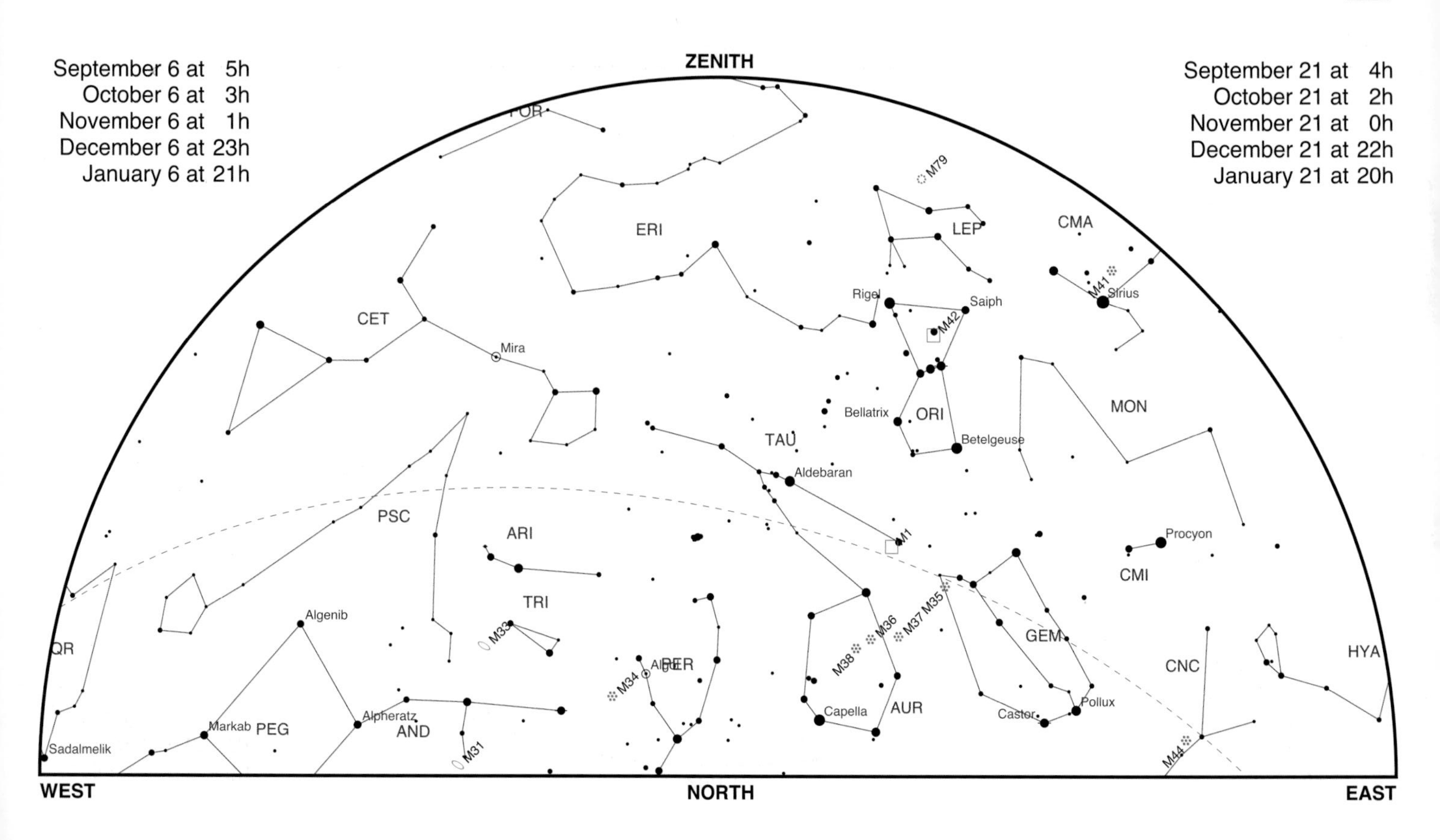
12N
September 6 at 5h
October 6 at 3h
November 6 at 1h
December 6 at 23h
January 6 at 21h
September 21 at 4h
October 21 at 2h
November 21 at 0h
December 21 at 22h
January 21 at 20h
ZENITH
WEST
NORTH
EAST
ERI
CET
Mira
PSC
ARI
TRI
M33
PEG
Markab
Sadalmelik
Algenib
Alpheratz
AND
M31
M34
Algol
PER
TAU
Aldebaran
M1
M79
LEP
Rigel
Saiph
M42
Bellatrix
ORI
Betelgeuse
CMA
M41
Sirius
MON
Procyon
CMI
Capella
AUR
M38
M36
M37
M35
GEM
Castor
Pollux
CNC
M44
HYA

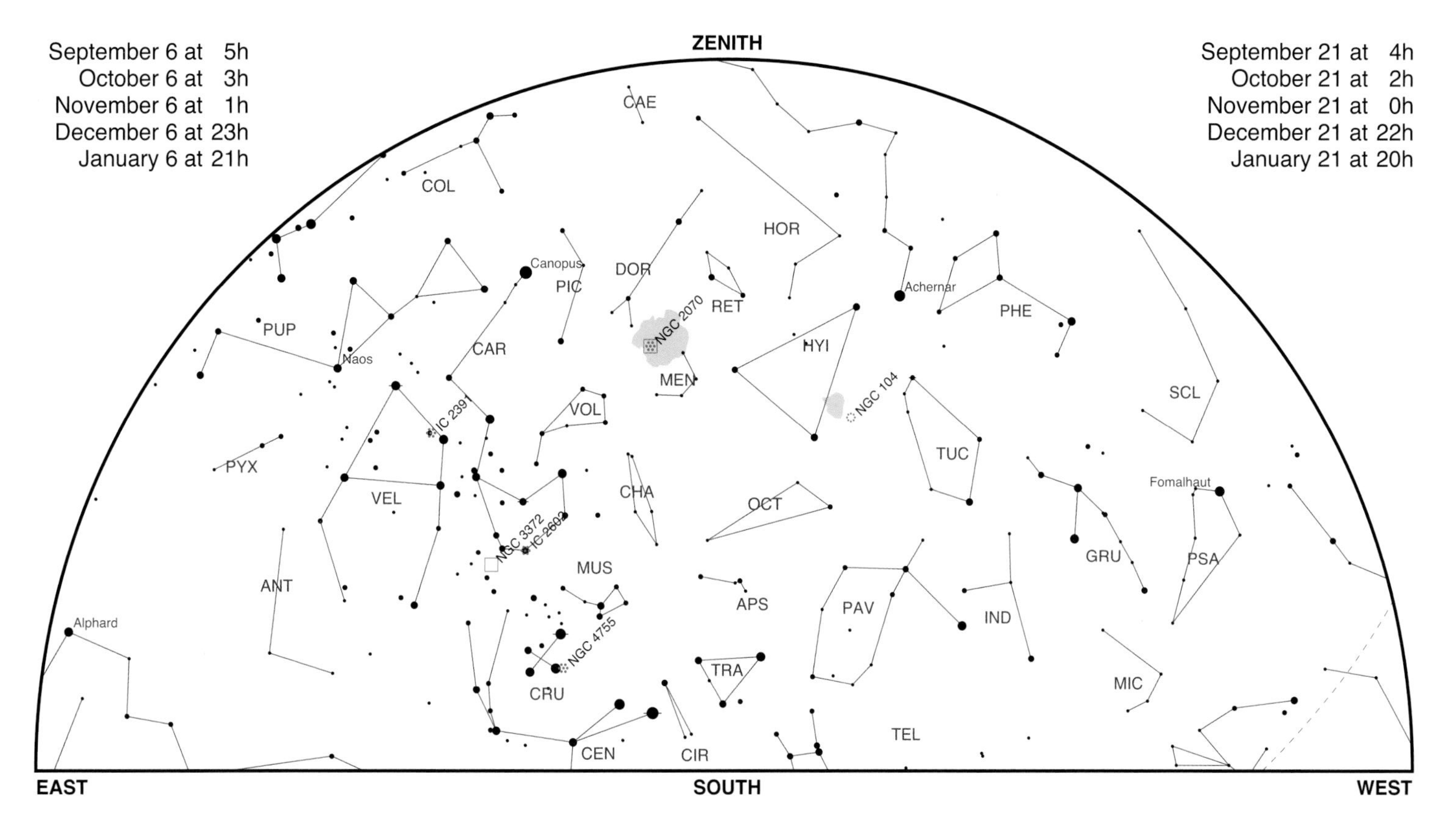
September 6 at 5h
October 6 at 3h
November 6 at 1h
December 6 at 23h
January 6 at 21h
September 21 at 4h
October 21 at 2h
November 21 at 0h
December 21 at 22h
January 21 at 20h
ZENITH
EAST
SOUTH
WEST
COL
CAE
HOR
DOR
RET
PIC
Canopus
Achernar
PHE
PUP
Naos
CAR
NGC 2070
HYI
MEN
NGC 104
SCL
VOL
IC 2391
TUC
PYX
VEL
CHA
OCT
Fomalhaut
GRU
PSA
NGC 3372
IC 2602
MUS
ANT
APS
PAV
IND
Alphard
NGC 4755
TRA
MIC
CRU
CEN
CIR
TEL

12S

The Planets in 2019

Lynne Marie Stockman

Mercury is difficult to spot as it is never far from the Sun. Northern latitude observers will have their best chance of seeing Mercury in the evening twilight during June, whilst November and December offer the best morning apparition. Planet watchers in the southern hemisphere should look for Mercury in the morning twilight in March and April, and in the evening in October. The inferior conjunction on 11 November is particularly noteworthy because Mercury will transit the face of the Sun.

Venus is at greatest elongation west on 6 January, when it is 47° from the Sun. It is a brilliant object in the morning sky between January and July, and is very well placed for southern hemisphere observers during this period. From northern temperate latitudes, however, this will be a disappointing apparition because Venus is always close to the horizon in the pre-dawn sky. It slowly draws toward to the Sun until it passes through superior conjunction in August. It then reappears in September after sunset to take its place as the evening star, gradually increasing in brightness through to the end of the year.

Mars does not come to opposition in 2019 so it is a quiet year for the red planet. It spends the first half of the year in the evening sky, passing through the constellations of Pisces, Aries, Taurus, Gemini, Cancer and Leo before the Sun finally catches up with it at conjunction in September. The following month, it reappears in the morning sky in Virgo and ends the year in Libra. Its next opposition occurs in October 2020.

Jupiter is a morning sky object early in the year. It is in Ophiuchus and Sagittarius during 2019, so it is best placed for viewers in the tropics and the southern hemisphere. From mid-March it rises before midnight in southern latitudes, and reaches opposition in June. Shining brightly in the evening sky from then until its December conjunction, it is found 1.4° north of Venus on 24 November (Venus being two magnitudes brighter).

Saturn is at conjunction at the beginning of 2019 and does not become visible in the morning sky until February. It rises before midnight from April when observed from the southern hemisphere, but viewers in northern temperate latitudes must wait until June before the ringed planet can be seen in the evening sky. Saturn reaches opposition in early July. It remains in Sagittarius all year, and from northern temperate latitudes, it never rises very high above the horizon. Saturn's rings were at their most open in 2017 when they were tilted almost 27° with respect to Earth. They have now begun to close but the tilt still exceed 24° for most of the year, reaching 25° in September and October. Saturn remains an impressive sight in small- or medium-sized telescopes. It undergoes a series of lunar occultations this year, beginning in January and ending in November.

Uranus is just visible to the naked eye under perfect seeing conditions. It is an evening sky object at the start of 2019, undergoing conjunction in April and reappearing in the morning sky afterwards. Uranus begins rising before midnight in August and opposition occurs in October. Beginning the year in Pisces, it moves to Aries in February where it stays for the rest of the year.

Neptune is an eighth-magnitude object and a telescope is always necessary to observe it. Like Uranus, Neptune is an evening sky object at the start of 2019 but is quickly lost in the Sun's glare, undergoing conjunction in March. It moves into the morning sky, rising ever earlier ahead of the Sun until June or July when it begins rising before midnight. Neptune reaches opposition in September and remains in Aquarius for the entire year.

Phases of the Moon in 2019

Month	New Moon	First Quarter	Full Moon	Last Quarter
January	6	14	21	27
February	4	12	19	26
March	6	14	21	28
April	5	12	19	26
May	4	12	18	26
June	3	10	17	25
July	2	9	16	25
August	1 and 30	7	15	23
September	28	6	14	22
October	28	5	13	21
November	26	4	12	19
December	26	4	12	19

Eclipses in 2019

On 5/6 January there will be a partial solar eclipse visible in parts of eastern Asia and the northern Pacific Ocean and the extreme west of Alaska including the Aleutian Islands. The eclipse begins at 23.34 UT on 5 January and ends at 03.49 UT on 6 January with maximum eclipse at 01:41 UT.

There will be a total lunar eclipse on 21 January visible throughout North America, South America, the eastern Pacific Ocean and western Atlantic Ocean, Greenland and Iceland, far western and north western Europe (including the UK), extreme northern Asia and westernmost parts of Africa. The eclipse begins at 02:36 UT and ends at 07:48 UT. The total phase begins at 04:41 UT and ends at 05:43 UT.

The path of totality of the total solar eclipse occurring on 2 July will only be visible from parts of the southern Pacific Ocean, central Chile and central Argentina, although a partial eclipse will be visible from most of the southern Pacific Ocean and western South America. The eclipse commences at 16:55 UT and ends at 21:50 UT, the total phase beginning at 18:01 UT and ending at 20:44 UT with maximum eclipse occurring at 19:23 UT.

On 16/17 July there will be a partial lunar eclipse with some if not all of the eclipse visible throughout southern and eastern North America, South America, most of Europe and Asia, Africa, Australia, the Indian Ocean and Antarctica. The penumbral eclipse begins at 18:43 UT on 16 July and ends at 00:17 UT on 17 July. The partial eclipse takes place between 20:02 UT and 23:00 UT on 16 July, with maximum eclipse at 21:31 UT.

The path of the annular solar eclipse taking place on 26 December will begin in Saudi Arabia and move eastwards across southern India and northern Sri Lanka, parts of the Indian Ocean, and Indonesia before ending in the eastern Pacific Ocean. A partial eclipse will be visible throughout most of Asia, far eastern Africa and northern Australia. The eclipse begins at 02:30 UT and ends at 08:05 UT with the full eclipse lasting from 03:34 UT to 07:01 UT and maximum eclipse at 05:17 UT.

Some Events in 2019

Month	Date	Object	Event
January	2	Saturn	Conjunction
	3	Earth	Perihelion (0.983 AU)
	3/4	Earth	Quadrantid Meteor Shower (ZHR 120)
	5/6	Earth/Moon	Partial Solar Eclipse
	6	Venus	Greatest Elongation West (47°)
	19	Uranus	East Quadrature
	21	Earth/Moon	Total Lunar Eclipse
	30	Mercury	Superior Conjunction
	31	Venus	Lunar Occultation
February	2	Saturn	Lunar Occultation
	5	Moon	Farthest Apogee of the Year (406,556 km)
	19	Moon	Closest Perigee of the Year (356,762 km)
	27	Mercury	Greatest Elongation East (18°)
March	1	Saturn	Lunar Occultation
	7	Neptune	Conjunction
	14	Jupiter	West Quadrature
	15	Mercury	Inferior Conjunction
	20	Earth	Equinox
	29	Saturn	Lunar Occultation
April	10	Saturn	West Quadrature
	10	2 Pallas	Opposition in Boötes
	11	Mercury	Greatest Elongation West (28°)
	22/23	Earth	Lyrid Meteor Shower (ZHR 18)
	23	Uranus	Conjunction
	25	Saturn	Lunar Occultation
May	6/7	Earth	Eta Aquarid Meteor Shower (ZHR 30)
	21	Mercury	Superior Conjunction
	22	Saturn	Lunar Occultation
	28	1 Ceres	Opposition in Ophiuchus
June	9	Neptune	West Quadrature
	10	Jupiter	Opposition in Ophiuchus
	19	Saturn	Lunar Occultation
	21	Earth	Solstice
	23	Mercury	Greatest Elongation East (25°)

Month	Date	Object	Event
July	2	Earth/Moon	Total Solar Eclipse
	4	Mars	Lunar Occultation
	4	Earth	Aphelion (1.017 AU)
	9	Saturn	Opposition in Sagittarius
	14	134340 Pluto	Opposition in Sagittarius
	16	Saturn	Lunar Occultation
	16/17	Earth/Moon	Partial Lunar Eclipse
	21	Mercury	Inferior Conjunction
	28/29	Earth	Delta Aquarid Meteor Shower (ZHR 20)
	29	Uranus	West Quadrature

Month	Date	Object	Event
August	9	Mercury	Greatest Elongation West (19°)
	12	Saturn	Lunar Occultation
	12/13	Earth	Perseid Meteor Shower (ZHR 80)
	14	Venus	Superior Conjunction

Month	Date	Object	Event
September	2	Mars	Conjunction
	4	Mercury	Superior Conjunction
	8	Saturn	Lunar Occultation
	8	Jupiter	East Quadrature
	10	Neptune	Opposition in Aquarius
	23	Earth	Equinox

Month	Date	Object	Event
October	5	Saturn	Lunar Occultation
	7	Saturn	East Quadrature
	20	Mercury	Greatest Elongation East (25°)
	21/22	Earth	Orionid Meteor Shower (ZHR 20)
	28	Uranus	Opposition in Aries

Month	Date	Object	Event
November	2	Saturn	Lunar Occultation
	5/6	Earth	Taurid Meteor Shower (ZHR 5)
	11	Mercury	Inferior Conjunction – Transit
	12	4 Vesta	Opposition in Cetus
	17/18	Earth	Leonid Meteor Shower (ZHR varies)
	28	Mercury	Greatest Elongation West (20°)
	28	Jupiter	Lunar Occultation
	29	Saturn	Lunar Occultation

Month	Date	Object	Event
December	8	Neptune	East Quadrature
	13/14	Earth	Geminid Meteor Shower (ZHR 75+)
	22	Earth	Solstice
	26	Earth/Moon	Annular Solar Eclipse
	27	Jupiter	Conjunction
	29	Venus	Lunar Occultation

The entries for meteor showers state the date of peak shower activity (maximum). The figures quoted in brackets in column 4 alongside each meteor shower entry are the expected Zenith Hourly Rate (ZHR) for that particular shower at maximum. For a more detailed explanation of ZHR, and for further details of the individual meteor showers listed here, please refer to the article *Meteor Showers in 2019* located elsewhere in this volume.

Monthly Sky Notes and Articles

January

New Moon: 6 January
Full Moon: 21 January

MERCURY begins the year at magnitude −0.4 low in the east before sunrise. Heading towards superior conjunction on 30 January, this tiny planet is lost to view early in the month. It reaches aphelion, the first of five this year, on 12 January and on the following day is only 1.7° south of Saturn. However, this event will likely go unobserved as both planets are very near to the Sun at that time.

VENUS is the brilliant morning star, magnitude −4.5, located 1.3° south of the waning crescent Moon on the first day of the year. Greatest elongation west occurs on 6 January when the planet is 47° from the Sun. This is also the time when Venus goes from its waxing crescent phase to waxing gibbous. It is highest in the sky when seen from southern latitudes and increasing in altitude every morning, rising three hours or more ahead of the Sun. Northern hemisphere observers are not so lucky at this apparition, with Venus remaining close to the pre-dawn eastern horizon until superior conjunction in August. On 22 January, Venus is located 2.4° north of Jupiter. It has an even closer encounter on the last day of the month with our satellite the Moon. Assorted islands in the western Pacific will actually witness Venus being occulted by the waning crescent Moon, beginning around 15:00 UT.

EARTH reaches the point in its orbit where it is closest to the Sun (perihelion) on 3 January. Our planet also enjoys two eclipses, a partial solar eclipse on 5–6 January and a total lunar eclipse on 21 January.

MARS starts 2019 in the constellation of Pisces, near the asterism known as the "Circlet". It is quite bright at the beginning of the year (magnitude +0.5) and easily visible in the evening sky from both hemispheres. On 15 January, Mars reaches the ascending node in its orbit, passing from south of the ecliptic to the north where it will remain for the rest of the year.

JUPITER spends the first half of the year growing brighter, starting out at magnitude −1.8 this month. Found in the constellation of Ophiuchus, the largest planet in the solar system is a morning sky object, rising just ahead of the Sun. It has two encounters with the waning crescent Moon, once on 3 January and again on 30 January. In between these two events, it pairs up with bright Venus, the morning star, on 22 January, when the two planets are less than 3° apart. Because of Jupiter's location against the background stars, southern hemisphere observers will get much the best views of this planet this year.

SATURN is at conjunction on the second day of the month and is occulted by the Moon three days later in what is the first of 13 such events this year. However, the ringed planet is too close to the Sun to be seen for most of January so this lunar occultation will go unobserved. Saturn appears very near Mercury on 13 January in the morning sky but only observers in the southern hemisphere may be able to view it. Saturn spends the year in the constellation of Sagittarius.

URANUS begins the year in Pisces, barely visible to the naked eye at magnitude +5.8. It reaches a stationary point on 7 January, when it finishes retrograde movement and returns to direct motion. East quadrature occurs on 19 January. It is reasonably placed in the evening sky for convenient viewing from both hemispheres.

NEPTUNE continues to reside in the constellation of Aquarius this year, always around eighth-magnitude. It is an evening sky object and is best seen from northern latitudes where the Sun sets early.

A Closer Look at Lepus: Hungry Hares or Thirsty Camels?

Brian Jones

One of the 48 constellations listed by the 2nd century astronomer Ptolemy in around 150 AD, the comparatively small but nonetheless distinct pattern of stars forming Lepus (the Hare) is visible immediately to the south of the

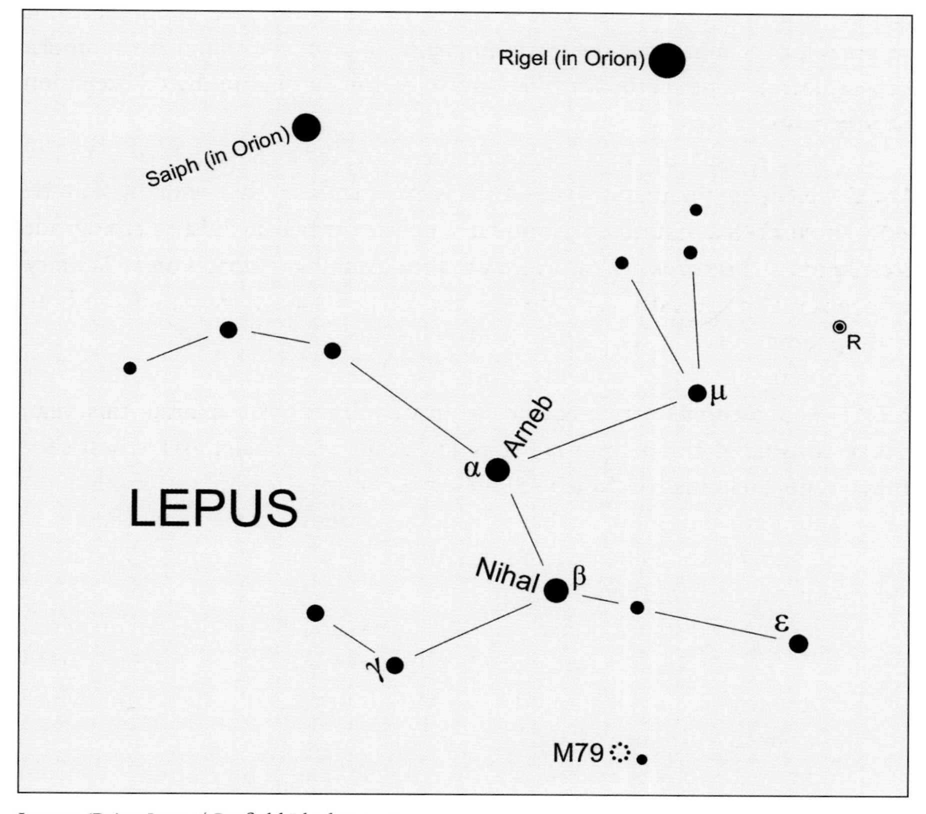

Lepus. (Brian Jones/Garfield Blackmore)

prominent constellation Orion. The two bright stars Rigel and Saiph, located at the southern part of Orion, are shown on the chart, and act as useful pointers to the group. Lepus can be viewed in its entirety from latitudes south of 63°N putting it within the visual reach of observers in central Canada, northern Europe and central Russia and from anywhere further south of these regions.

In his book *Star Lore – Myths, Legends and Facts* the American writer and amateur astronomer William Tyler Olcott described Lepus as '... the timid Hare fleeing before the Hounds of Orion'. This somewhat vivid depiction reflects the proximity of the celestial hare to the mighty Orion and his two canine companions Canis Major and Canis Minor, and is borne out by the most popular of the numerous accounts explaining the origins of Lepus, which came from the Greeks who lived on Sicily. In early times there appears to have been a great deal of crop devastation wrought by the local hare population, the response of the inhabitants of the island being to place the animal up in the sky quite close to Orion, in the expectation that the mighty hunter could keep their numbers under control!

To the Egyptians, the stars in Lepus represented the legendary Boat of Osiris, the powerful Egyptian god whose form was depicted by the group of stars we now know as Orion. Arabic astronomers, on the other hand, likened the group to a herd of thirst-slaking camels who were drinking from the nearby Milky Way.

The brightest star in Lepus is the magnitude 2.58 supergiant Arneb (α Leporis). Deriving its name from the Arabic ***'al-arnab'*** meaning 'the Hare', the light from Arneb reaches us from a distance of around 2,200 light years. Somewhat closer to us is the yellow-white giant Nihal (β Leporis), the magnitude 2.81 glow of which set off on its journey towards us around 160 years ago. Reflecting how Arabic astronomers identified the constellation as a whole, Nihal takes its name from the Arabic ***'al-nihal'*** meaning 'the Camels Beginning to Quench Their Thirst'.

Shining from a distance of just over 200 light years, the orange-red giant star Epsilon (ε) Leporis glows at magnitude 3.19 making it slightly brighter, and a little more remote, than magnitude 3.29 Mu (μ) Leporis, the light from which reaches us from a distance of 186 light years.

Gamma (γ) Leporis is a yellowish-white star shining at magnitude 3.59 from a distance of only 29 light years. Closer examination will reveal that Gamma has a magnitude 6.28 companion star with a distinctly yellow-orange tint. The pair can be split with powerful binoculars or a small telescope and presents an attractive colour contrast to the observer.

R Leporis – Hind's Crimson Star

The Mira-type variable star R Leporis is located near the western boundary of Lepus. Also known as Hind's Crimson Star, the variability of R Leporis was discovered in 1845 by the English astronomer John Russell Hind after whom the star gained its popular name. Shining from a distance in excess of 1,000 light years, R Leporis varies in brightness between magnitudes 5.5 and 11.5 over a period of around 430 days, and observation under dark skies either with binoculars or a small telescope should enable you to follow it through its entire period of variability. If you search for R Leporis and have difficulty in finding it, this may well be because it is at or near minimum magnitude. With a little patience, and given the fact that the star is distinguished by its strong red colour, Hind's Crimson Star should slowly but surely come into view as it brightens again.

John Russell Hind. (Wikimedia Commons/ Henry Joseph Whitlock)

Messier 79 – A Fine Globular Cluster

Lepus plays host to the beautiful globular cluster Messier 79 (NGC 1904) which can be detected as a hazy star-like object lying roughly on a line taken from Arneb, through Nihal, and projected for approximately the same distance again. One of around 25 deep sky objects discovered by the French astronomer Pierre François André Méchain, M79 first came to light on 26 October 1780, although evidence seems to suggest that it may have been recorded by the Italian astronomer Giovanni Batista Hodierna prior to 1654.

Messier 79. (Wikimedia Commons / NASA / HST / Fabian RRRR)

Observation has shown that M79 contains some 150,000 stars spread over a diameter of around 100 light years. Lying at a distance of a little over 40,000 light years, M79 can be detected in binoculars, which will show it as a fuzzy star like object. Astronomers at mid-northern latitudes may have trouble tracking this object down if there is moonlight or horizon mist or glow present, which will tend to hide it from view, although for observers at more southerly latitudes it will be situated higher in the sky and should be picked up fairly easily. Described by the Rev. Thomas William Webb in his *Celestial Objects for Common Telescopes* as being '… tolerably bright … blazing in centre …' M79 merits the attentions of the backyard astronomer.

February

New Moon: 4 February
Full Moon: 19 February

MERCURY reappears in the western sky after sunset in what is a moderately good evening apparition for northern hemisphere observers. A lunar occultation takes place on 5 February but Mercury is too near to the Sun for this event to be observable. Mercury appears just 0.8° north of Neptune on 19 February and attains greatest elongation east on the penultimate day of the month. This elongation is the shallowest of the year, with Mercury straying only 18° away from the Sun. Beginning February at magnitude −1.5, Mercury fades to −0.2 by the end of the month.

VENUS continues to rule the pre-dawn skies above the eastern horizon. Never brighter than it was on the first day of the year, it dims slightly from magnitude −4.3 to −4.1 over the course of this month. The phase is increasing, from 62% to 72%, but at the same time, the planet's apparent diameter is diminishing as Venus gets farther from Earth. After a close approach to Jupiter last month, the morning star is found 1.1° north of the considerably fainter Saturn on 18 February. The best views of Venus are to be found from the tropics and the southern hemisphere, where it continues to climb higher in the sky for the first half of the month before starting its slow descent back toward the Sun.

EARTH sees a so-called "Super Moon" this month. The Full Moon on 19 February occurs less than 7 hours after the closest perigee of the year. This follows the farthest apogee of the year on 5 February, less than a day after New Moon.

MARS is in direct motion throughout 2019 and passes eastwards from Pisces to Aries on 13 February. On the same day, the first-magnitude planet is found just 1° away from faint Uranus. Mars is well aloft in the evening sky, not setting until around midnight.

JUPITER is the bright object shining at magnitude −2.0 in the otherwise unremarkable constellation of Ophiuchus. It is visible in the pre-dawn heavens and is best seen from southern latitudes where the ecliptic rises high in the sky. On 27 February, it appears near the waning crescent Moon.

SATURN appears at magnitude +0.6 low in the east just before sunrise and is best viewed from southern latitudes. The planet spends all of 2019 in the constellation of Sagittarius, criss-crossing the sky between the asterisms of the "Teapot" and the "Teaspoon". This year also sees Saturn repeatedly occulted by the Moon. Observers in western Europe and north western Africa will get their chance on 2 February, beginning at 05:00 UT, to see the very old crescent Moon pass in front of the ringed planet. Venus joins Saturn on 18 February in their closest appulse of the year.

URANUS leaves Pisces for Aries on 6 February where it spends the rest of the year. It comes within a degree of the much brighter Mars on 13 February and like Mars, sets around or just before midnight.

NEPTUNE is getting closer to the Sun and more difficult to see as it heads toward conjunction next month. On 19 February it has a close encounter with Mercury in the west after sunset but a telescope will be needed to see this eighth-magnitude planet amongst the stars of Aquarius.

ICON Explores the Ionosphere

Richard Pearson

We are all familiar with the beautiful displays of the aurora caused by solar wind emitted from solar flares and CMEs (Coronal Mass Ejections) on the Sun. Planetary scientists want to learn more about the processes which produce the aurora in our ionosphere along with other activity high above the Earth's surface.

Charged particles in Earth's atmosphere – which make up the ionosphere – create bands of colour above Earth's surface, known as airglow. ICON, depicted in this artist's concept, will study the ionosphere from a height of about 350 miles to understand how the combined effects of terrestrial weather and space weather influence this ionised layer of particles. (NASA's Goddard Conceptual Image Lab / B. Monroe)

The Ionospheric Connection Explorer, or ICON, is a 2-year mission that will study the boundary between our atmosphere and space: the dynamic zone high in our atmosphere where planetary weather and space weather meet. Originally scheduled for launch on 8 December 2017 aboard an Orbital ATK Pegasus XL rocket from the Reagan Test Site on Kwajalein Atoll in the Marshall Islands, the ICON mission has been postponed until 2018 with no definite launch date arranged (at the time of writing).

The ICON spacecraft is designed to study and analyse the Earth's atmosphere and the immediate space environment in several different ways, including tracking the movement, composition and temperature of both the high-altitude winds and the charged gas (plasma) in the ionosphere. As well as helping us to understand what is happening in the Earth's ionosphere during space weather events, ICON will lead the way in our understanding of the instability that can lead to severe interference with communications and GPS signals.

ICON will orbit the Earth at a 27-degree inclination and at an altitude of around 580 kilometres (360 miles), thereby placing it in a position to observe

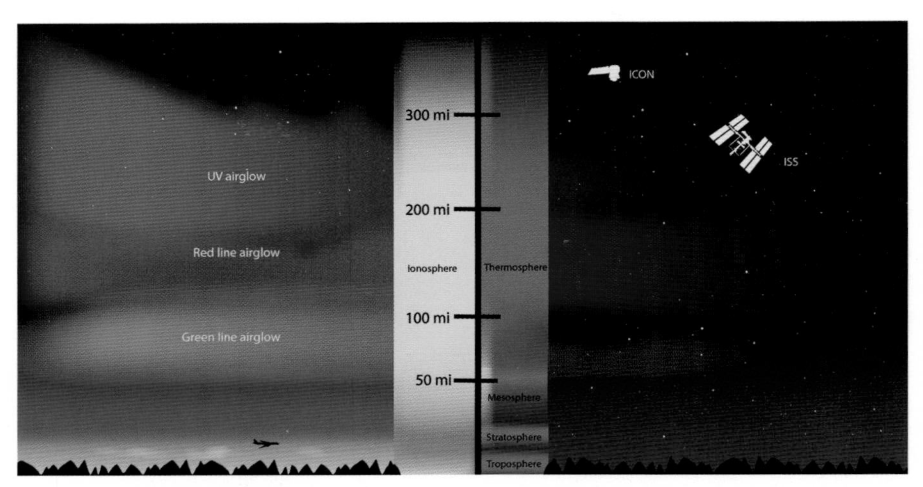

NASA's ICON mission will orbit the upper atmosphere, through the bottom edge of near-Earth space. From this vantage point, ICON will be able to observe both the upper atmosphere – made of neutral particles – and a layer of charged particles called the ionosphere, which extends from about 50 to 360 miles above the surface of Earth. Processes in the ionosphere also create bright swaths of colour in the sky, known as airglow. ICON will observe how interactions between terrestrial weather and the ionosphere create such shimmering airglow as well as other changes in the space environment. (NASA's Goddard Space Flight Center/ICON)

the ionosphere around the equator. Targeting its instruments for a view of what is happening at the lowest boundary of space at about 90 km (55 miles) up to 580 km (360 miles), ICON is equipped with four instruments designed to collect images of the ionosphere and to directly measure features of the space environment through which it flies. This range of instruments offers a perspective that would otherwise require two or more orbiting spacecraft and will provide the first comprehensive look at this important region to help scientists understand – and some day predict – what drives disturbances in the ionosphere.

The **Michelson Interferometer for Global High-resolution Thermospheric Imaging** (MIGHTI) observes the temperature and speed of the neutral atmosphere. These winds and temperature fluctuations are driven by weather patterns closer to Earth's surface, the neutral winds, in turn, driving the motions of the charged particles in space.

The **Ion Velocity Meter** (IVM) observes the speed of the charged particles in response to the push of the high altitude winds and the electric fields they generate.

The **Extreme Ultra-Violet** (EUV) instrument captures images of oxygen glowing in the upper atmosphere, to measure the height and density of the daytime ionosphere, helping to monitor the response of the space environment to weather in the lower atmosphere.

Moreover, the **Far Ultra-Violet** (FUV) instrument captures images of the upper atmosphere in the far ultraviolet. By day, FUV measures changes in the chemistry of the upper atmosphere – the source of the charged gases found higher up in space. At night this device measures the density of the ionosphere by tracking how it responds to weather in the lower atmosphere.

The ICON mission will provide us with incredible new views of the ionosphere, helping scientists determine the physics of our space environment, thus paving the way for mitigating its effects on our technology, communications systems and society. It will also help us to prepare more accurate space weather forecasts which are the dream of physicists at the moment.

March

New Moon: 6 March
Full Moon: 21 March

MERCURY is still visible early this month in the evening sky for those in northern temperate latitudes but it soon vanishes as it approaches the Sun and inferior conjunction on 15 March. Mercury's apparent path in the sky undergoes a loop, changing direction on 5 March from direct to retrograde and then resuming direct motion on 27 March. Shortly after conjunction, Mercury reappears in the east before sunrise in what is the best morning apparition this year for southern hemisphere observers. They will have the greatest chance of seeing Mercury and Neptune 3.4° apart on 22 March. Mercury's brightness varies greatly during the time around inferior conjunction and in March, this tiny planet goes from magnitude −0.1 to +6.0 at inferior conjunction and then back to +1.1.

VENUS continues to dazzle as the morning star, especially in the skies as seen from the southern hemisphere, where it rises three hours ahead of the Sun. The planet dims slightly from magnitude −4.1 to −4.0 by the end of the month. The phase continues to increase, from 72% to 81%, but Venus appears ever smaller in the telescopic eyepiece (shrinking from 15.7 arc-seconds to 13.3 arc-seconds by the end of the month) as it moves away from Earth and toward superior conjunction in August. It continues to lose altitude throughout the month and is already rather low for observers in northern temperate latitudes. On 2 March, the waning crescent Moon passes 1.2° south of Venus.

EARTH is at an equinox on 20 March. This is the day that the Sun crosses the celestial equator from south to north, marking the beginning of astronomical spring in the northern hemisphere and astronomical autumn in the south. Another equinox occurs in September.

MARS is nearly a magnitude fainter than it was at the beginning of the year and is now shining at magnitude +1.3. The planet moves out of Aries and into the neighbouring constellation of Taurus on 23 March, and passes 3.1° south of M45, the Pleiades, on the last day of the month. Observers in northern latitudes have slightly better views of Mars where it sets just before midnight.

JUPITER continues to brighten as it pulls farther from the Sun. Now at magnitude −2.1, Jupiter reaches west quadrature on 14 March. This is the time when the planet's disk has a slightly gibbous appearance when viewed through a telescope, leading to some interesting visual effects as its large satellites disappear in the planet's shadow. The nearly Last Quarter Moon glides close by on 27 March. Jupiter is in Ophiuchus, rising around midnight or just after, and is particularly well-placed for viewing from equatorial and southern latitudes.

SATURN is best seen from southern and equatorial regions, where it is gaining some helpful altitude above the eastern horizon. A morning sky object located in Sagittarius, it is occulted by the Moon twice this month. On the first day of March, some islands in the south Pacific will see the waning crescent Moon eclipse the first-magnitude object in an event beginning around 16:00 UT. Then, on 29 March, another occultation will begin at approximately 02:30 UT when southern Africa sees the planet disappear behind the disk of the Moon.

URANUS is barely visible to the naked eye (magnitude +5.9) in Aries. The best viewpoint for this faint object is the northern hemisphere, from where Uranus doesn't set until mid-evening. It may already be too deep into evening twilight for observers in tropical and southern latitudes to see.

NEPTUNE reaches conjunction on 7 March, rendering it invisible to observation this month. It moves into the morning sky along with Mercury where the two planets have another close encounter on 22 March. However, they may both be too close to the Sun for this appulse to be observable.

A Closer Look at Sextans:
The Unimpressive Triangle of Johannes Hevelius

Brian Jones

One of a number of scientific instruments that decorate the southern sky and taking the form of a small triangle of faint stars, the dim and somewhat unimpressive constellation Sextans (the Sextant) was introduced by the Polish astronomer Johannes Hevelius in the 17th century. Intended to commemorate the sextant, an instrument used by Hevelius to measure star altitudes, this group of stars was described by the Rev. Thomas William Webb in his *Celestial Objects for Common Telescopes* as: "A modern asterism … one of the minor constellations, formed by Hevelius out of unclaimed stars lying between the ancient ones." It can be safely said that none of the stars in Sextans are particularly bright and any form of moonlight or light pollution will tend to hide the group from view although the constellation itself can, with a little imagination, be seen to at least vaguely resemble the object it depicts.

A portrait of Johannes Hevelius by the Polish painter Daniel Schultz the Younger. (Wikimedia Commons)

Sextans lies on the celestial equator, and the entire constellation can be seen from almost every inhabited region of our planet. To find it, first of all locate the prominent constellation Leo (the Lion) which should be easy to pick out, its overall shape indeed resembling that of a large, crouching lion. Looking a little way to the south of Regulus, and providing the sky is dark and clear, you might be able to pick out Sextans with the naked eye, although a pair of binoculars would definitely be an asset

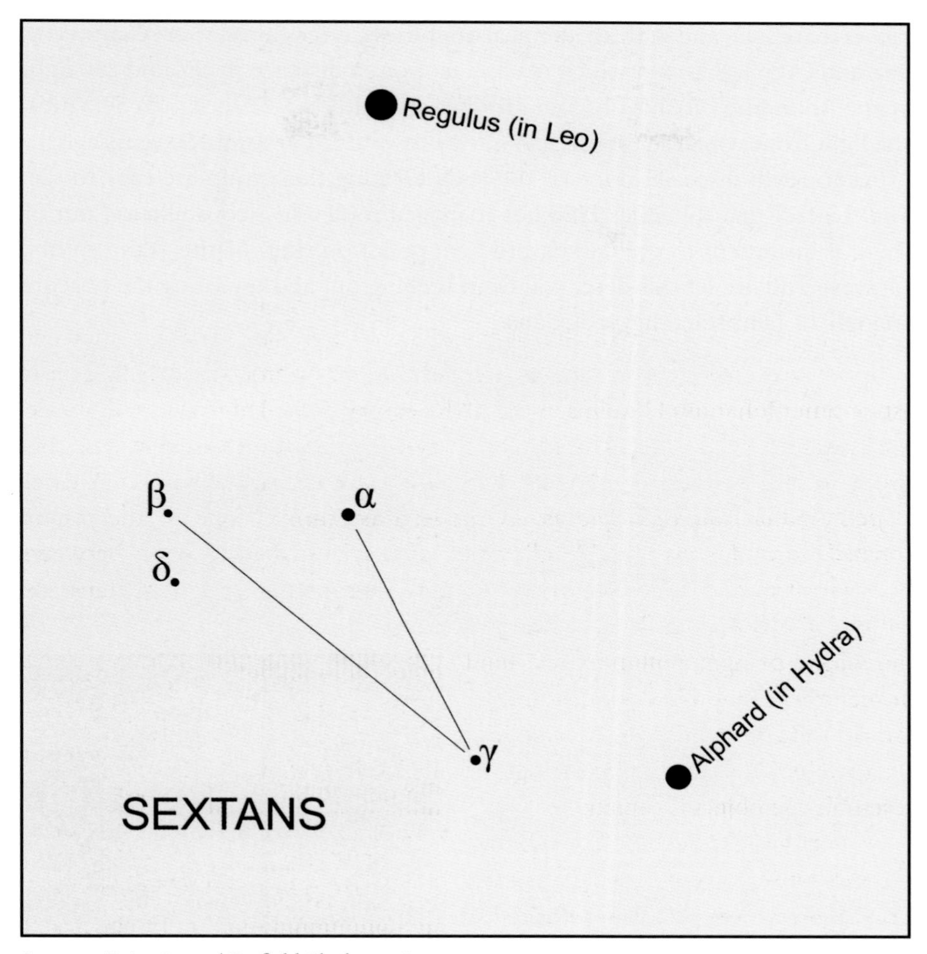

Sextans. (Brian Jones / Garfield Blackmore)

to your search. The constellation is depicted here with Regulus and, a little to the west, the bright star Alphard in Hydra, both of which can be used as guides to locating the group.

As far as the individual stars in Sextans are concerned, the brightest of these is the blue giant Alpha (α) Sextantis, the magnitude 4.49 glow of which reaches us from a distance of around 280 light years. Slightly fainter is magnitude 5.07 Beta (β) Sextantis which lies at a distance of 400 light years. Somewhat closer than Beta, and with an identical brightness, is magnitude 5.07 Gamma (γ) Sextantis, the light from which reaches us from a distance of around 280 light years. Rounding off this obscure group is magnitude 5.25 Delta (δ) Sextantis, the light from which set off on its journey towards us around 320 years ago.

As you will discover, none of the stars forming this group are easy to see, and the fact that Johannes Hevelius managed to devise a constellation out of them is testament to the darker, pre-light-pollution skies of the 17th century. However, do not let this deter you from seeking out and exploring the obscure triangle of faint stars that is Sextans.

April

New Moon: 5 April
Full Moon: 19 April

MERCURY is well-placed for viewing in the pre-dawn sky for those living in southern regions of Earth. On 2 April, it is only 0.4° north of Neptune and 3.6° north of the waning crescent Moon. Greatest elongation west (which, at 28°, is the largest of the year) occurs on 11 April. Five days later, Mercury and Venus are less than 5° apart. Although Mercury is brightening throughout the month (from magnitude +1.0 to −0.3), it is still around four magnitudes fainter than Venus.

VENUS has yet another close encounter with the waning crescent Moon, appearing 2.7° north of our satellite on the second day of the month. It has a much closer appulse on 10 April when it is only 0.3° south of eighth-magnitude Neptune in the morning sky. Mercury and Venus come together briefly on 16 April, and two days later Venus reaches aphelion, the point in its orbit where it is farthest from the Sun. As viewed from the southern hemisphere, the morning star, shining at magnitude −3.9, is still quite high in the eastern sky before sunrise, but is slowly losing altitude as the month progresses, rising less than three hours before the Sun by the end of April. When viewed through a telescope, Venus has a gibbous appearance, with 88% of its disk illuminated at month's end.

MARS is the bright reddish object passing eastwards through Taurus in April. By mid-month, the red planet is moving past the Hyades (C41), and Aldebaran, the brightest star (magnitude +0.8) in the constellation. Mars continues to dim as it travels farther away from the Earth, ending the month around magnitude +1.6, and is best seen from the northern hemisphere where it doesn't set until just before midnight.

2 PALLAS comes to opposition on 10 April. A small telescope is necessary to observe this eighth-magnitude object in the constellation of Boötes, not far from the bright star Arcturus.

JUPITER is located in the non-zodiacal constellation of Ophiuchus, brightly shining at magnitude −2.3. Observers from southern and equatorial regions have the best views of this planet which now rises before local midnight. Those in northern temperate latitudes are not so lucky, with Jupiter remaining fairly low to the horizon and not rising until after midnight. On 10 April, Jupiter reaches a stationary point and reverses direction across the sky, turning from direct to retrograde motion. The waning gibbous Moon comes within 2° of the planet on 23 April.

SATURN reaches west quadrature on 10 April. This is when Saturn-Earth-Sun makes a right angle and is an ideal time for observing the planet telescopically. The Moon once again occults Saturn this month. On 25 April, observers in eastern Australia, New Zealand, and islands in the south Pacific will have a chance to see the ringed planet disappear behind the limb of the waning gibbous Moon in an event starting around 12:15 UT. Saturn is shining at magnitude +0.5 in Sagittarius and is high in the morning sky for viewers in southern and equatorial latitudes. Unfortunately for those in the northern temperate regions, even though Saturn reaches its maximum declination north late this month, the gas giant remains low in the sky. Saturn reverses course and begins retrograde motion on the last day of April.

URANUS is at conjunction on 23 April and is lost in the glare from the Sun this month.

NEPTUNE is visible in the morning sky in Aquarius and is found in the company of Venus on 10 April. A telescope will be necessary to see both planets as Neptune, at magnitude +8.0, is below naked-eye visibility. Neptune is best viewed from southern latitudes as the sky is already brightening in the northern hemisphere when Neptune rises.

Movable Feasts and Leaping Moons: Astronomy and the Date of Easter

David Harper

Most ancient calendars were based on the phases of the Moon, and this is still the case with the Islamic, Jewish and Hindu religious calendars and the Chinese traditional calendar. The Gregorian calendar, which is the civil calendar for most of the world, is predominantly a solar calendar. Its months are independent of the waxing and waning of the Moon, and its leap year rules aim to keep the calendar in step with the seasons. But one link to the Moon remains: the date of Easter.

For Christians, Easter commemorates the crucifixion and resurrection of Jesus. These events occurred, according to the Bible, at the Jewish festival of Passover, which begins on the 15th day of the month of Nisan. In the Jewish calendar, this is the night of the Full Moon following the Spring Equinox.

The first Christians observed Easter at the same time that their Jewish neighbours were celebrating Passover, and they relied on the Jewish authorities to tell them when Passover would occur. As Christianity sought to distance itself from its Jewish roots, it became desirable for the early Church to find a way to calculate the date of Easter. The link to the Full Moon and the Spring Equinox was kept, but this led to a new problem: how to predict the phases of the Moon far into the future. The mathematics required to calculate the position of the Moon accurately would not be developed for more than a thousand years.

A little astronomical ingenuity came to the rescue. The average length of the lunar month was well known, and so were cycles such as the 8-year octaeteris and the 19-year Metonic cycle, which approximate a whole number of solar years to a whole number of lunar months. For several centuries, Easter tables were constructed using an 84-year cycle which contained 1039 lunar months. This was not as accurate as the Metonic cycle, but it had the advantage that it also contained a whole number of weeks, which was significant because Easter must be a Sunday.

Image of the Full Moon taken on 22 October 2010. (Wikipedia/Gregory H. Revera)

Eventually, a 532-year cycle was adopted. This combined the 19-year Metonic cycle of the Moon with the 28-year cycle of days of the week. The rule for Easter had also been formalised: it was the Sunday after the Full Moon on or after the Spring Equinox, which was taken to be 21 March.

To simplify the calculation of the date of the Easter Full Moon, the concept of the epact was introduced. This is the age of the Moon on 1 January. Now 12 lunar months are about 11 days shorter than a calendar year, so the epact increases by 11 from one year to the next, and since the Moon can never be more than 30 days old, the epact is always reduced by 30 whenever it exceeds the maximum. The table below shows the epacts for a complete Metonic cycle in the Julian calendar.

YEAR	1	2	3	4	5	6	7	8	9	10	11	12	13	14	15	16	17	18	19
EPACT	0	11	22	3	14	25	6	17	28	9	20	1	12	23	4	15	26	7	18

To make the epact return to the same value at the start of each 19-year Metonic cycle, it must be increased by 12, rather than 11, in passing from year 19 of one cycle to year 1 of the next. This is known as the saltus or the leap of the moon.

Over the course of many centuries, a new problem emerged. The leap year rule introduced into the Roman calendar by Julius Caesar yielded an average calendar year of 365.25 days, which was slightly longer than the true length of the year as defined by the seasons. This is called the tropical year and it is 365.24219 days. The astronomical Spring Equinox drifted steadily away from 21 March by almost a day every century. By the 16th century, it was falling on 11 March, and this caused consternation within the Church, because it meant that Easter was no longer being celebrated on the correct date.

Pope Gregory XIII sought advice from astronomers, and in 1582 he issued a papal bull called Inter Gravissimus which introduced a reformed calendar. Ten days were omitted from October that year, to bring the equinox back to 21 March. The leap year rule was modified slightly, so that century years were no longer leap years unless they were multiples of 400. And the method for calculating the date of Easter was changed to ensure that the phases of the "ecclesiastical" Moon defined by the epact did not drift too far from the real Moon. The result is a cycle of dates of Easter that does not repeat itself until 5,700,000 years have passed.

Pope Gregory XIII. (Wikimedia Commons)

There is a common misconception that the date of Easter is difficult to calculate, and that it jumps around the calendar unpredictably. Neither of these is true. The calculation of the date of Easter is quite straightforward, involving a sequence of simple steps that can be followed by any bright ten-year-old with a calculator. The interested reader can try this at home by following the steps given at **dateofeaster.com** (the reader will have to supply their own bright ten-year-old and calculator).

The date of Easter also follows a predictable pattern. For example, since Easter falls on 21 April in 2019, then in 2020 there are only two possible dates for Easter: 5 April (with a 34.2% probability) or 12 April (with a 65.8% probability). In fact, in 2020, Easter is on 12 April.

There are occasional calls for the rules defining the date of Easter to be changed, so that it cannot fall on such a wide range of dates. In 1928, the British Parliament passed the Easter Act, which proposed that Easter should be the Sunday following the second Saturday in April. Easter would always fall between 9 and 15 April. The act was never implemented, and Easter remains the only public holiday in the U.K. which follows the phases of the Moon.

May

New Moon: 4 May
Full Moon: 18 May

MERCURY is gone from the morning sky for northern hemisphere viewers soon after the month begins but those from more southerly observation points have a few extra days in which to planet watch. In particular, they may be able to see the nearly New Moon pass 2.9° south of Mercury on 3 May and possibly even glimpse the pairing of Mercury and Uranus five days later. However, Mercury vanishes into the Sun's glare before mid-month, undergoing superior conjunction on 21 May and reaching perihelion on 24 May. Mercury reaches its theoretical maximum magnitude of −2.3 at superior conjunction but is not actually visible at that time.

VENUS continues its tenure as the morning star, still best seen from southern latitudes where it precedes the Sun by two hours or so. The nearly New Moon passes less than 4° away from the planet on the second day of the month. Later, on 18 May, Venus is found only 1.2° south of Uranus. Although Venus is shining brightly at magnitude −3.9, faint Uranus will be difficult to spot in the dawn sky.

EARTH gets a traditional "Blue Moon" on 18 May when the Moon becomes Full at 21:11 UT. A traditional Blue Moon is the third Full Moon in an astronomical season containing four. Before that, however, on 11 May, the waxing crescent Moon passes through Praesepe (also known as the Beehive Cluster) in the constellation of Cancer.

MARS has an encounter with the Moon this month as the waxing crescent Moon appears close to the small planet on 7 May. On 16 May, Mars moves out of Taurus into Gemini and attains its most northerly declination of the year, spending the rest of the month at magnitude +1.7 and setting before midnight.

1 CERES comes to opposition on 28 May in Ophiuchus. It is only seventh magnitude so a small telescope will be necessary to view it.

JUPITER still resides in Ophiuchus, at magnitude −2.5 by far the brightest object (apart from the Moon, of course) in that part of the sky. It rises before midnight and is best observed from southern and equatorial latitudes. On 20 May, it is 1.7° south of the Moon.

SATURN continues to brighten as it approaches opposition later this summer, going from magnitude +0.5 to +0.3 in Sagittarius. Southern hemisphere viewers continue to be favoured, with Saturn now rising before midnight. It is still strictly a morning sky object for observers farther north, remaining low to the horizon. On 22 May, Saturn is occulted by the waning gibbous Moon as seen from southern Africa and the islands of the southern Indian Ocean. The occultation begins around 20:00 UT and finishes early the next day.

URANUS is now a morning sky object, rising just ahead of the Sun and possibly just about visible by the end of the month. It has two close encounters with the inferior planets this month, one with Mercury on 8 May and the other with Venus ten days later. However, Uranus is only magnitude +5.9 and may be lost in the dawn sky.

NEPTUNE continues to reside in Aquarius, shining at magnitude +7.9, and thus necessitating the use of a telescope to see it. Southern and equatorial regions get much the best of views of this distant object which is well aloft before sunrise.

When Astronomy Meets the Law: A Personal Perspective on Twilight

David Harper

A fascination with astronomy can lead you to some unexpected places: observatories atop remote volcanoes, eclipse-chasing in exotic locations … or the witness stand at the Old Bailey.

For a couple of years in the late 1990s, I had the privilege of working in Her Majesty's Nautical Almanac Office (HMNAO), which was then part of the Royal Greenwich Observatory in Cambridge. One of my duties was to prepare legal statements at the request of solicitors or the police in cases where the position of the Sun or the level of natural light was a factor. A driver involved in a car accident might claim that he was dazzled by the setting Sun, for example, or the reliability of an eyewitness might be called into question in twilight conditions.

Astronomers recognise three levels of twilight. Evening civil twilight, when the Sun is 6° below the horizon, is the time when it is has become too dark to play sports or read a newspaper by natural light. At nautical twilight, when the Sun is 12° below the horizon, the sea horizon can no longer be discerned, making observations with a sextant impossible. Astronomical twilight, when the Sun is 18° below the horizon, is when sixth-magnitude stars become visible at the zenith, and the sky is dark enough for serious astronomical observations. At the latitude of the British Isles, astronomical twilight does not occur between mid May and late July, as the Sun never dips far enough beneath the horizon.

In one memorable case, we were asked by the police to compile times of nautical twilight for the south coast of England on a particular range of dates, as part of a murder enquiry where the suspect was an amateur yachtsman. The police believed that he had killed his business partner and disposed of the body at sea from his yacht, since the remains of the unfortunate victim had subsequently been caught up in a trawler's nets. The suspect was not a good enough sailor to navigate his yacht in complete darkness, hence the police's interest in the times of nautical twilight, which could be used to test his alibi.

This unusual twilight view of the Very Large Telescope (VLT), taken from the north slope of Paranal Mountain, shows two of the four 8.2-metre Unit Telescopes (UTs), with the small but distinctive shape of the Southern Cross clearly visible just to the upper right of the closer UT. (G. Hüdepohl (atacamaphoto.com)/ESO)

When such cases came to trial, the written expert witness statements provided by members of HMNAO were normally accepted without question by both sides. It was rare for an astronomer to be called to testify in person. Indeed, we hoped it would not happen, because according to HMNAO folklore, a very senior colleague had once been questioned so aggressively by a barrister in court that he felt as if he was 'the accused' rather than a witness.

The reader will therefore understand my trepidation when, two years after leaving HMNAO, I received a summons to appear as a witness at the Old Bailey in connection with a written statement I had made concerning the time of sunset. I put on my best suit, caught the train to London, and entered the imposing building that is the Central Criminal Court. I was briefed by the police officers attached to the case: speak clearly and slowly, direct your answers towards the jury, and always address the judge as 'my lord'. The usher called me into the courtroom and I took the oath. What manner of inquisition was I about to undergo?

The entrance to the Central Criminal Court, better known as the Old Bailey. The court was founded in the 16th century, although the present building dates from 1902. The most serious criminal cases in England have been tried here for almost two centuries. (Wikimedia Commons/Tbmurray)

DEFENCE BARRISTER: In your written statement, Dr Harper, you said that the time of sunset was 6:24 p.m. Is that correct?

ME: Yes.

DEFENCE BARRISTER: The incident in question took place at 6:35 p.m. Would it be correct to say that the incident took place eleven minutes after sunset?

ME: Yes.

DEFENCE BARRISTER: No further questions.

THE JUDGE: Thank you, Dr Harper, you may step down.

And having served British justice by confirming that 35 minus 24 is indeed eleven, my extremely brief career as an expert astronomical witness was over.

June

New Moon: 3 June
Full Moon: 17 June

MERCURY moves into the evening sky this month, providing everyone with quite good observing opportunities. On 18 June, Mercury and Mars are found only 0.2° apart. Greatest elongation east (25° from the Sun) occurs on 23 June, by which time Mercury is descending toward the horizon. Over the course of the month, Mercury fades from −1.2 to +1.0, a change of over two magnitudes.

VENUS is overtaken by the very old crescent Moon on 1 June. Up until this month, the best views of Venus have been from southern latitudes but from mid-month until August, no part of Earth will have a clear advantage in observing this bright planet in the morning twilight. Venus, still shining at magnitude −3.9, appears nearly full in a telescope, with 98% of the planet's disk (now only 10 arc-seconds in diameter) illuminated. It continues its descent toward the eastern horizon and conjunction with the Sun in August.

EARTH reaches solstice on 21 June. On this day, the Sun is at its most northerly declination, heralding the start of astronomical summer in the northern hemisphere and astronomical winter in the south. The next solstice is in December.

MARS continues its eastward journey across the sky, moving from Gemini to Cancer on 28 June. Earlier in the month, the young crescent Moon passes less than 2° away from Mars on 5 June, and Mercury and the red planet undergo a close pairing on 18 June. Found in the west after sunset, Mars continues to dim slightly and sets mid-evening.

JUPITER is at opposition on 10 June which is when the planet is at its brightest. Blazing away at magnitude −2.6 in Ophiuchus, Jupiter is visible all night. Observers from northern latitudes do not get the excellent views of this planet afforded to those in the southern hemisphere. This is because the ecliptic is fairly low to the horizon when seen from northern temperate regions. The nearly Full Moon is in close attendance to Jupiter on 16 June.

SATURN is found between the asterisms of the "Teaspoon" and the handle of the "Teapot" in the constellation of Sagittarius this month, brightening to magnitude +0.1. The planet remains close to the horizon when viewed from northern temperate regions but is well-placed for observation from latitudes farther south where it rises by early evening. Saturn is occulted by the waning gibbous Moon on 19 June when people in southern South America, southern Africa and the islands of the south Atlantic witness the ringed planet disappearing behind the Moon's disk. This event begins around 01:30 UT.

URANUS is still lost in morning twilight for northern hemisphere observers but early risers in more southerly latitudes will be able to spot this sixth-magnitude planet amongst the fainter stars in Aries.

NEPTUNE reaches west quadrature on 9 June. Later in the month, on 22 June, it changes from direct to retrograde motion across the sky. Neptune never gains much altitude when viewed from northern latitudes but it is well-placed for observation the farther south you are. This ice giant is still a morning sky object, magnitude +7.9, in the watery constellation of Aquarius.

Lunar Occultations and Conjunctions in 2019

Richard Pearson

The Moon appears so brilliant in our skies that we often fail to realise that it is a small body with a diameter of just 3,475 km (as against the 12,742 km of Earth). The time taken for the Moon to complete one journey around the Earth (or, more correctly, around the barycentre) is 27.3 days.

This beautiful image of a triple near-conjunction above the round domes of the telescopes at ESO's La Silla Observatory in northern Chile was captured on Sunday 26 May 2013 when the three planets Jupiter (top), Venus (lower left) and Mercury (lower right) were engaged in a spectacular cosmic dance. (Y. Beletsky (LCO)/ESO)

As the Moon travels across the sky, it occasionally passes in front of stars. During these events, which are known as lunar occultations, the star is temporarily hidden from view. Occultations of fainter stars are common enough, although there are a number of bright stars that lie on or near the ecliptic, and can therefore be occulted by the Moon, including Aldebaran (α Tauri) and Regulus (α Leonis). Amateurs carry out valuable work in observing occultations, using nothing more elaborate than a telescope and an accurate stop-watch to carefully record the exact time when the star is hidden by the onward-moving limb of the Moon.

All the planets in the solar system lie close to the plane of the ecliptic and are often occulted by the Moon, or positioned close to the Moon in the sky. A conjunction occurs when two objects are lined up with each other (or nearly so) as seen from Earth and, in the case of Saturn, the planet is frequently lined up with the Moon during 2019 (although only one of these events will be observable from the UK).

The six most favourable occultations of Saturn in 2019 are detailed here, the remainder being visible only from the Atlantic Ocean and Antarctica.

Date	Time (UT)	Area of Visibility
2 February 2019	07:05	UK, Europe, North Africa, India, Pakistan and Russia
25 April 2019	14:27	Eastern Australia
19 June 2019	03:46	South America
16 July 2019	07:14	South America
8 September 2019	13:42	Philippines, New Zealand and Australia
5 October 2019	20:36	South Africa

In addition, Jupiter will be just 0.7° south of the Moon on 28 November 2019 at 10:49 UT, although this event will be a challenge to see from mid-northern latitudes.

Planets appear as tiny discs when seen from Earth, unlike stars, which are seen as points of light, resulting in occultations being virtually instantaneous. Before occultation, the star is seen shining steadily, although as the lunar disc covers it up, the light from the star disappears suddenly. The reappearance of the star from behind the opposite limb of the Moon is equally sudden. Because the positions of the stars in the sky are known much more accurately than that of the ever-shifting Moon, the successful timing of occultations gives the exact position of the Moon at that precise moment.

Occultations of stars by the Moon are particularly impressive when the Moon is waxing since the occultation takes place at the dark limb, which can not be seen unless lit up by earthshine. When the Moon passes through a star cluster such as the Beehive (M44) in Cancer (the Crab), half a dozen naked-eye stars are occulted over a relatively short period. The Moon passes near the Beehive during most months of the year. During 2019 the occasions when the Moon approaches to within 0.2° of the cluster will be on 7 June, 4 July and 28 August, while the conjunction of 13 April may be particularly interesting. These events provide excellent visual and photographic opportunities for those who venture outdoors to view them.

July

New Moon: 2 July
Full Moon: 16 July

MERCURY is getting lower in the west after sunset but is still visible early in the month when, on 4 July, it has a close encounter with the Moon and is also 2.5° south of Praesepe (M44) in the constellation of Cancer. Two days later, Mercury is within 4° of Mars, but this may be difficult to observe due to the low altitude of both objects. As in March, Mercury undergoes a short period of retrograde motion, beginning on 7 July and ending on the last day of the month. Coincidentally, Mercury also arrives at aphelion on 7 July. The closest planet to the Sun is lost to view by early or mid-month as it reaches inferior conjunction on 21 July, reappearing before the end of July in the east before sunrise. Once again, Mercury rapidly dims around inferior conjunction, with magnitudes varying from +1.1 at the beginning of July to +6.0 at inferior conjunction and quickly back to +2.4 by the end of the month.

VENUS is getting increasingly difficult to observe in the dawn sky, with the northern hemisphere possibly having slightly better views. Although still a bright magnitude −3.9, it all but vanishes from sight by the end of the month.

EARTH is at aphelion, the point in its orbit farthest from the Sun, on 4 July. A total solar eclipse is visible from our planet two days before that and a partial lunar eclipse takes place on 16/17 July.

MARS has been dodging the Moon all year but this month, on 4 July, it is actually occulted by the young crescent Moon during daylight hours. Two days later, it is a few degrees away from the planet Mercury in the western sky. Mars passes within half a degree of Praesepe, on 13 July, and on 30 July, moves from Cancer to Leo.

JUPITER is now past opposition and an evening sky object in the large and nondescript constellation of Ophiuchus. It dims slightly this month, down 0.1 of a magnitude to −2.5, but still dominates its section of sky. It is best seen from the southern hemisphere as it remains low to the horizon when viewed from northern temperate latitudes. The waxing gibbous Moon passes a little over 2° north of the planet on 13 July.

SATURN finally reaches opposition on 9 July, six months after its January conjunction, and is visible all night in Sagittarius. Now at magnitude +0.1, this gas giant is high in the sky when viewed from tropical and southern latitudes. The rings are tilted at more than 24° with respect to Earth, down from the maximum of 27° in 2017, but they are still a lovely sight in a small- to medium-sized telescope. The ecliptic is close to the horizon as seen from northern temperate regions so Saturn remains low in the sky. Another lunar occultation takes place this month when, on 16 July, the Full Moon passes in front of Saturn as seen from many of the islands of the south Pacific and much of South America. This event begins at approximately 05:00 UT.

URANUS reaches west quadrature on 29 July in Aries. The faintest of the naked-eye planets is only magnitude +5.8 which makes it difficult to see in the semi-twilight of the northern hemisphere early morning skies, but southern observers get excellent views of Uranus before sunrise.

NEPTUNE rises in the evening for observers in tropical and southern latitudes but makes a much later arrival for northern hemisphere planet watchers, not appearing until midnight. It is best seen from the southern hemisphere where Aquarius is high in the sky during the early morning hours. A telescope is always necessary to observe this eighth-magnitude planet.

134340 PLUTO arrives at opposition on 14 July. A large telescope and detailed finder chart is necessary to spot this faint dot (magnitude only +14.5) amongst the crowded star fields of Sagittarius.

Georges Henri Joseph Édouard Lemaître

David M. Harland

Georges Lemaître was born at Charleroi in Wallonia, Belgium, on 17 July 1894. After a classical education at a Jesuit secondary school, the Collège du Sacré-Coeur in his home town, he went to the Catholic University of Louvain to become a civil engineer. His studies were interrupted by the Great War in 1914.

Albert Einstein with Georges Lemaître (right) at the California Institute of Technology in 1933. (Associated Press)

After the war, Lemaître returned to Louvain to study physics and graduated in 1920. He then entered a seminary and was ordained as a priest in 1923. Next was a year's research at the University Cambridge in England, where he met Arthur Eddington, the expert on the structure of stars.

Meanwhile, in 1915 Albert Einstein had published the Theory of General Relativity, in which he showed that gravity could be interpreted in terms of geometry. At that time, astronomers believed the universe to be eternal, and static, so Einstein added a term to his equations to represent the force which would be required to prevent the universe from collapsing in on itself under gravity. He naturally presumed that the solution he had identified would be unique. But in 1917 the Dutch physicist Willem de Sitter found a solution in which the universe expanded exponentially; then five years later Aleksandr Friedmann in Russia found a family of solutions, some expanding universes and others contracting ones.

While working for his doctorate Lemaître spent a year attending Harvard and the Massachusetts Institute of Technology where he became aware of the work of Edwin Hubble, who was proving that spiral nebulae were not only far outside our own galaxy but systems of similar size whose recession indicated the universe was expanding.

On returning to Belgium in 1925, Lemaître became a lecturer at his *alma mater*, where in 1927 he became Professor of Astrophysics. On independently discovering Friedmann's solutions to Einstein's equations he readily accepted that the universe was expanding but chose to avoid the issue of an 'origin' by suggesting that, after spending a long time in the static state that Einstein had envisaged, an instability of some form had initiated a runaway expansion. In 1927 he published this idea in *Annals of the Scientific Society of Brussels* in a paper entitled 'A homogeneous Universe of constant mass and growing radius accounting for the radial velocity of extragalactic nebulae.' In 1930 Eddington arranged for it to be reissued in the *Monthly Notices of the Royal Astronomical Society*. By this point however, Einstein had rejected the special term that he had added to his equations to hold the universe static.

Given that the universe was expanding, Lemaître realised that if one could wind the clock back, so to speak, there must have been a time when all of the matter in the universe had been a single, super dense entity that he named the 'primeval atom'. In an analogy with the spontaneous process of radioactive decay, he suggested this had exploded and what we observed was the debris scattering. He published this in 1931 in a paper 'The beginning of the world from the point of view of Quantum Theory', this time in *Nature* to ensure the widest possible audience.

In 1932 Einstein and de Sitter jointly authored a paper that embraced the work of Friedmann and Lemaître and explained the illusion that galaxies are moving away from us as if we were at the centre of the universe by pointing out that the galaxies aren't receding *through* space at the speeds indicated by their radial velocities, it was *space* that was expanding and the galaxies were simply being carried along. It was a geometrical effect of the expansion that made those further away *appear* to recede exponentially.

In January 1933 Lemaître visited the California Institute of Technology, where Hubble was based, and gave a seminar on his ideas. Einstein, who was on an extended sabbatical in America, stood up afterwards and said, "This is the most beautiful and satisfactory explanation of creation to which I've ever listened."

Nevertheless, the hypothesis of the 'primeval atom' was largely ignored by astronomers. But then in 1946 George Gamov, who had moved from Russia to the United States, expanded on the theme and predicted that there should be a 'black body' radiation field with its peak intensity in the microwave region of the electromagnetic spectrum. At that time there were no instruments capable of testing this prediction. And in any case, in 1950 Fred Hoyle at Cambridge rejected the idea that the universe was "… created in one big bang at a particular time in the remote past."

But then in the 1960s a theory group at Princeton headed by Robert Dicke independently realised that if the universe originated in a hot fireball then we should be able to prove that by measuring the cosmic microwave background. Just as they were about to build a detector, they heard 'on the grape vine' that this radiation had been detected by radio astronomers Arno Penzias and Robert Wilson with a larger antenna at the nearby Bell Telephone Laboratories. Papers by the two teams were published simultaneously in May 1965. By then Lemaître had turned to mathematical research, but this evidence in favour of his hypothesis, by then known as the Big Bang origin of the universe, must have been deeply satisfying.

Georges Henri Joseph Édouard Lemaître died at Louvain on 20 June 1966.

Further reading:

The Day Without Yesterday: Lemaître, Einstein, and the Birth of Modern Cosmology by John Farrell (Basic Books, 2006)

August

New Moon: 1 August
Full Moon: 15 August
New Moon: 30 August

MERCURY is a morning sky object this month, with somewhat better views from the northern hemisphere than the south. It reaches greatest elongation (19° west of the Sun) on 9 August and begins to get lower in the eastern sky shortly afterwards, vanishing before the end of the month. It makes its closest approach to Praesepe (M44) on 17 August and reaches perihelion three days later. Mercury brightens by almost four magnitudes this month, beginning at +2.1 and ending at −1.7.

VENUS undergoes a daytime lunar occultation on the first day of the month. It reaches perihelion on 8 August and superior conjunction six days later. Venus reappears very low in the west at sunset by the end of the month, ready to assume the mantle of the evening star. However, it is too close to the Sun to be visible during most of August.

EARTH saw two Full Moons in both January and March of last year, at least in most time zones. The second Full Moon in a calendar month is popularly known as a "Blue Moon" so 2018 was a double Blue Moon year. This year, August sees two New Moons in a calendar month. The second New Moon is sometimes referred to as a "Black Moon".

MARS has been moving steadily eastwards across the sky all year but the Sun is slowly overtaking it and the red planet is lost to view, low in the west, well before the end of the month. It reaches aphelion on 26 August whilst in the constellation of Leo.

JUPITER is an evening sky object, dominating the sky at magnitude −2.3 in the non-zodiacal constellation of Ophiuchus. The waxing gibbous Moon passes by on 9 August. Two days later, Jupiter returns to direct motion, having been in retrograde since early April. For those in northern temperate latitudes, Jupiter sets by midnight but southern hemisphere observers get an extra hour or two before the gas giant vanishes below the western horizon.

SATURN is now past opposition and is an evening sky object, fading slightly to magnitude +0.3. It continues to stay low to the horizon in Sagittarius when viewed from the northern hemisphere, whilst observers farther south enjoy superior views of the ringed planet. The waxing gibbous Moon occults Saturn on 12 August in an event beginning about 07:30 UT. It will be visible from Micronesia and much of Australia.

URANUS reverses course for the last time this year, reaching a stationary point on 12 August and changing from direct to retrograde motion across the background stars in Aries. It brightens slightly from magnitude +5.8 to +5.7 as it approaches opposition in October. Uranus now rises before midnight and can be observed from both northern and southern hemispheres.

NEPTUNE is heading toward opposition next month and is becoming easier to see even from northern temperate latitudes where it is finally rising before midnight. However, at magnitude +7.8, a small telescope will be required to see the planet in Aquarius.

First Light for Gravitational Waves

Richard Pearson

The detection of gravitational waves is a new and exciting branch of science which will enable scientists to learn more about the sources producing them and about gravity itself, and thereby increasing our knowledge of the nature of the universe.

When Albert Einstein published his general theory of relativity in 1915, he predicted the existence of gravitational waves, picturing them as ripples in the fabric of space-time produced by massive, accelerating bodies, such as black holes in orbit around each other. Gravitational waves are continually passing Earth, yet even the strongest have a minimal effect and their sources generally lie at great distances from us. Scientists have been trying to track down gravitational waves for many years although their detection requires the use of extremely sensitive equipment.

The Laser Interferometer Gravitational-Wave Observatory (LIGO) is a multi-kilometre-scale project designed to detect the tiny ripples in space-time caused by passing gravitational waves through the use of laser interferometry. It consists of two widely separated interferometers located in the United States, based in Livingston, Louisiana (LIGO Livingston) and Hanford, Washington (LIGO Hanford). Separated by a distance of around 3,000 kilometres, they are operated in unison to collect data simultaneously.

LIGO detected gravitational waves from a signal on 14 September 2015 of two black holes of 29 and 36 solar masses merging about 1.3 billion light-years away. The signal came from the southern celestial hemisphere, in the approximate direction of the Magellanic Clouds. Announced in February 2016, this first historic discovery was followed by three more events attributed to mergers of black holes. The last of these signals in August 2017 (image 1) was also observed by the Virgo Collaboration, the addition of the third instrument allowing a much more accurate location of the source. Virgo is a large interferometer, located near Pisa in Italy, also designed to detect gravitational waves, and named for the Virgo Cluster of galaxies.

In October 2017, the LIGO project was recognised by the award of the Nobel Prize in Physics. One half of the value was awarded to Rainer Weiss, the other half jointly to Barry Clark Barish and Kip Stephen Thorne 'for decisive contributions to the LIGO detector and the observation of gravitational waves.'

Soon after the presentation of the Nobel Prize, a new and significant discovery was aired by astronomers. For the first time, both gravitational waves and electromagnetic radiation arising from the same event have been observed, thanks to a combined global effort.

On 17 August 2017 LIGO in the United States, working with the Virgo Interferometer, detected a gravitational wave signal passing the Earth. The

Image 1. This artist's impression shows two tiny but very dense neutron stars at the point at which they merge and explode as a kilonova. The cataclysmic aftermaths of this kind of merger disperse heavy elements such as gold and platinum throughout the Universe. Such rare events were expected to produce both gravitational waves and a short gamma-ray burst, both of which were observed on 17 August 2017. (ESO/L. Calçada/M. Kornmesser)

signal was named **GW 170817** (deriving from 'Gravitational Wave' and the date of observation 20**17-08-17**) and was the first detection of a collision of two neutron stars (previous detections being of colliding black hole pairs).

GW 170817 was also the first gravitational wave event observed to have a simultaneous electromagnetic signal. Around two seconds after the signal was recorded, two space observatories, NASA's Fermi Gamma-ray Space Telescope (FGST) and ESA's INTErnational Gamma-Ray Astrophysics Laboratory (INTEGRAL), detected a short gamma-ray burst (named GRB 170817A) coming from the same area of sky. (image 2)

The LIGO–Virgo observatory network positioned the source within a vast region of the southern sky. As night fell in Chile, many telescopes peered at this patch of sky, searching for new sources. The 1.0-metre Swope Telescope at Las Campanas Observatory in the southern Atacama Desert of Chile was

Image 2. This image from the VIsible MultiObject Spectrograph (VIMOS) instrument on ESO's Very Large Telescope at the Paranal Observatory in Chile shows the galaxy NGC 4993. Although the galaxy itself is not unusual, it contains something never before witnessed – the aftermath of the explosion of a pair of merging neutron stars in an event known as a kilonova (indicated with the arrow). (ESO)

the first to announce a new point of light. It appeared very close to NGC 4993, a 12th magnitude lenticular galaxy in the constellation of Hydra, and VISTA observations pinpointed this source at infrared wavelengths almost at the same time. Distance estimates from both the gravitational wave data and other observations agree that GW 170817 is located at the same distance as NGC 4993, around 130 million light years from Earth.

Further discoveries of this kind will help change our perspective of the Universe as we know it, a change which has its roots in Albert Einstein's prediction of gravitational waves made just over a century ago.

September

Full Moon: 14 September
New Moon: 28 September

MERCURY is invisible at the beginning of the month, undergoing superior conjunction on 4 September and reappearing by mid-month in the west after sunset in what is the best evening apparition for southern hemisphere observers. Mercury and Venus come within 0.3° of each other on 13 September and on the penultimate day of the month, the closest planet to the Sun is found only 1.2° north of the first-magnitude star Spica in the constellation of Virgo. Mercury is fading in brightness after superior conjunction and is magnitude −0.2 at the end of the month.

VENUS continues to favour observers in southern latitudes as it begins to climb higher above the western horizon after sunset. It has a very close appulse with Mercury on 13 September but the two planets may be too low to the horizon for this event to be visible. Venus remains at magnitude −3.9 this month, appearing in its waning gibbous phase (although still nearly full).

EARTH reaches equinox on 23 September. It is on this day that the Sun crosses the celestial equator from north to south, heralding the beginning of astronomical spring in the southern hemisphere and astronomical autumn in the north. In North America, the Full Moon nearest to the autumnal equinox is traditionally called the "Harvest Moon". This year the Harvest Moon occurs on 14 September. Coincidentally, the Harvest Moon is also this year's "Micro Moon", the Full Moon occurring closest to apogee.

MARS is at conjunction on 2 September and officially moves from the evening to the morning sky, although the planet is too close to the Sun to be observed before the end of the month. On 24 September, Mars crosses from Leo into Virgo.

JUPITER continues to outshine every other (non-lunar) celestial body in the constellation of Ophiuchus although it dims slightly from magnitude −2.2 to −2.1. The First Quarter Moon glides past on 6 September and two days later, Jupiter reaches east quadrature. As at west quadrature in March, this is an ideal time to observe the interplay of the planet, its Galilean satellites and their respective shadows which are cast noticeably off to one side. Northern hemisphere observers lose the planet early in the evening but those in southern and equatorial latitudes have until midnight before Jupiter sets.

SATURN undergoes yet another lunar occultation, this one occurring on 8 September when, at approximately 11:15 UT, the waxing gibbous Moon eclipses the planet as seen from places including south eastern Africa, Madagascar, Micronesia and many islands in the Indian Ocean. Retrograde motion ends on 18 September when Saturn reaches a stationary point and ten days later, the gas giant reaches its maximum southerly declination for the year. Saturn is an evening sky object, located in the constellation of Sagittarius between the asterisms of the "Teapot" and the "Teaspoon", and dimming from magnitude +0.4 to +0.5. The tilt of the rings briefly exceeds 25° this month and next. It is best seen from southern and equatorial regions where it doesn't set until the early morning hours.

URANUS continues its sojourn through Aries, shining dimly at magnitude +5.7. Although it rises before midnight, it is still best observed early in the morning when it gains some useful altitude above the horizon.

NEPTUNE passes the fourth-magnitude star Phi (φ) Aquarii in early September on its way to opposition on the tenth. Never brighter than magnitude +7.8 this year, you will need a small telescope to observe this faint planet. Neptune is visible all night in Aquarius.

Humboldt and the Gegenschein

Richard Pearson

On crisp, clear evenings during the 17th century, when the absence of street lighting ensured really dark skies, sky glows such as the zodiacal light could be seen far more easily than they can be today. Its presence had probably been first reported by the English academic Joshua Childrey (1623–1670) as far back as 1661. However, in 1683 the French astronomer Giovanni Domenico Cassini (1625–1712) referred to the zodiacal light as a celestial light in the zodiac and predicted the presence of an aura of material surrounding the Sun.

Portrait of Friedrich Wilhelm Heinrich Alexander von Humboldt in 1843 by the German painter Joseph Karl Stieler (1781–1858). (Wikimedia Commons/Joseph Karl Stieler)

The zodiacal light is visible as a faint and diffuse white glow, roughly triangular in shape and extending from the vicinity of the Sun along the zodiac in the plane of the ecliptic. Decreasing in intensity with distance from the Sun the zodiacal light is caused through the scattering of sunlight by interplanetary dust and is best seen from the tropics when the zodiac is at the steepest angle to the horizon, either during twilight after sunset in spring or before sunrise in autumn. Under exceptional seeing conditions and very dark skies it is discernible as a band (known as the zodiacal band) extending completely around the ecliptic, although the glow is so faint that any form of moonlight and/or light pollution will outshine it, rendering it invisible.

Another phenomenon, which takes the form of a faint (but slightly brighter) oval glow seen directly opposite the Sun, is the Gegenschein. In 1730, what we now term the Gegenschein was first described by the French Jesuit astronomer and mathematician Esprit Pezenas (1692–1776), although he did not give it the name by which we know it today. That privilege went to the renowned German botanist and scientist Friedrich Wilhelm Heinrich Alexander von Humboldt.

Alexander von Humboldt was born on 14 September 1769 into a wealthy, upper-class Berlin family. In his early years Alexander was tutored privately at his family estate in Tegel, then a wooded suburb 16 kilometres north of Berlin, after which he studied for a short time in 1787 at the University of Frankfurt (Oder) before attending the renowned university in Göttingen where he matriculated in 1789.

Alexander's father Alexander Georg von Humboldt had died in 1779, and the subsequent death of his mother Maria Elisabeth in 1796 left Alexander

The skies above the European Southern Observatory's Paranal Observatory are so dark that the often difficult to observe Gegenschein can be seen in its full glory. This image shows the Gegenschein as a band running diagonally from upper left to lower right. At the time the picture was taken in October 2007 the weather conditions at Paranal were excellent and sky transparency close to perfect. This allowed the photographer to capture many of the very faint details within the Gegenschein, thereby revealing its fine structure. (ESO / Y. Beletsky)

with an inheritance which allowed him to fulfil his desire to travel and explore. He seized the opportunity and began planning expeditions with the French explorer and botanist Aimé Jacques Alexandre Bonpland (1773–1858).

Following a visit to Madrid to obtain the necessary permission to explore Spain's realms in the Americas, Aimé Bonpland and Alexander von Humboldt set sail aboard the ship Pizarro on 5 June 1799, and on 16 July they finally landed at Cumaná, Venezuela. After calling at the town of Caripe where they visited the Guácharo cavern, they returned to Cumaná from where Alexander observed a remarkable meteor shower (the Leonids) on the night of 11/12 November 1799.

Bonpland and Humboldt continued their explorations of South America. Taking a trip to the Andes, their ascent of the stratovolcano Pichincha near the city of Quito was followed by a climb to the top of Chimborazo (a currently-inactive stratovolcano in the Cordillera Occidental range of the Andes). They continued to Lima, Peru, from where Humboldt observed the transit of the planet Mercury on 9 November 1802 (there is also a transit of Mercury on 11 November 2019 as detailed elsewhere in this Yearbook).

It was during his Spanish-American expedition of 1799–1804 with Aimé Bonpland that Alexander von Humboldt carried out observations of the phenomenon we now know as the Gegenschein. This is the very faint patch of light, roughly oval in shape, which is sometimes visible near to the ecliptic directly opposite the Sun. The name derives from the German for 'counterglow' which was an excellent description of what he and Aimé Bonpland occasionally observed during their expedition.

As is the case with the zodiacal light, the Gegenschein is caused by the scattering of sunlight back towards the Earth from tiny dust particles in the plane of the solar system. Although the major axis of the Gegenschein in the ecliptic plane generally extends for around 20° and the minor axis for around 10°, its overall size can be somewhat more significant when viewed in the tropics. It was in 1854 that the Danish astronomer Theodor Johan Christian Ambders Brorsen (1819–1895) published the first thorough investigations of the Gegenschein, although it was in his notebooks, compiled half a century before during his Spanish-American expedition, that Humboldt used the word 'Gegenschein' for the first time, and the term remains with us to this day.

Alexander von Humboldt worked with and influenced great men such as the English naturalist, geologist and biologist Charles Darwin, the American essayist, philosopher and naturalist Henry David Thoreau and the German

An oblique view taken by Lunar Orbiter 4 showing most of Mare Humboldtianum. (Wikimedia Commons/NASA/James Stuby)

biologist, naturalist and philosopher Ernst Haeckel. When Humboldt died on 6 May 1859 at the age of 89, the news reached London via telegraph from Berlin within hours of his death: 'Berlin is plunged in sorrow!' As the sad news spread over the world, people were shocked that the 'age of Humboldt' had come to an end, although his name lives on in the fact that many geographical features, places, universities and schools, plants and animals are named for him, as is the 273 km diameter Mare Humboldtianum, a lunar mare located along the northeastern limb of the Moon a little to the east of Mare Frigoris.

October

Full Moon: 13 October
New Moon: 28 October

MERCURY is high in the evening sky for those living in equatorial and southern latitudes. On 20 October, it reaches greatest elongation east, appearing 25° from the Sun, and begins to lose altitude. It has another close encounter with Venus on the last day of the month, just before Mercury enters into retrograde motion. This evening apparition is very poor for those in northern temperate latitudes and the planet may be largely unobservable.

VENUS passes a few degrees north of Spica, the brightest star in the constellation of Virgo, on the third day of the month, but Venus, at magnitude −3.9, is much the more conspicuous object. Mercury and Venus are again found together in the sky on the last day of October. For observers in southern latitudes, the evening star is beginning to attain some useful altitude, setting about an hour after the Sun, but it remains low to the western horizon when viewed from northern temperate latitudes. Planet watchers armed with telescopes can watch the phase of Venus slowly decrease to 94% by the last day of the month.

MARS reappears low in the east after last month's conjunction, rising just ahead of the Sun, and shining at magnitude +1.8 in Virgo.

JUPITER is growing fainter as the year progresses, down to a still-brilliant magnitude −2.0. Found in Ophiuchus, Jupiter has two close encounters with the waxing crescent Moon, on 3 October and again on the last day of the month. Never very high this year when viewed from northern latitudes, Jupiter is getting harder to observe as it sets mid-evening. The best views are from the southern hemisphere where the planet is still conveniently situated for observing, setting just before midnight.

SATURN reaches east quadrature on 7 October in Sagittarius. This is an excellent opportunity for astrophotographers to catch interesting shadow effects in the Saturnian system as the shadows of the planet, rings and satellites are cast somewhat to one side. However, the best chance to see the planet is from southern latitudes where the planet is fairly high in the sky during evening hours. Two days before quadrature the Moon gets in on the act when it occults Saturn. The occultation event begins around 18:15 UT and may be seen from southern South America. Saturn continues to fade in brightness, decreasing slightly in magnitude from +0.5 to +0.6.

URANUS finally reaches opposition on 28 October and is visible much of the night. It is found shining at magnitude +5.7 in Aries where it is well-placed for observers in both hemispheres.

NEPTUNE was at opposition last month and is an evening sky object in Aquarius, setting as the sky begins to lighten in the east. Although best seen from southern and tropical latitudes, telescopic observers in northern temperate regions can also glimpse this elusive eighth-magnitude object.

A Closer Look at Equuleus: The Little Horse that Trots across the Night Sky

Brian Jones

The tiny constellation Equuleus (the Little Horse) takes the form of a small irregular pentagon of faint stars visible immediately to the west of the much larger constellation Pegasus. This little group of stars was introduced nearly two thousand years ago by the Greek astronomer Ptolemy on his list of 48 constellations drawn up during the 2nd century AD. Equuleus can be tracked down fairly easily provided the sky is dark, clear and moonless. If you have any difficulty then a pair of binoculars will help.

Such is the position of Equuleus in the sky that it can be seen from anywhere north of latitude 77°S, making it accessible to observers in almost any inhabited

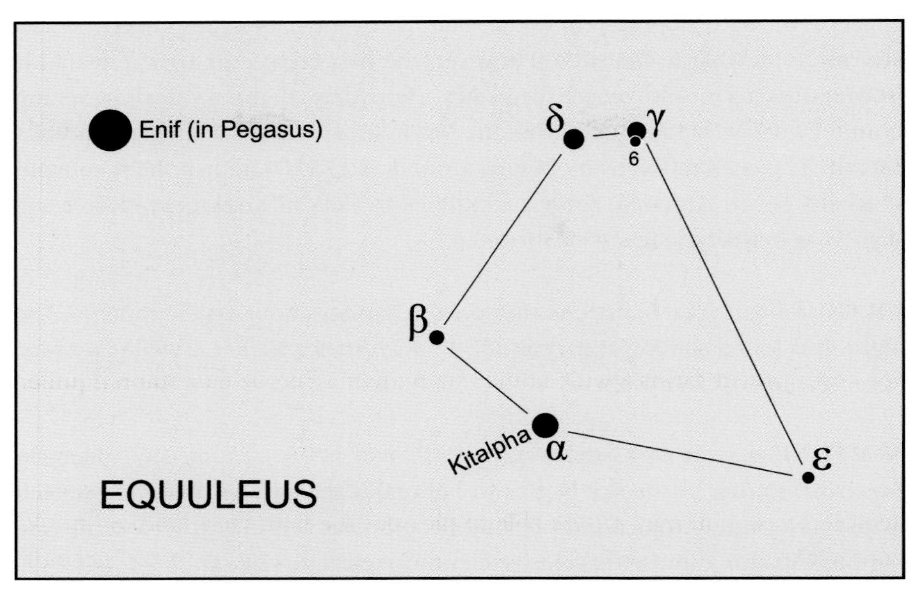

Equuleus. (Brian Jones/Garfield Blackmore)

part of the world. To find it, first of all seek out the large pattern of stars in Pegasus known as the Square of Pegasus, formed from the four bright stars Alpheratz (α Andromedae), Scheat, Markab and Algenib. For those of you observing from mid-northern latitudes this pattern of stars will be visible high in the southern sky, although from mid southern latitudes you will need to look a short way above the northern horizon (and also bear in mind that it will appear 'inverted' in relation to how it is depicted on this chart). Now follow an imaginary line from Alpheratz diagonally across the Square of Pegasus, through Markab and on roughly as far again, until you reach the relatively bright star Enif from where the small but distinctive pattern formed from the stars in Equuleus is easily tracked down.

The brightest star in Equuleus is magnitude 3.92 Kitalpha (α Equulei), its name being (appropriately) derived from the Arabic ***'qit at al-faras'*** meaning 'the Section of the Horse'. Kitalpha shines from a distance of around 190 light years which means the light we are seeing today from this star actually set off towards us during the reign of William IV.

The second brightest star in the group is Delta (δ) Equulei, which lies much closer to us, the light from this magnitude 4.47 star having taken just 60 years to reach our planet.

Much further away than either of these is faint Beta (β) Equulei which, shining at magnitude 5.16 from a distance of 330 light years, lies at roughly twice the distance of magnitude 5.30 Epsilon (ε) Equulei, the light from which reaches us from 175 light years away.

The magnitude 4.70 white giant star Gamma (γ) Equulei completes the main outline of Equuleus, the light from Gamma having set off towards us 118 years ago, just before the end of the reign of Queen Victoria. A closer look at Gamma will reveal that it forms a wide double with the magnitude 6.07 star 6 Equulei. However, the relationship between these two stars is nothing more than a line of sight effect, 6 Equulei being located at a distance of around 440 light years, nearly four times that of Gamma. Provided the sky is exceptionally dark, clear and moonless, you may just be able to pick out the faint glow from 6 Equulei and so resolve this pair with the naked eye, although you will probably need the help of a pair of binoculars which will bring out both stars quite well.

November

Full Moon: 12 November
New Moon: 26 November

MERCURY is an evening sky object shining at magnitude +0.6 at the beginning of the month. However, it is soon lost to view as it dims and undergoes inferior conjunction on 11 November. This is a particularly special conjunction as the planet is at its ascending node (passing through the ecliptic from south to north) on the same day, meaning that the Sun, Mercury and Earth are in perfect alignment. The result is a planetary transit across the face of our star, full details of which can be found in the article *The Transit of Mercury* elsewhere in this volume. The last Mercury transit was in 2016 but the next one won't occur until 2032. (It is interesting to note that the Mars rover Curiosity observed a Mercury transit from Mars in 2014!) Mercury reaches its fourth and last perihelion of the year on 16 November by which time it should be just about visible in the morning sky. Retrograde motion ceases on 20 November. It is within 2° of the Moon on 25 November and reaches greatest elongation west (20°) three days later. By the end of the month, Mercury brightens to magnitude −0.5.

VENUS is found in the vicinity of Antares, the lucida of the constellation of Scorpius, on 9 November. It has a rather closer encounter with bright Jupiter, appearing just 1.4° south of the gas giant on 24 November. Venus is the more brilliant of the two planets, shining at magnitude −3.9 to Jupiter's −1.9. Four days after that, the waxing crescent Moon passes less than 2° north of the evening star. Best seen from the southern hemisphere, Venus is finally beginning to show itself during evening twilight to observers in northern temperate latitudes. With Venus now in its waning gibbous phase, the illuminated fraction of the planet's disk reduces from 94% to 89% this month.

MARS brightens slightly this month, from magnitude +1.8 to +1.7, as it traverses Virgo. It appears less than 3° away from Spica, the alpha star of that constellation, on 9 November. Look for Mars in the east, just ahead of the Sun.

4 VESTA is the brightest member of the main asteroid belt between Mars and Jupiter. It attains opposition on 12 November in the constellation of Cetus and is nearly visible to the naked eye at magnitude +6.5.

JUPITER finally leaves Ophiuchus to enter Sagittarius on 16 November. The gas giant is drawing closer to the Sun and conjunction late next month, so it is not surprising that it is found in the vicinity of an inferior planet on 24 November. Venus and Jupiter, only 1.4° apart at their closest (approximately 12:15 UT), make a dazzling sight in the west after sunset. Four days after this planetary appulse, Jupiter is occulted by the very young crescent Moon as seen from central and southern Asia. This occultation event begins around 10:00 UT.

SATURN concludes its 2019 series of lunar occultations with two of them this month. The first occurs on 2 November when the waxing crescent Moon glides across the planet as seen from New Zealand. The event begins about 07:30 UT. The final occultation takes place on the penultimate day of the month but is visible only from Antarctica. Saturn shines at magnitude +0.6 in Sagittarius.

URANUS is well aloft by the time the Sun sets, a sixth-magnitude object south of the ecliptic in the constellation of Aries.

NEPTUNE reverses direction this month, reaching a stationary point on 27 November and resuming direct motion after five months in retrograde. Look for it (with a telescope as the planet is only magnitude +7.9) in the evening sky in Aquarius.

The Transit of Mercury

Richard Pearson

Mercury is the closest planet to the Sun, orbiting our parent star once every 88 days at a mean distance of just 48 million km. With a diameter of just 4,880 km, it is also the smallest planet in our solar system. This combination of small size and proximity to the Sun makes Mercury quite difficult to see. Often swamped by the Sun's glare, the planet always appears very close to the Sun, the only times we can actually spot Mercury being either during the early morning before sunrise or late evening after sunset, when the Sun's

During the transit of Mercury on 7 May 2003, the planet could be seen as a small, black dot on the surface of the Sun. (ESO)

glare will not interfere too much and we can find the planet hovering over the horizon.

An excellent opportunity to see Mercury will be presented on 11 November 2019 when the planet will transit the Sun, and will appear as a tiny black dot silhouetted against the solar disk. Transits are quite rare events, and in the case of Mercury there are only around 13 or 14 per century. Lasting around five and a half hours, the 2019 transit of Mercury offers an excellent opportunity

In this photograph, taken during the transit of 9 May 2016, the planet Mercury can be seen towards the lower left of image, appearing as a tiny disc silhouetted against the bright backdrop of the Sun. (NASA/Bill Ingalls)

to astrophotographers to capture the event, as well as providing a chance for astronomical societies to arrange a viewing with the public and to share the transit experience with them.

The entire event will be visible from South America, Central America, the eastern United States and Canada, and the far west of Africa. Observers in Europe, the rest of Africa and the Middle East will see the start of the transit, which will still be in progress at sunset. For observers in the central and western United States and Canada (except for Alaska) and in New Zealand, the transit will already be in progress at sunrise. The event will not be visible at all from Australia, Alaska, Asia and the far north of Norway, Sweden and Finland.

The transit begins (first contact) at 12h 35m (UT) and is at maximum extent (the time at which the planet passes closest to the centre of the Sun as seen from the centre of Earth) at 15h 20m. From the UK the event is still in progress when the Sun sets in the west, the transit of Mercury ending (last contact) at 18h 04m.

When viewed during a transit, Mercury can be seen as a tiny black dot moving slowly across the face of the Sun. Because the planet will appear so small you will need a telescope to see it. The best method to observe the event is to project an image of the Sun onto a suitable screen, such as a piece of white card, held at a convenient distance behind the eyepiece. *On no account must you attempt to look directly at the Sun through a telescope or binoculars without proper protection, even when it is either obscured by mist or fog or located low in the sky. If you do, then severe and permanent eye damage, even blindness, can result.* You should also bear in mind that modern telescopes often have plastic components inside the draw tube and eyepieces. To avoid the possibility of melting these, attention should be paid to keeping the beam of sunlight central in the telescope.

As we have seen, Mercury is often difficult to see due to its proximity to the Sun and is often lost in the twilight sky. Details of the best times to observe the planet are given in the Monthly Sky Notes which will confirm its position throughout the year. Mercury is at its most accessible for a few days either side of the times it reaches its greatest separation (elongation) from the Sun. At these times, a pair of binoculars will help you to track the planet down and, once identified, you may be able to spot Mercury with the naked eye. *Please remember not to carry out a search for the planet until the Sun is below the horizon, to avoid any possibility of the solar glare causing damage to your eyes.*

December

Full Moon: 12 December
New Moon: 26 December

MERCURY is best seen in the northern hemisphere during its last morning apparition of the year but it isn't visible for long, vanishing by mid-month. On 30 December, it reaches aphelion for the fifth and final time in 2019. Mercury is at magnitude −0.5 for most of the month, brightening somewhat toward the end of the year.

VENUS reaches its peak altitude as the evening star for those observing it from southern latitudes but the view from the northern hemisphere continues to improve into next year as the bright planet climbs ever higher above the western horizon. Setting two hours after sunset, it also begins to brighten slightly after spending the last few months at magnitude −3.9. The decrease in phase (89% to 82%) is more than offset by the decrease in distance between Venus and Earth and subsequent increase in apparent disk diameter (11.7 arc-seconds to 13.1 arc-seconds over the course of the month). It is found 1.8° south of Saturn on 11 December and finishes off the year with a lunar occultation on 29 December. Only those living in the southernmost parts of South America will see this event which begins around 02:00 UT.

EARTH is at solstice on 22 December. The Sun is at its most southerly declination, resulting in the longest day of the year in the southern hemisphere and the start of astronomical summer. Meanwhile, the northern hemisphere endures the shortest day of the year and the onset of astronomical winter. The final eclipse of the year, an annular solar eclipse, takes place on 26 December.

MARS arrives in the constellation of Libra on the first day of December. The old crescent Moon passes less than 4° away from the red planet on 23 December. Mars is found in the east before sunrise, slowly brightening from magnitude +1.7 to +1.6 by month's end.

JUPITER reaches its maximum declination south for the year on 8 December whilst in Sagittarius. It is difficult to observe this month and soon vanishes in the west, with a lunar occultation occurring on 26 December and conjunction with the Sun the day after. It returns to the morning sky next year.

SATURN is drawing closer to the Sun as it approaches conjunction in January. Still in Sagittarius, it is an evening sky object shining at magnitude +0.6 and is difficult to see from northern latitudes where it sets not long after the end of twilight. Saturn makes a pretty picture in the west with Venus on 11 December when the two planets are less than 2° apart. The very young crescent Moon pays a visit on 27 December but is not quite close enough to occult the planet this time.

URANUS is an evening sky object in Aries, setting in the hours just before dawn. At magnitude +5.7, it is difficult to spot except in the darkest of skies.

NEPTUNE attains east quadrature on 8 December in Aquarius. It sets around or just before midnight and is best viewed from the southern hemisphere where it is high above the western horizon at sunset. As Neptune is only magnitude +7.9, a small telescope will be necessary to observe it.

William Frederick Denning

Richard Pearson

The Ursids is a minor meteor shower which runs annually from 17 to 25 December and produces between 5 and 10 meteors per hour. The shower peaks on the night of 21/22 December 2019, meaning that the waning crescent moon should not interfere too much and skies should be dark enough for what could be a good show. Best viewing will be just after midnight from a dark location far away from city lights.

The Ursids is associated with comet 8P/Tuttle which was first spotted on 9 January 1790 by the French astronomer Pierre François André Méchain. The last observation of the comet was on 1 February of that year, resulting in an observed arc of just 24 days. This was insufficient to allow astronomers to identify the object as being periodic, and so no accurate orbit was calculated. However, the comet was accidentally rediscovered by the American astronomer Horace Parnell Tuttle on 5 January 1858 and has been observed on every return since, apart from during the unfavourable apparition of 1953.

William Frederick Denning. (Courtesy of the Library and Archives of the Royal Astronomical Society)

During the period in which the Earth passes through the dusty tail of 8P/Tuttle, the resulting meteors appear to radiate from a point close to the star Beta (β) Ursae Minoris (Kochab), although they can appear anywhere in the sky.

The Ursids were first identified by the English astronomer William Frederick Denning who spent a lot of his time studying meteor showers. Born on 25 November 1848 at the village of Redpost, near Radstock, in Somerset, William Denning was the eldest of four children born to Isaac Poyntz and Lydia (née Padfield) and was destined to become one of the leading astronomers of his time. Little is known of William Denning's early childhood although in 1856, when William was just eight years old, he and his family moved from the rural settings of Redpost to the city of Bristol. Here his father Isaac established the accountancy partnership of Denning, Smith and Company, working in this business until his death in 1884. The Denning family had many relatives in the Bristol area, and William lived in the city for the remainder of his life.

William turned his attention to astronomy at the age of 17, and in 1866 purchased a 4¼-inch refractor. However, following his observation of the great Leonid meteor shower in 1866 (which put on a particularly impressive display at the time), William decided to devote most of his attention to meteors, a

William Denning owned and used several telescopes, including this 10-inch alt-azimuth mounted Browning reflector with which he carried out many of his cometary searches. (Courtesy of the Library and Archives of the Royal Astronomical Society)

branch of astronomy in which no telescope is necessary. His most celebrated work was devoted both to the determination of meteor shower radiant points and to the main characteristics of the showers themselves.

In 1867 he became a journalist, writing a series of astronomy articles for the local newspapers, and sent his first paper (on the Andromedids meteor shower) to the Royal Astronomical Society (RAS) in 1872. Elected a Fellow of the RAS in 1877, he continued to pinpoint the radiants of many more meteor showers, including that for the Lyrids, following his observations of the shower in 1873 and 1874. His first list, published in 1876, contained 27 radiants derived from

Lunar Orbiter 1 image of Denning lunar crater. (James Stuby/NASA)

his observations at Bristol. This was followed by the publication of further lists in 1890, 1912 and 1923. It was his list of 918 radiants, deduced from 9,177 observed paths between 1873 and 1890 that led to the committee of the RAS awarding him their Gold Medal in 1898. His work on meteors also led to him receiving the Valz Prize from the French Academy of Sciences on 23 July 1890.

Denning's enthusiasm for observing the night sky is considered the motivation for his discovery of several comets, including 1881 V on 4 October 1891 in Leo (this being the short-period comet 72P/Denning-Fujikawa which was lost until its accidental rediscovery in October 1978 by Shigehisa Fujikawa). His other comet discoveries include those of 1890 VI on 23 July 1896 in Ursa Minor; 1892 II on 18 March 1892 near Delta Cephei; and 1894 I on 26 March 1894 in Leo Minor. He is also credited with the independent discovery, on 31 March 1891, of the comet 1891 I (Barnard-Denning), first seen by American astronomer Edward Emerson Barnard the previous evening. William is also recognised as being one of the discoverers of Nova Aquilae in 1918 and with the discovery of Nova Cygni in 1920.

William Denning wrote just one book, *Telescopic Work for Starlight Evenings*, which was published in 1891. In 1895, he summed up his enthusiasm for the night sky when he wrote that he had "... been engaged, as an amateur astronomer, in observing celestial objects or in exploring the heavens, since 1865."

William Frederick Denning died at his home in Bristol on 9 June 1931. He had never married and had no children. Craters on both the Moon and Mars have been named in his honour. It can indeed be said that today William Frederick Denning remains an icon for amateur astronomers who sit in their deck-chairs and recliners, holding a red torch and patiently recording shooting stars as they flash across the clear night sky.

Comets in 2019

Neil Norman

If comets are your astronomical passion and you long for another impressive display like that of Comet Hale-Bopp (which passed perihelion on 1 April 1977 and was one of the brightest and most widely-observed comets of the 20th century), then 2019 is not destined to be your year ...

Of the 69 comets that are due to reach perihelion this year, 54 are short-period comets belonging to the Jupiter family; three are defunct (being asteroidal in appearance or even lost); four are long-period comets with orbit periods of 200 years or more; and the remaining eight are objects of asteroidal appearance that will display a coma at times, thereby indicating their cometary nature. The comets that are of potential interest to those without expensive CCD devices or very large telescopes are all at their visual best between January and February of 2019 and are as follows.

46P/Wirtanen

The original target for the highly successful Rosetta mission, this comet passed perihelion in late-2018, approaching to within 11,620,000 kilometres (7,220,000 miles) of Earth in December of that year. Discovered by the American astronomer Carl Alvar Wirtanen on 17 January 1948, 46P/Wirtanen has an orbital period of 5.4 years and begins 2019 as a 4th magnitude object in the constellation of Camelopardalis. Although visible to the naked eye, it will be fairly large and diffuse, and you may need a pair of binoculars to help you see it. The comet passes through Lynx and enters Ursa Major in mid-January, dropping in magnitude as it does so. 46P/Wirtanen remains well placed as we move into March, although beginning to fade out of binocular range.

DATE	R.A.	DEC	MAG	CONSTELLATION
1 Jan 2019	07 08 04	+59 23 44	4.6	Lynx
15 Jan 2019	08 57 41	+60 00 23	6.2	Ursa Major
1 Feb 2019	09 31 17	+55 22 55	7.9	Ursa Major
15 Feb 2019	09 37 12	+50 48 16	9.3	Ursa Major
1 Mar 2019	09 41 30	+45 57 34	10.6	Ursa Major

Comet 46P/Wirtanen imaged on 6 February 2008. (C. Rinner, F. Kugel - Observatoire de Dauban, Banon, France)

38P/Stephan-Oterma

Discovered in January 1867 by the French astronomer Jérôme Eugène Coggia, this Halley-type comet has an orbital period of 38 years and was at perihelion in late-2018. 38P/Stephan-Oterma commences 2019 at around magnitude 10 which means that a small telescope will be required to see it. The comet passes through Lynx during January and early-February, reducing in magnitude as it moves ever further away from the inner reaches of the Solar System, so do take every opportunity to go out and view it.

DATE	R.A.	DEC	MAG	CONSTELLATION
1 Jan 2019	08 37 22	+40 53 29	10.0	Lynx
15 Jan 2019	08 36 05	+44 57 21	11.0	Lynx
1 Feb 2019	08 30 17	+47 23 20	11.5	Lynx
15 Feb 2019	08 28 03	+47 27 55	12.0	Lynx

64P/ Swift-Gehrels

Originally discovered on 16 November 1889 by Lewis A. Swift, who described it as a '... faint, round, nebulous mass, without a tail ...', this comet was missed on future returns until Tom Gehrels accidentally rediscovered it as a 19th magnitude object on 8 February 1973 during a search for Apollo-type minor planets. Orbital calculations revealed the sighting to be a return of 64P, and the comet subsequently received the designation 64P/Swift-Gehrels. Its orbital period has been determined as 9.23 years.

64P/Swift-Gehrels has a current perihelion distance of 1.95 AU, which is far enough away from the Sun to ensure that the volatile ices within the comet do not receive enough solar radiation to fully sublimate, the result being that the nucleus does not produce much of a show for the casual observer. The comet returned to perihelion in November 2018 and remains well placed for

Comet 64P/Swift-Gehrels imaged on 1 October 2009. (C. Rinner, F. Kugel - Observatoire de Dauban,Banon, France)

observation going into 2019, when it can be found in the constellation of Aries at the beginning of the year. 64P/Swift-Gehrels crosses into Taurus during early- to mid-January, although because it is quite faint, a small to medium sized telescope will be required in order to detect it.

DATE	R.A.	DEC	MAG	CONSTELLATION
1 Jan 2019	02 57 43	+29 54 47	10	Aries
15 Jan 2019	03 31 06	+28 16 56	11	Taurus
1 Feb 2019	04 10 04	+26 49 06	12	Taurus
15 Feb 2019	04 40 54	+25 54 20	13	Taurus

29P/ Schwassmann-Wachmann

This comet was discovered by German astronomers Arnold Schwassmann and Arno Arthur Wachmann on photographic plates taken at the Hamburg Observatory on 15 November 1927. A search for precovery images revealed that the comet had been photographed as a 12th magnitude object on 4 March 1902.

29P/Schwassmann-Wachmann undergoes regular outbursts, during which its brightness can vary by anything between 1 and 4 magnitudes. At the time of discovery the comet was in outburst, resulting in a discovery magnitude of around 13, somewhat brighter than its normal magnitude of around 16 which it displays when in a 'stable' state. Outbursts occur at an average rate of 7.3 per year, and generally fade over a couple of weeks or so. Observation has shown that, during particularly strong outbursts, the brightness of the comet can range from 9th to 19th magnitude, which suggests that its surface must be undergoing highly changing processes and perhaps playing host to some fascinating geology.

This comet is believed to be a member of the class of objects known as 'centaurs', of which around 80 are currently known, and all of which travel around the Sun between the orbits of Jupiter and Neptune. Orbital integrations indicate that this comet and other like it are escapees from the Kuiper Belt, their migrations being due to the gravitational perturbations of Neptune. The gravitational effects of Jupiter will probably result in 29P/Schwassmann-Wachmann eventually migrating either inward or outward of the Solar System.

For the past few years 29P/Schwassmann-Wachmann has been located below the celestial equator, thereby favouring observers in the southern

Morphology of 29P/Schwassmann-Wachmann from 7 August 2016 to 22 October 2017 taken from Swan Hills, New South Wales, Australia. (Justin Tilbrook)

hemisphere. However, following perihelion on 7 March 2019, the comet will reach opposition in October, situated north of the celestial equator in Pisces and so favouring observers in the northern hemisphere.

DATE	R.A.	DEC	MAG	CONSTELLATION
1 Nov 2019	00 31 32	+14 22 36	14?	Pisces
15 Nov 2019	00 27 24	+13 45 12	14?	Pisces
1 Dec 2019	00 25 06	+13 11 43	14?	Pisces
15 Dec 2019	00 25 28	+12 54 38	14?	Pisces

Minor Planets in 2019

Neil Norman

We currently know of more than 700,000 minor planets, with around 450,000 or so having been allotted a permanent number due to their orbits being well defined. Of these, over 16,000 have been given names ranging from musicians to scientists and from places and plants to animals and mythological characters.

The vast majority of these objects lie within the main asteroid belt, located between the orbits of Mars and Jupiter. However, some of these rocky travellers, known as Potentially Hazardous Asteroids (PHAs), travel around the Sun in larger elliptical orbits that often bring them into close encounters with the planets. To date there are 1,786 known PHAs with around 150 of these believed to have diameters of more than a kilometre. To qualify as a PHA these objects must have the capability to pass within 8 million km (5 million miles) of Earth and to be larger than 100 metres across.

A large number of smaller asteroids pass close to Earth on a regular basis and can range in size from a couple of meters to tens of meters in size. Of course, much smaller ones routinely enter the Earth's atmosphere and burn up harmlessly as meteors.

The dedicated asteroid observer, or more serious astronomer interested in this particular branch of astronomy, should go to the home page of the Minor Planet Center, whose job it is to catalogue and keep track of these objects as they are discovered. This page can be accessed by visiting **www.minorplanetcenter.net** Once here you will see a table of newly discovered minor planets and Near Earth Objects (NEOs). This is the main page and should be consulted on a regular basis as it is updated daily. At the top of the page is a search box that you can use to find information on any object that you are interested in, and from this you can obtain ephemerides of the chosen subject. Requests for observations are often requested by the Minor Planet Center, and backyard astronomers can play a useful role in monitoring these objects.

We begin our look at 2019 with an object designated (89959) 2002 NT7 and with a diameter thought to be in the range of 2 km. Discovered on 9 July 2002, by the

Lincoln Near Earth Asteroid Research team (LINEAR) in Socorro, New Mexico, this object caused some initial concern once its orbit had been determined. (89959) 2002 NT7 was placed with a positive numbered rating on the Palermo Technical Impact Hazard Scale at +0.06, indicating a higher threat of future Earth impact. (The Palermo Scale is a logarithmic scale used by astronomers to rate the future potential impact risk of an object with Earth. Any object with a value between −2 and 0 warrants careful monitoring, with objects rated 0 to 2 being the most likely to impact Earth within 100 years). The risk posed by (89959) 2002 NT7 was downgraded on 1 August 2002 after further observations enabled astronomers to refine the orbit. (89959) 2002 NT7 now has an observational arc of over 60 years with a number of precovery images located. Orbiting the Sun over a period of 835 days, with a perihelion at 0.81 AU and aphelion at 2.65 AU, (89959) 2002 NT7 will make a close passage to Earth on 13 January 2019 when it will approach to within 0.40 AU (37,910,000 miles) of our planet.

During its forthcoming close approach to Earth, this object will be exceedingly faint, and observers will require large telescopes to locate it. (89959) 2002 NT7 can be located from mid-northerly latitudes at the following positions:

DATE	R.A.	DEC	MAGNITUDE	CONSTELLATION
12 Jan 2019	14 02 42	−06 09 38	17.1	Virgo
13 Jan 2019	14 05 10	−04 24 52	17.1	Virgo
14 Jan 2019	14 07 37	−02 39 53	17.1	Virgo

The following minor planets are well placed for observation during 2019 and make ideal targets for backyard astronomers equipped with only moderate optical aid.

2 Pallas

The second asteroid to be found, Pallas was discovered on 28 March 1802 by the German astronomer Heinrich Wilhelm Matthias Olbers. With a diameter of 512 km (318 miles) and an orbital period of 4.61 years, its path

Lithograph portrait of the German astronomer Heinrich Wilhelm Matthias Olbers by the German portrait painter and lithographer Rudolf Suhrlandt. (Wikimedia Commons)

around the Sun is highly eccentric and quite steeply inclined to the main plane of the asteroid belt.

DATE	R.A.	DEC	MAGNITUDE	CONSTELLATION
1 Jan 2019	13 23 07	−06 03 36	9.0	Virgo
1 Feb 2019	13 58 09	−02 13 10	8.6	Virgo
1 Mar 2019	14 12 08	+04 55 12	8.2	Virgo
1 Apr 2019	14 03 01	+15 18 58	7.9	Boötes
1 May 2019	13 40 46	+22 30 02	8.2	Boötes
1 Jun 2019	13 28 38	+23 53 46	8.9	Coma Berenices
1 Jul 2019	13 36 36	+21 24 09	9.4	Boötes

3 Juno

Discovered by the German astronomer, Karl Harding on 1 September 1804, and with a diameter of around 230 km (143 miles), Juno is the eleventh largest asteroid and one of the two largest stony (S-type) asteroids. Its mean distance from the Sun is 2.6 AU, its journey around our star taking 4.36 years to complete.

DATE	R.A.	DEC	MAGNITUDE	CONSTELLATION
1 Jan 2019	03 32 45	−02 37 14	8.2	Eridanus
1 Feb 2019	03 51 30	+02 38 07	8.8	Taurus
1 Mar 2019	04 27 46	+07 41 18	9.3	Taurus
1 Apr 2019	05 20 55	+12 07 15	9.7	Orion

4 Vesta

Vesta as imaged by the Dawn spacecraft in 2011. (NASA/JPL-Caltech/UCLA/MPS/DLR/IDA)

Another discovery by Heinrich Wilhelm Matthias Olbers, Vesta was first seen on 29 March 1807 and is one of the largest asteroids, with a diameter of 525 km (326 miles) and an orbital period of 3.63 years.

DATE	R.A.	DEC	MAGNITUDE	CONSTELLATION
1 Aug 2019	03 12 15	+10 30 38	8.0	Aries
1 Sep 2019	03 40 40	+11 03 16	7.7	Taurus
1 Oct 2019	03 48 07	+10 31 47	7.2	Taurus
1 Nov 2019	03 29 26	+08 55 53	6.6	Taurus
1 Dec 2019	02 59 40	+08 00 09	6.8	Cetus

5 Astraea

Another S-type object, Astraea was discovered by the German amateur astronomer Karl Ludwig Hencke on 8 December 1845 and has a diameter of 119 km (74 miles), orbiting the Sun once every 4.12 years. Astraea has a highly reflective surface, which suggests that it is composed of a mixture of nickel-iron with silicates of magnesium and iron.

DATE	R.A.	DEC	MAGNITUDE	CONSTELLATION
15 Sep 2019	06 53 39	+18 21 06	11.7	Gemini
15 Oct 2019	07 41 18	+16 34 44	11.3	Gemini
15 Nov 2019	08 17 01	+14 36 42	10.8	Cancer
15 Dec 2019	08 29 31	+13 56 24	10.0	Cancer

6 Hebe

Discovered on 1 July 1847 by Karl Ludwig Hencke, this asteroid has a mean diameter of 186 km (116 miles) and orbits the Sun once every 3.78 years. Hebe has a high bulk density and contains around 50% of the mass of the asteroid belt, indicating that it is a very solid body which has not undergone impacts or collisions. This is somewhat unusual for an asteroid of this size for which multiple impacts are the norm. Hebe is most likely to be the parent body of the H type ordinary chondrites, the most common type of meteorite and which account for about 40% of all meteorites hitting Earth.

DATE	R.A.	DEC	MAGNITUDE	CONSTELLATION
1 Jan 2019	06 18 56	+05 47 40	8.5	Orion
1 Feb 2019	05 57 02	+10 52 52	9.2	Orion
1 Mar 2019	06 02 18	+15 14 21	9.8	Orion
1 Apr 2019	06 31 18	+18 32 26	10.4	Gemini
1 May 2019	07 12 59	+20 00 35	10.8	Gemini

1L/2017 U1 'Oumuamua

A remotely-taken image of 1L/2017 U1 'Oumuamua (previously A/2017 U1) built up from 25, 150-second exposures taken on 26 October 2017 with the 16-inch, f/3.75 robotic unit ('Pearl') of Tenagra Observatories in Arizona. The object is extremely faint (magnitude 22.2) and marked here with two red lines. The telescope tracked the apparent motion of the object across the stars. (Gianluca Masi/The Virtual Telescope Project and Michael Schwartz/Tenagra Observatories)

An unusual designation (and one not seen before) this object was discovered on 18 October 2017 by the Panoramic Survey Telescope and Rapid Response System (Pan-STARRS) when located 0.2 AU (30,000,000 km) from Earth. Initially believed to be a comet, and previously designated C/2017 U1 (PANSTARRS), it was reclassified as an asteroid on 26 October 2017, with a designation of A/2017 U1 (later named 1L/2017 U1 'Oumuamua).

Although not sounding particularly worthy of attention, 1L/2017 U1 has the highest known orbital eccentricity of any object in the Solar System. Its hyperbolic orbit took it past perihelion on 9 September 2017 and to within 0.16 AU (24,000,000 km) of the Earth on 5 October 2017. This object appears to have come from the direction of the star Vega in the constellation Lyra, and is the first confirmed example of an interstellar object entering the inner Solar System.

Large variations in the light curves of 1L/2017 U1 ‘Oumuamua reveal it to be a highly elongated object measuring approximately 180m × 30m × 30m (assuming it to have an albedo of 10%). At time of writing, the object is already at a distance of around 2 AU from the Sun and receding from us at a rate of 38 km/s.

An article on 1L/2017 U1 ‘Oumuamua and interstellar objects will appear in the 2020 edition of the Yearbook of Astronomy.

Meteor Showers in 2019

Neil Norman

On any given night of the year one can reasonably expect to see several meteors dash across the sky. Quite often these will be 'sporadic' meteors, that is to say they can appear at any time and from any direction. These are usually nothing more than pieces of space debris, ranging from grain-sized pieces of rock to materials lost during rocket launches or space walks. However, at certain times of the year the Earth encounters more organised streams of debris that produce meteors over a regular time span and which seem to emerge from the same point in the sky. These are known as meteor showers.

These streams of debris are the scattered remnants of comets that have made repeated passes through the inner solar system. The ascending and descending nodes of their orbits lie at or near the plane of the Earth's orbit around the Sun, the result of which is that at certain times of the year the Earth encounters and passes through a number of these swarms of particles. The term 'shower' must not be taken too literally. Generally speaking, even the strongest annual showers will only produce one or two meteors a minute at best, this depending on what time of the evening or morning that you are observing and the lunar phase at the time.

Noting down the colours of meteors will tell you something about their composition. For example red is nitrogen/oxygen, yellow is iron, orange is sodium, purple is calcium and turquoise is magnesium. To avoid confusion with sporadic meteors, be sure to have a star map with you, so you can compare the trajectories against it and, hopefully, trace their origins back to the shower radiant.

When observing meteors it is always good to make notes of what you see. For example, how many meteors you actually observe; how many of these are either shower or sporadic meteors; the appearance of any bolides (bright sustained meteors); any observable colour of a meteor; and whether there was any noise associated with meteors seen. The thermal stress of entering the atmosphere at such large velocities can cause some bolides to break apart, an event which can produce the audible 'crackling' sound that has often been documented by observers.

A table of the principal meteor showers for 2019 is given here, the information it provides including the name of the shower; the period over which the shower is active; the ZHR (Zenith Hourly Rate); the parent object from which the meteors originate; the date of peak shower activity; and the constellation in which the radiant of the shower is situated. Most of this is self-explanatory, although the ZHR may need a little explanation. The Zenith Hourly Rate is the number of meteors you may expect to see if the radiant (the point in the sky from where the meteors appear to emerge) is at the zenith (or overhead point) and if the observing conditions were perfect and included dark, clear and moonless skies with no form of light pollution whatsoever. However, the ZHR should not to be taken as gospel, and you should not expect to actually see the quantities stated, although 'outbursts' can occur with significant activity being seen.

Meteor Showers in 2019

SHOWER	DATE	ZHR	PARENT	PEAK	CONSTELLATION
Quadrantids	1 Jan to 5 Jan	120	2003 EH1 (asteroid)	3/4 Jan	Boötes
Lyrids	16 Apr to 25 Apr	18	C/1861 G1 Thatcher	22/23 Apr	Lyra
Eta Aquarids	19 Apr to 28 May	30	1P/Halley	6/7 May	Aquarius
Delta Aquarids	12 Jul to 23 Aug	20	96P/Machholz	28/29 Jul	Aquarius
Perseids	17 Jul to 24 Aug	80	109P/Swift-Tuttle	12/13 Aug	Perseus
Orionids	16 Oct to 27 Oct	20	1P/Halley	21/22 Oct	Orion
Taurids	10 Sep to 20 Nov	5	2P/Encke	5/6 Nov	Taurus
Leonids	15 Nov to 20 Nov	Varies	55P/Tempel/Tuttle	17/18 Nov	Leo
Geminids	7 Dec to 17 Dec	75+	3200 Phaethon	13/14 Dec	Gemini

Quadrantids

The Quadrantids is a significant shower and at its peak, which takes place on the night of 3/4 January, observed rates can reach over 100 meteors an hour. The radiant lies a little to the east of the star Alkaid in Ursa Major and the meteors emerging from it are rapid, reaching speeds of 40 km/s or more. A drawback to this shower is that the period of maximum occurs over a period of just 2 or 3 hours. The parent object has been tied down to the Amor asteroid 2003 EH which is likely to be an extinct comet. The moon will be a thin crescent this year and should not interfere with what could be a good show.

Lyrids

These meteors move quite rapidly and can clock up speeds of up to 48 km/s. The rates vary, but typical values of 10 to 15 per hour are recorded. The maximum falls on the night of 22/23 April with the radiant lying near the prominent star Vega in the constellation Lyra. The parent of the shower is the long-period comet C/1861 G1 Thatcher, which last came to perihelion on 3 June 1861 having passed Earth at a distance of almost 50 million kilometres (31 million miles) in May of that year. The period of the comet is 415 years and it will next approach perihelion in 2280. This year, many of the fainter Lyrids will be blocked out by the waning gibbous moon, although clear skies should still allow you to catch a few of the brightest ones.

Eta Aquarids

This is one of the two showers associated with 1P/Halley and is active for a full month between 19 April and 21 May. The radiant lies just to the east of the star Sadalmelik in Aquarius and maximum activity will occur pre-dawn on 7 May with up to 30 meteors per hour expected. These will be travelling at 65 km/s, the high speed due to the parent comet being in a retrograde orbit resulting in the debris entering the atmosphere head-on. The thin crescent moon will set early in the evening leaving dark skies for what should be a good show.

Delta Aquarids

The Delta Aquarids are a long, drawn-out meteor shower which was not officially recognised until the 1950s. The shower coincides with the much more prominent Perseids, although Delta Aquarid meteors are generally much dimmer than those associated with the Perseids, making identification much easier for the observer.

The peak of the shower occurs during the early morning of 29 July each year, and in 2019 a waning crescent moon will not present too much of a problem. The radiant lies close to the star Skat in Aquarius, situated to the south of the Square of Pegusus and a little to the north of the bright star Fomalhaut in Piscis Austrinus. Although never very high above the horizon as seen from mid-northern latitudes, the radiant is well placed for those observers situated in the southern hemisphere.

The parent object was once thought to be a member of the Marsden/Kracht sungrazer group, but now it is strongly suspected that the Delta Aquarids are linked to the short-period sungrazing comet 96P/ Machholz, which was discovered on 12 May 1986 by the American amateur astronomer Donald Machholz.

Perseids

A (potentially) gorgeous sight to see! I have personally noted a variation in yearly amounts, some displays in recent years being rather under par, this following the 2012 and 2013 displays which were more in line with what can be reasonably expected. The Perseids are also fast moving, clocking in at 58 km/s, due to the parent comet (109P/Swift-Tuttle) being in a retrograde orbit. Up to

A Perseid meteor captured by Guy Wells in August 2015. (Northolt Branch Observatories, England)

80 meteors per hour can be seen at their peak, which occurs on 12/13 August, and large fireballs are often observed, with some even seen to cast shadows. A note of caution though – there are a few other showers active at around this time, so inexperienced observers should ensure they follow the trajectories of any meteors seen back to the radiant point in the northern reaches of Perseus. The nearly full moon will block out most of the fainter meteors this year, but the Perseids are so bright and numerous that it could still be a good show.

Orionids

This is the second of the meteor showers associated with 1P/Halley, occurring between 16 and 27 October with a peak on the night of 21/22 October. The radiant of this shower is situated a little way to the north of the red super giant star Betelgeuse in the shoulder of Orion the Hunter, and the Orionids are best

An impressive Orionid fireball photographed by David Rankin, Utah, U.S.A. on 21 October 2017. (David Rankin)

viewed in the early hours leading up until dawn when the constellation is at its highest. The velocity of the meteors entering the atmosphere is a speedy 67 km/s. Although a number of the fainter meteors may be blocked out by the third quarter moon, the Orionids tend to be fairly bright so the 2019 display could still be a good show.

Taurids

This shower is associated with periodic comet 2P/Encke which was last at perihelion in early-2017 and which is the comet with the shortest known orbital period (3.3 years). The stream of debris left by the comet is truly vast, and is by far the largest in the inner Solar System, and with particles of a larger mass than the other showers, fireballs are widely reported with this shower. The reason for this is believed to be that a much larger comet fragmented some 20,000 to 30,000 years ago, leaving behind a large piece that we know today as 2P/Encke, together with a huge amount of smaller debris. The Tunguska event of 1908 may well be related to this stream, though this remains a debatable issue.

The Taurid radiant lies a little to the south of the Pleiades open star cluster. The ZHR is low, and only five or so can be realistically expected, although they are worth waiting for as they appear to glide across the sky at a distinctly sedate pace of around 27 km/s per hour. The moon will be at first quarter for the 2019 maximum, setting shortly after midnight and leaving dark and hopefully-clear skies for viewing.

Leonids

This is another fast moving shower with atmosphere impact speeds of 72 km/s and containing a lot of larger sized pieces of debris, with diameters in the order of 10mm and masses of around half a gram. These can create lovely bright meteors that occasionally attain magnitude −1.5 or better. It is interesting to note that each year around 15 tonnes of material is deposited over our planet from the Leonid stream. The shower peaks on the night of 17/18 November, although because the moon will be approaching third quarter phase at the time of the 2019 maximum, its light will drown out many of the fainter meteors.

The parent of the Leonid shower is the periodic comet 55P/Tempel Tuttle which orbits the Sun every 33 years and which was last at perihelion in 1998 and is due to return in late-May 2031. The radiant is located a few degrees to the north of the bright star Regulus in Leo. The Zenith Hourly Rates vary due to the Earth encountering material from different perihelion passages of the parent comet. For example, the storm of 1833 was due to the 1800 passage, the 1733 passage was responsible for the 1866 storm and the 1966 storm resulted from the 1899 passage.

Geminids

The Geminids were first recorded in 1862 and the debris stream has clearly been perturbed into its current orbit by Jupiter. The parent of the shower is the object 3200 Phaethon, an asteroid that is in many ways behaving like a comet. Discovered in October 1983, this rocky 5-kilometre wide object is classed as an Apollo asteroid and has an unusual orbit that takes it closer to the Sun than any other named asteroid. Classified as a potentially hazardous asteroid (PHA), 3200 Phaethon made a relative close approach to Earth on 10 December 2017, when it came to within 0.069 AU (10.3 million kilometres / 6.4 million miles) of our planet.

Geminid meteors travel at speeds of 35 km/s and disintegrate at heights of around 40 kilometres above the Earth's surface. The shower radiates from a point close to the bright star Castor in Gemini and peaks on the night of 13/14 December. This is considered by many to be the best shower of the year, and it is interesting to note that the number of observed meteors appears to be increasing each year, although as far as the 2019 shower is concerned, the nearly full moon will block out many of the fainter Geminids ensuring a less-than-spectacular display.

Article Section

Astronomy in 2018

Rod Hine

Life on Exoplanets?

There are few topics that have inspired so many interesting and ingenious research projects as exoplanets and the possibility of discovering evidence of extra-terrestrial life. This year has seen some remarkable advances in techniques for discovering new exoplanets and examining them for habitability, as well as theoretical studies of possible ways that life might be created and sustained in even the most unlikely circumstances.

The star system TRAPPIST-1, first reported in 2016, continues to inspire much interest. Consisting of a cool red dwarf star surrounded by no less than seven planets somewhat similar to Earth, it is a mere 40 light-years from Earth and thus within reach of the most powerful telescopes. Transit observations have given estimates of the sizes and masses of the individual planets and, based on this limited amount of information, Amy Barr of the Planetary Science Institute, Tucson, Arizona, has built a convincing picture of the composition and likely environment of each planet. The basic assumption is that the planets are composed of a mixture of iron, rock and water in correct proportions to give the observed size and mass. From the distance from the star and the star's temperature it was then possible to deduce the state of these materials and hence some idea of the conditions on the planet including any tidal effects. The planets closer to the star could well harbour liquid magma due to tidal heating while the low density of the outermost planets indicates the presence of water in the form of ice. Overall it seems that two of the planets, d & e, could provide a habitable environment with liquid water and an atmosphere.

Initial studies of TRAPPIST-1 pointed to the fact that small cool red dwarf stars are prone to outbursts of X-ray and UV flares that would be inimical to survival of life even within an otherwise habitable zone. However, it has also been suggested that the right amount of such energy could provide the stimulus for the reactions to produce the molecules necessary for life, and the long lifetime of red dwarf stars would give plenty of time for some kind

This is an artist's impression of the TRAPPIST-1 system, showcasing all seven planets in various phases. When a planet transits across the disk of the red dwarf host star, as two of the planets here are shown to do, it creates a dip in the light from the star, which can be detected from Earth. (NASA)

of life to emerge. A team led by Matt Tilley at University of Washington has proposed that a thick haze, such as found in the atmosphere of Titan, could survive solar flares and shield life on the surface. If such scenarios are possible then the abundance of red dwarf stars could provide even more opportunities for life in the universe.

Most Distant Quasar

Eduardo Bañados and his colleagues at the Observatories of the Carnegie Institution for Science in Pasadena, California, have reported the existence of a quasar with a red-shift $z = 7.54$. Not only does this make it the most distant quasar so far observed but it is also the brightest. It is 4×10^{13} times brighter than the sun and 800 million times more massive than the sun. Therein lies a significant problem for current theories on the formation of black holes, by which it should take billions of years for such large black holes to form by accretion. From the red-shift data this object must have existed a mere 690 million years after the formation of the universe – clearly there must be some mechanism that allows the direct formation of black holes of huge mass, perhaps three or four orders of magnitude more massive than the Sun. One possibility is the rapid collapse of massive gas clouds into black holes early in the life of the universe, after which many such black holes could merge to form objects such as this quasar. Alternative theories involving the merger of huge early stars require too long as there is a limit to the rate at which a star can grow due to the increasing radiation pressure impeding the inflow of gas.

Another study of this quasar, headed by Bram Venemans of the Max Planck Institute for Astronomy, Heidelberg, Germany, showed that the galaxy that hosts the quasar is rich in highly luminous dust, including metals. This implies a high rate of star formation at a very early stage in the life of the universe. A final curious feature of this observation is that the spectrum of light from the quasar is missing some wavelengths, which suggests that the environment around the host galaxy was rich in neutral hydrogen. In other words, the period of reionisation may have been in progress. In our epoch, most gas in interstellar space is already ionized by the radiation from stars and galaxies. It seems therefore that the early universe was very different from our own.

Hubble took this photograph of 3C273 in 2013. Discovered in the constellation of Virgo in 1962, this object lies at a distance of 2.44 billion light years and is one of the nearest quasars to the Solar System. It is also the brightest, its apparent magnitude of 12.9 making it the most distant celestial object bright enough to be seen by amateur astronomers. This superb image clearly shows a jet of charged particles moving at relativistic speeds. Such fast-moving particles travel in a spiral path due to the local magnetic fields thus producing synchrotron radiation with a distinctive non-thermal spectrum and polarisation. (ESA/Hubble & NASA)

Galaxies without Stars?

The Hubble Deep Field observation caused amazement in 1995 when it revealed that a tiny patch of apparently empty sky was full of around three thousand galaxies. The work was repeated with the newer ACS (Advanced Camera for Surveys) instrument in 2012 as the Hubble eXtreme Deep Field, with 5,500 galaxies going back some 13.2 billion years.

Recent work using the Multi Unit Spectroscopic Explorer instrument (MUSE) at the Very Large Telescope (VLT) in Chile has examined the same region in Fornax with even greater depth. MUSE is described as "a panoramic integral-field spectrograph operating in the visible wavelength range". In other words,

The Hubble Ultra Deep Field is a tiny but much-studied region in the constellation of Fornax, and is seen here as observed with the MUSE instrument on ESO's Very Large Telescope. However, this image only gives a very partial view of the riches of the MUSE data, which also provide a spectrum for each pixel in the picture. This data set has allowed astronomers not only to measure distances for far more of these galaxies than before – a total of 1,600 – but also to find out much more about each of them. Surprisingly 72 new galaxies were found that had eluded deep imaging with the NASA/ESA Hubble Space Telescope. (ESO/MUSE HUDF collaboration)

it is a high resolution camera providing a spectrum of each pixel. The complete instrument weighs eight tons and is mounted at the Nasmyth platform of the VLT. It works by slicing the incoming wide-field image into 24 sections and then each section is imaged by one of 24 identical spectroscopic imaging units. All the data is then processed and recombined to form the final image – around 3.2 Gigabytes per exposure.

The team, led by Floriane Leclerq of the University of Lyon, France, has studied some 1600 faint galaxies including 72 that were not visible by Hubble. These faint galaxies appear to shine almost entirely at the wavelength known as Lyman-alpha, a result of glowing hydrogen alone with little evidence of stars. The researchers concede that they do not yet have enough data to decide what is happening. One possibility is that they are seeing a halo of dense ionized gas around a star-forming region. Other possibilities are some kind of fluorescence, cooling radiation from cold gas accretion or emission from satellite galaxies.

Dark Matter and Dark Energy – Still Missing!

Every year brings new attempts by theorists to explain the twin mysteries of Dark Matter and Dark Energy, and every new theory makes the experimentalists nervous in case they have missed something in the data. It is precisely an experimental discrepancy that has led Dan Hooper of Fermi National Accelerator Laboratory (Fermilab), Illinois, to propose a novel idea for the origin of dark matter. It is well known that free neutrons decay with a mean lifetime of about 14½ minutes. The neutron turns into a proton and releases an electron and an antineutrino. So far, so good. Fire a beam of neutrinos into a suitable detector and the mean lifetime can be measured at 888 seconds with all protons, electrons and antineutrinos accounted for. However, trap some neutrons near absolute zero in a "neutron bottle" and just let them decay and the mean lifetime comes out at 879.6 seconds. Bartoz Fornal and Benjamin Grinstein of UCSD, California, have speculated that perhaps neutrons can decay in some other way, with maybe around one in a hundred of the ultracold neutrons decaying into a "dark particle" that sneaks out of the bottle without interacting with other matter.

To be a candidate for dark matter the particle would need to have a mass just less than that of a neutron, and the difference in mass would be shown up as a photon with an energy of the order of one MeV. Such a photon could be detected easily enough in a suitable experiment. The idea of dark matter being composed of "dead neutrons" may be bizarre but it is no less bizarre than some of the other suggestions and it may be amenable to experimental verification. It may even tie up with a theory called "asymmetric dark matter" in which neutrons and dark matter particles are produced by similar processes.

Theories of Dark Energy also have plenty of problems to keep researchers awake at night. Adam Reiss of Johns Hopkins University, Baltimore, Maryland, has been working on the measurement of Hubble's Constant for more than ten years. Continual refinement of results from stars and galaxies, using "standard candles" has allowed him to calculate the value at 73.2 kilometres per second per megaparsec with an error of about +/−1. An alternative method derived from the Planck satellite observations gives the value of 67.3 +/−0.5 units.

Other studies by the Kilo-Degree Survey group (KiDS) have also given results that differ from the Planck observations. KiDS is a wide-field imaging survey using the VLT Survey Telescope (VST) at Paranal in the Atacama Desert, Chile, to map the large-scale distribution of matter in the universe. One of the researchers, Catherine Heymans at University of Edinburgh, has mapped dark matter using weak lensing techniques, whereby light bends around concentrations of mass. She finds that the dark matter is distributed much more smoothly than predicted by Planck. Another member of KiDS is Shahab Joudaki of University of Oxford who suggests that some of these discrepancies can be reconciled by allowing the density of dark energy to vary over the lifetime of the universe. Unfortunately there are implications for the standard Lambda-CDM model and the idea that dark energy could increase with time leads to predictions of ordinary matter being ripped apart long before now.

Fortuitously, the LIGO and Virgo gravitational wave observations of the collision of two neutron stars may give new insights into the speed of cosmic expansion. The ripples in space-time were observed by the two LIGO detectors in the USA and the Virgo detector in Italy so by triangulation it was possible to direct optical telescopes to the correct location while the event was still visible. Within 1.7 seconds of the neutron star merger, the Fermi Gamma Ray Space Telescope detected a short gamma ray burst, also associated with the merger. A combination of optical and gravitational observations of the fireball may allow an independent measurement of the rate of expansion of the universe without the need to rely on other methods.

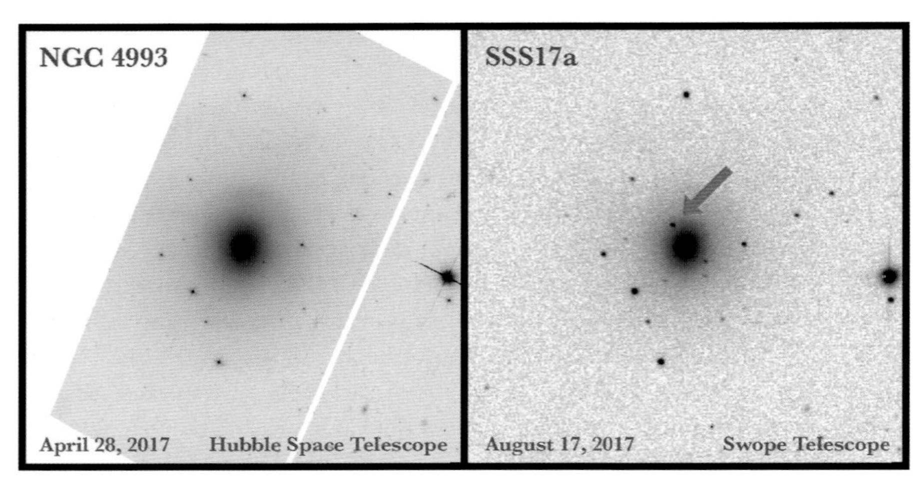

The first image of an 'optical transient' resulting from the merger of two neutron stars, and the first image of an optical counterpart to a gravitational wave detection. The box at left shows the host galaxy NGC4993, 130 million light years distant, as it appeared in a Hubble Space Telescope image taken on 28 April 2017. At right is the same galaxy imaged by the Swope Telescope just a few hours after the gravitational wave and gamma ray detections on 17 August 2017. The arrow points to the short-lived visible fireball that resulted from the merger of two neutron stars in that galaxy. (Credit: Swope Supernova Survey via UC Santa Cruz)

Another result of the LIGO / Virgo observation was that the speed of light and the speed of gravitational waves have been confirmed to be in agreement to one part in a million billion. Some of the wilder theories of dark energy and modified gravity implied that gravitational waves would travel either faster or slower than light and so now these can be ruled out. Just how the Dark Matter / Dark Energy problem will be resolved remains to be seen. Perhaps the long-sought combination of general relativity and quantum mechanics into a single coherent theory will provide the answer after all.

Popular Thesis

It has been said that Stephen Hawking's book *A Brief History of Time* is the most widely unread best-seller, since it has sold over 10 million copies but anecdotally many people have admitted not finishing it. This does not do justice to the book which explains very clearly the state of the art in physics in 1988. History seems to be repeating itself with the news that Stephen Hawking's PhD thesis is the most frequently downloaded thesis from the University of Cambridge Open Access Repository.

The thesis is typewritten with numerous hand-drawn formulae and equations so will come as quite a culture shock to younger readers, and the text is highly technical but clearly expressed. Many of the ideas Hawking explores are now commonly accepted and it is fascinating to imagine the young PhD student making such confident statements in 1966.

However, most of the equations require thorough understanding of tensor notation so, sadly, even fewer people who download it will finish reading it. Perhaps some will return to *A Brief History of Time* and the sequel *The Universe in a Nutshell* to see into the mind of the incomparable Professor Hawking.

Solar System Exploration in 2018

Peter Rea

This article was written in the winter of 2017 / 2018 before some of the missions discussed here have been launched, so the author will keep his fingers crossed at launch time! As all the other missions mentioned are active, some details may change for operational reasons. It is hoped that 2018 will continue to be another exciting year in Solar System exploration. A new NASA mission to Mars will study the interior of the planet, an area of science not fully explored before. The European Space Agency in collaboration with the Japan Aerospace Exploration Agency will send two spacecraft on a single carrier rocket to the innermost planet of the solar system. The USA will also launch an ambitious mission to send a spacecraft extremely close to the Sun.

An Insight into the Martian Interior

We have learned so much about Mars since the first successful flyby mission of Mariner 4 in 1965. We have mapped the surface of Mars to great accuracy; taken its temperature, studied its atmosphere and analysed the surface rocks. Yet little is known of the interior. NASA hopes to put this right when they launch their next mission to Mars. InSight (Interior Exploration using Seismic Investigations, Geodesy and Heat Transport) is a NASA Discovery Programme mission. By studying the size, thickness, density and overall structure of the Red Planet's core, mantle and crust, as well as the rate at which heat escapes from the planet's interior. The InSight mission will let us delve into the evolutionary processes of Mars. The launch window to Mars opens on 5 May 2018. The Atlas 5 launch vehicle will leave from Vandenberg Air Force Base on the west coast of the USA. This is a first for a planetary mission as all other American missions to the planets have taken off from Cape Canaveral on the east coast.

The structure of InSight is based on the highly successful Mars Phoenix lander to the Martian polar regions. The main payload consists of the Seismic Experiment for Interior Structure (SEIS), shown on the left in this picture (figure 1) which depicts the Insight spacecraft on the surface of Mars. SEIS will

Figure 1. (NASA/JPL-Caltech)

listen for Mars quakes and listen to meteorite impacts that send shockwaves through the mantle to better understand the internal structure of Mars. The second main payload is the Heat Flow and Physical Properties Package (HP3). This is a self-penetrating heat flow probe which will burrow up to 5 metres below the surface to measure how much heat comes from Mars' core, revealing a lot about the thermal history of Mars.

Using the spacecrafts own X band radio signal to Earth the Rotation and Interior Structure Experiment (RISE) will precisely measure the planets rotation and wobble to accurately measure the size and density of the Martian core and mantle. Wind and temperature sensors plus a magnetometer make up the rest of the experiments. The SEIS and HP3 experiments need to be accurately placed onto the surface of Mars. A robotic arm fitted with a camera will perform this task. The landing is scheduled for 26 November 2018 in the Elysium Planitia region.

Mission website can be found at **http://insight.jpl.nasa.gov**

A Mission to Touch the Sun

You only need to look at the Aurora to appreciate the effects our Sun has on the Earth. We live inside the Sun's atmosphere and are constantly being bombarded by the solar wind. NASA is about to launch an ambitious mission to study the Sun up close. Originally called simply Solar Probe the mission was renamed Parker Solar Probe after solar astrophysicist Eugene Parker. The spacecraft will go close enough to the sun to watch the solar wind speed up from subsonic to supersonic, and it will fly though the birthplace of the highest-energy solar particles. Launching in the summer of 2018 on a 7-year journey, Parker Solar Probe needs to use the gravitational field of Venus 7 times to obtain the correct orbit for very close passes of the Sun, these phasing orbits being shown in this diagram (figure 2). After nearly 7 years the first of the very close flybys of the sun will occur bringing the probe to less than 6.4 million kilometres above the photosphere of the Sun, when orbital velocity will reach ~725,000 kilometres per hour (~200 kilometres per second), faster by far than any other manmade spacecraft. To perform these unprecedented investigations,

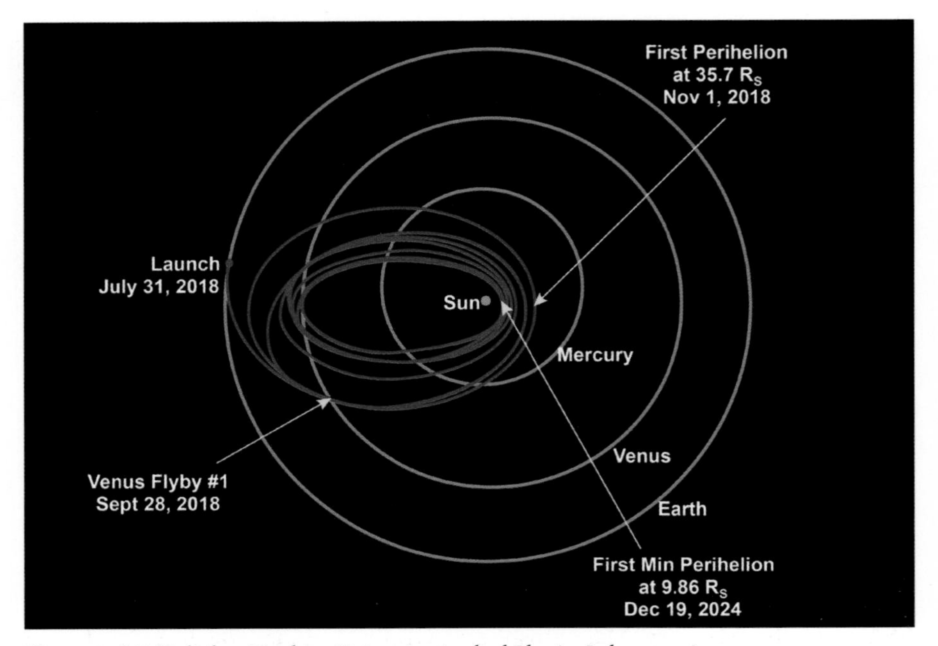

Figure 2. (NASA/Johns Hopkins University Applied Physics Laboratory)

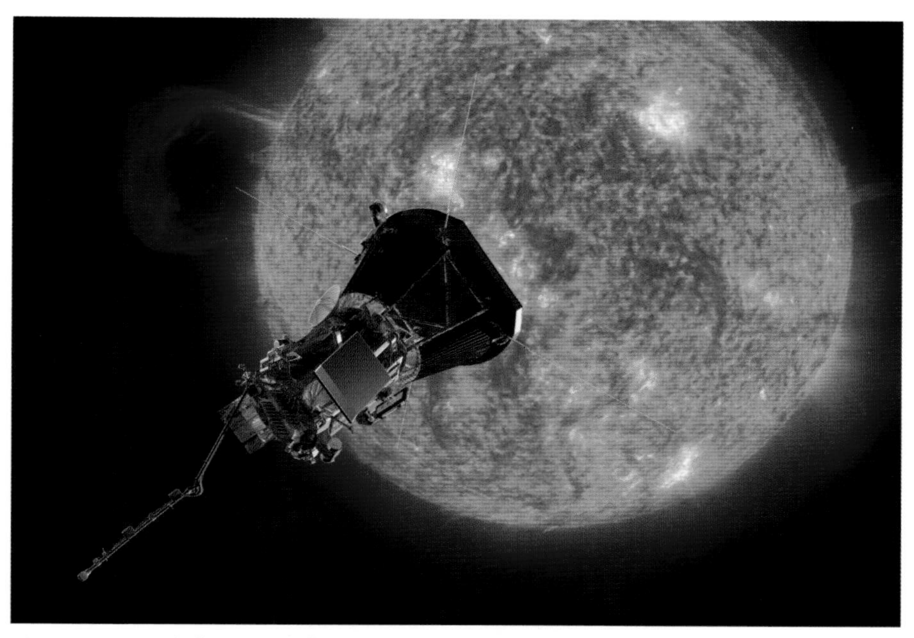

Figure 3. (NASA/Johns Hopkins University Applied Physics Laboratory)

the spacecraft and instruments will be protected from the Sun's heat by an 11.43cm (4.5-inch) thick carbon-composite shield, which will need to withstand temperatures outside the spacecraft that reach more than 1,300 degrees Celsius. This artist impression (figure 3) shows the Parker Solar Probe near the Sun.

Mission website can be found at **http://parkersolarprobe.jhuapl.edu**

Return to Mercury with BepiColombo

The European Space Agency (ESA) and the Japan Aerospace Exploration Agency (JAXA) are sending a new mission to Mercury in 2018. The mission is named for Italian scientist Giuseppe 'Bepi' Colombo (1920–1984) noted for his work on multiple planetary flybys using the gravitational 'slingshot'. The first mission to Mercury, Mariner 10, was made possible by his work. BepiColombo is due for launch in October 2018 by an Ariane 5 launcher. The mission comprises two spacecraft, the Mercury Planetary Orbiter (funded by ESA) and the Mercury Magnetospheric Orbiter (funded by JAXA). During

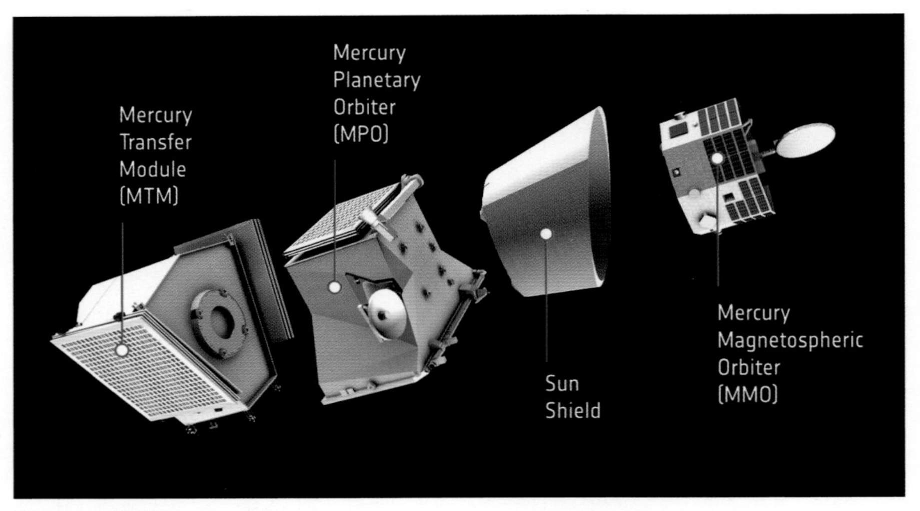

Figure 4. (ESA)

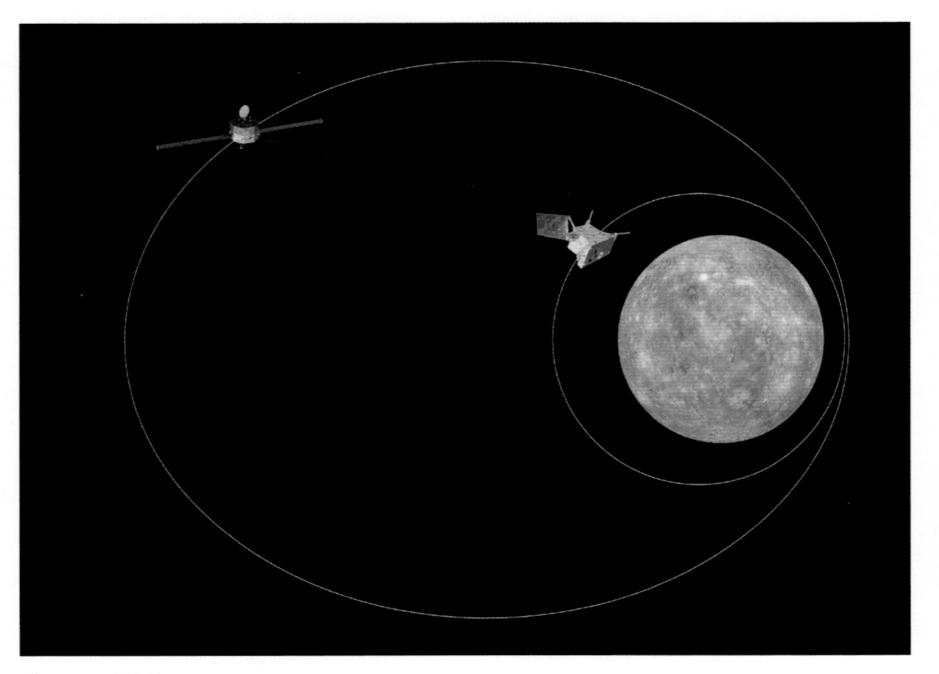

Figure 5. (ESA)

the cruise to Mercury both spacecraft are attached to the Mercury Transport Module which contains the propulsion systems necessary to reach the inner solar system. These elements are shown here (figure 4). Getting to Mercury is not easy due to the intense gravity field of the Sun. The journey will take around 7 years and will include 9 planetary flybys before entering a polar orbit around Mercury in 2025. These flybys are Earth once, Venus twice and Mercury six times. BepiColombo has been given a suite of instruments to provide the measurements necessary to study and understand the composition, geophysics, atmosphere, magnetosphere and history of Mercury. To better understand the planet the Mercury Planetary Orbiter (MPO) and Mercury Magnetospheric Orbiter (MMO) will be placed into the different orbits (figure 5) shown here. The MPO will be in a 480 × 1500 km, 2.3 hour period polar orbit and the MMO in a 590km × 11,640 km, 9.3 hour period polar orbit. A nominal mission would last around a year.

Mission website can be found at **http://sci.esa.int/bepicolombo**

All Lined Up for Bennu Encounter

The OSIRIS-REx (who thinks of these names?) mission was launched on 8 September 2016 toward the C type or carbonaceous asteroid 101955 Bennu, discovered in September 1999 by the Lincoln Near-Earth Asteroid Research (LINEAR) project situated at White Sands, New Mexico. This diagram (figure 6) shows that the orbit of Bennu is inclined at 6 degrees relative to the orbit of Earth. To rendezvous with Bennu, the spacecraft must be in the same orbital plane. Plane changes using the onboard propulsion system would require substantial amounts of propellant, increasing the mass and launch energy requirements with all its associated cost increases. However, the mission can achieve this plane change for free by using the Earth's gravity. Almost a year after launch the orbit of OSIRIS-REx brought it back to the vicinity of Earth for a flyby. On 22 September 2017 the spacecraft came within 17,240 kilometres of Antarctica as shown in this graphic representation (figure 7) of the Earth flyby. The flyby increased the orbit velocity of the spacecraft by 13,600 kilometres (8,450 miles) per hour and increased the orbital inclination putting the spacecraft on a course to rendezvous with Bennu about 11 months later.

After entering orbit around Bennu, OSIRIS-REx will begin mapping the asteroid for over a year before slowly lowering the orbit to allow the spacecraft

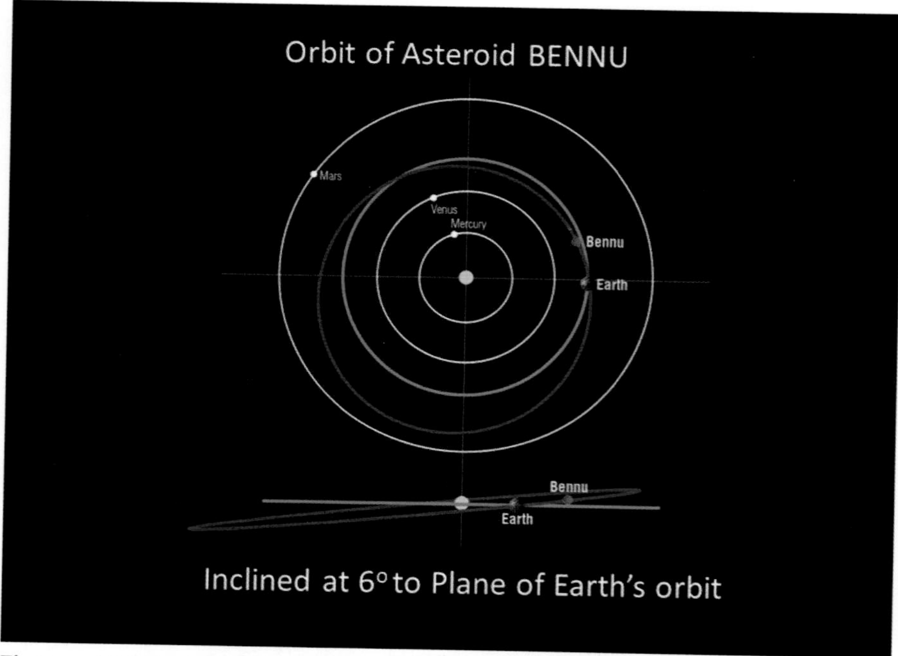

Figure 6. (NASA's Goddard Space Flight Center/University of Arizona)

Figure 7. (NASA's Goddard Space Flight Center/University of Arizona)

to gently touch the surface and collect a small sample. This sample will be placed into a special return capsule. After completing science observations, the spacecraft will depart Bennu around March 2021. This date is determined by the relative positions of Bennu and Earth. The return journey to Earth will take around two and a half years. In September 2023 the sample return capsule will separate from the spacecraft and enter the Earth's atmosphere. The capsule containing the sample will be collected at the Utah Test and Training Range in the USA.

For the record OSIRIS-REx is an acronym of Origins, Spectral Interpretation, Resource Identification, Security, Regolith Explorer.

Mission website can be found at **http://www.asteroidmission.org**

Continuing Science at Jupiter

The NASA spacecraft JUNO entered a polar orbit around Jupiter on 5 July 2016, having left Earth on 5 August 2011. This orbit brings Juno within 4,200 kilometres (2,600 miles) of the planet and out to 8.1 million kilometres (5.0 million miles) taking 53 days to complete one orbit. This orbit was intended to be only a 14-day orbit but issues with the onboard propulsion system deemed that JUNO would stay in the current 53-day orbit. This does not impact on science return, it just takes longer!

In under two hours on 1 September 2017 the spacecraft passed over Jupiter's north pole, through its closest approach (perijove), then over the south pole on its way back out. In this sequence of 11 colour-enhanced images (figure 8) from perijove 8, taken by an instrument called Junocam over a period of roughly 1 hour and 35 minutes, the north pole is on the right (first image in the sequence) and the south pole is on the left (11th image in the sequence).

Users of the mission website can download raw image data from Junocam and process the image with software of their choice. The processed image can

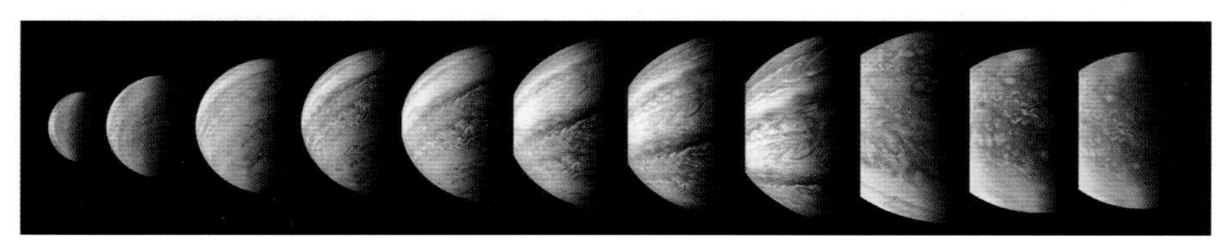

Figure 8. (NASA / SwRI / MSSS)

then be uploaded for all to see. A splendid example of Citizen Science. The end of the mission was scheduled for 2018 but as the journey around Jupiter is taking longer, the mission is likely to conclude around 2021 depending on funding.

The mission website can be found at **www.missionjuno.swri.edu**

Closing in on the Kuiper Belt

After the highly successful Pluto / Charon flyby in July 2015, the NASA New Horizons spacecraft has been heading toward its next target of Kuiper Belt Object 2014 MU69. In July of 2017 the KBO passed in front of (occulted) a star as seen from certain parts of the southern hemisphere. A team of New Horizon scientists observed this occultation from Argentina and based on their observations and those from the SOFIA airborne observatory, 2014 MU69

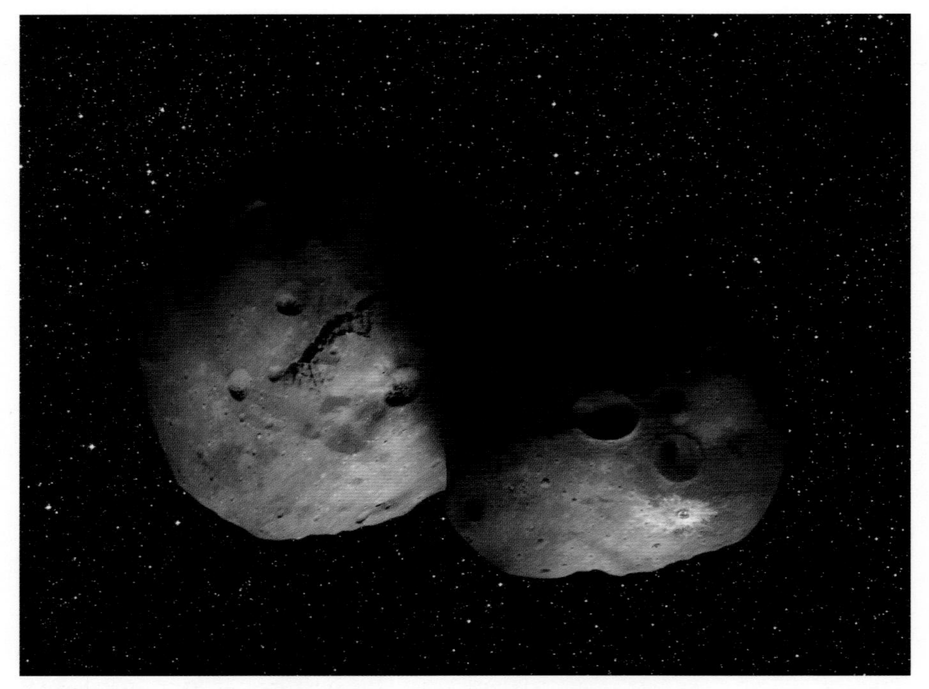

Figure 9. (NASA/Johns Hopkins University Applied Physics Laboratory/Southwest Research Institute)

may not be a single object. It could be either very irregular in shape or be two separate objects very close to each other or even touching (a so-called contact binary) as shown in this artist impression (figure 9) of the KBO based on these observations. However, we will have to wait until 1 January 2019 when the onboard cameras will reveal its true shape and structure.

Mission website can be found at **http://pluto.jhuapl.edu**

Asteroid Explorer

Hayabusa 2 is an asteroid sample return mission launched by JAXA, the Japan Aerospace Exploration Agency, on 3 December 2014 toward asteroid 162173 Ryugu. Hayabusa returned to the vicinity of Earth on 3 December 2015, exactly one year after launch, for a gravitational slingshot to increase orbital velocity and refine the trajectory toward Ryugu. The spacecraft should rendezvous with the asteroid around July 2018 for an 18-month survey. A major objective is to collect a sample of the asteroid and return this back to Earth. Hayabusa 2 is expected to leave the asteroid around the end of 2019 and return the capsule to

Figure 10. (JAXA)

Earth about a year later. This mission follows in the footsteps of the pioneering work of the original Hayabusa mission that was sent to asteroid 25143 Itokawa. This artist impression (figure 10) shows Hayabusa performing science at Ryugu.

Mission website can be found at **http://global.jaxa.jp/projects/sat/hayabusa2**

End of Aerobraking, Beginning of Science

The European Space Agency's Trace Gas Orbiter arrived at Mars on 19 October 2016 and was placed into a highly elliptical orbit measuring 250km x 98,000km orbit with a period of 4.2 days. TGO spent most of 2017 reducing the size of this orbit down to 400km x 400km with a period of two hours. Using onboard propulsion to achieve this would require substantial amounts of propellant making TGO too heavy to launch! However, the mission makes use of the thin Martian atmosphere. The low point in the orbit was reduced to the point where the spacecraft started to "feel" the atmosphere. This caused a drag on the spacecraft that robbed it of energy so that the next apoareion or high point

Figure 11. (ESA)

in the orbit was lower. Repeated passes through the upper atmosphere slowly reduced the orbital period.

Once the science orbit of 400km x 400km is achieved, full science operations can begin which should be in the spring of 2018. TGO is depicted at work above Mars in this image (figure 11). A key goal of this mission is to gain a better understanding of methane and other atmospheric gases that are present in small concentrations but could be evidence for possible biological or geological activity. The Trace Gas Orbiter will monitor seasonal changes in the atmosphere's composition and temperature to create and refine detailed atmospheric models.

Mission website can be found at **http://exploration.esa.int/mars/46475-trace-gas-orbiter**

Mars Summary

Mars continues to dominate Solar System exploration. At the time of writing there are eight active missions at Mars – five from the USA, two from Europe and one from India.

A NASA website on their Martian exploration can be found at **https://mars.nasa.gov** Information on the European Mars missions can be found at **http://exploration.esa.int/mars** Details on the Indian Mars mission can be found at **https://www.isro.gov.in/pslv-c25-mars-orbiter-mission**

Not Forgetting the Moon

To conclude this year's review of Solar System exploration I would like to pay homage to a remarkable mission which is still operating (as at time of writing) around the Moon. This is NASA's Lunar Reconnaissance Orbiter (LRO) which entered lunar orbit on 23 June 2009 and is shown in lunar orbit in this artist impression (figure 12). LRO carried a large suite of instruments for studying the Moon, but of note were the twin Lunar Reconnaissance Orbiter Cameras. With 195mm primary mirrors and focal lengths of 700mm they have taken thousands of high resolution images of the lunar surface including the image at left (figure 13) showing a view of the central peak of Tycho crater, and the extraordinary view at right of the Earth rising above the Moon's limb with Compton crater in the foreground (figure 14) (echoing a similar picture taken

Figure 12. (NASA/GSFC/Arizona State University)

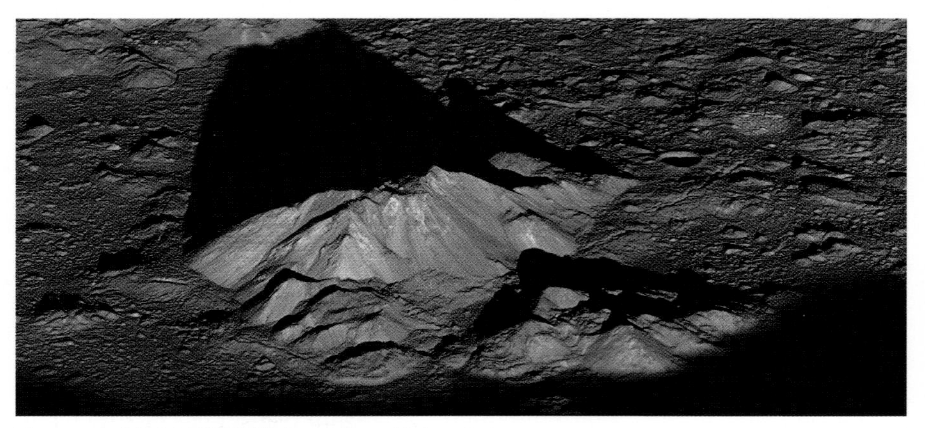

Figure 13. (NASA/GSFC/Arizona State University)

Figure 14. (NASA/GSFC/Arizona State University)

during the Apollo 8 mission to the Moon in December 1968). Readers of this publication are encouraged to view the vast array of LRO images, which can be found on the mission website given below. Visit and enjoy!

LRO Mission website can be found at **https://lunar.gsfc.nasa.gov**
The LRO camera has a dedicated website at **http://www.lroc.asu.edu**

Solar System exploration continues to excite and inspire, and next year promises to be no different.

Anniversaries in 2019

Neil Haggath

Jeremiah Horrocks

This year sees – probably – the 400th anniversary of the birth of a man who might have been one of the greatest of astronomers, had his life not been tragically cut short – Jeremiah Horrocks.

Horrocks was born in Toxteth, near Liverpool. The date of his birth is unknown, as no record exists; even the year is not certain. Some historians say he was born in 1618, others in 1619. He died on 3 January 1641, and it is

Jeremiah Horrocks observing the transit of Venus in 1639, depicted by English painter Eyre Crowe. (Wikimedia Commons)

commonly stated that he was 22 years old – which would date his birth as 1618. However, Allan Chapman notes that one of his contemporaries described him as "in his twenty-second year" at the time of his death, and therefore concludes that he was probably born in 1619.

Horrocks went to Emmanuel College, Cambridge, at the age of 13 or 14 – that was in fact quite common in those days, when the level of university education was somewhat lower than it is today – but for reasons unknown, he left without graduating. He was an early supporter of Copernicanism, and studied the works of Tycho Brahe, Johannes Kepler and others.

His achievements during his short life show that he had the potential to be a truly great scientist; he appears to have been a rather brilliant young man. He was the first to demonstrate that the Moon's orbit is elliptical, and proposed that comets also move in elliptical orbits, decades before Halley. His work on the motion of the Moon was acknowledged by Isaac Newton in the *Principia*.

Horrocks is best known, of course, as the first person to observe a transit of Venus. Johannes Kepler (1571–1630) was the first to realise that Mercury and Venus undergo transits; in 1627, he predicted that both planets would do so in 1631 – but sadly, he did not live to see them. The transit of Mercury in that year was observed; that of Venus was not, as it was not visible from Europe – and there were very few, if any, telescopes outside Europe at that time! However, Kepler failed to realise that transits of Venus occur in pairs, eight years apart; he actually predicted a "near miss" in 1639.

It was the 20-year-old Horrocks, in 1639, who realised that Kepler's "near miss" would in fact be another transit – and this one *would* be visible from Europe, including Britain. He made that discovery only a few weeks before the event, and so had very little time to publicise it!

The transit occurred on 24 November, according to the Julian Calendar (which was then still in use in Britain), or 4 December by the Gregorian Calendar (which had been adopted by most of Europe). It was observed by only two people – Horrocks himself, who by then lived in the Lancashire village of Much Hoole, and his friend William Crabtree (1610–1644) in Salford. Horrocks wrote a paper on the event, entitled in Latin *Venus in Sole Visa* (Venus in the Face of the Sun), which was not published until years after his untimely death.

A common misconception about Horrocks is that he was a clergyman; he is often referred to by later writers as "the Reverend Jeremiah Horrocks", and as the

Vicar or Curate of Much Hoole. He was in fact no such thing – though he probably would have been, had he lived long enough. The idea arose from Horrocks' own statement (in his paper on the transit) that he had to leave his observations to attend to his church duties. However, as Allan Chapman points out, it was not possible at the time for anyone to be ordained as a clergyman by the age of 21 or 22. He was probably a church assistant of some kind, such as a Bible clerk, and was probably studying for the clergy – but a Reverend, he was not.

Horrocks observed the transit, according to tradition, from Carr House, the home of a wealthy family; it is likely that he was employed as a tutor to that family's children.

While the date of Horrocks' death is known, the cause is not. It must have been sudden and unexpected, as he had been planning to travel 30 miles to visit Crabtree the following day. We can only wonder what he might have achieved, had he lived a full life!

Captain Cook and the Transit of Venus

3 June marks the 250th anniversary of the second of the pair of transits of Venus in the 1760s, which was of vital scientific importance, and was observed by among others, the great navigator Captain James Cook.

As is well known, transits of Venus occur in pairs eight years apart, with the intervals between the pairs alternating between 105.5 and 121.5 years. As is also well known, the transits of 1761 and 1769 were used to refine the measurement of the Astronomical Unit, the Earth's mean distance from the Sun. This was proposed by Edmond Halley (1656–1742); after observing a transit of Mercury in 1677, he realised that by observing a transit from widely different longitudes and accurately timing it, the distance of the Sun could be deduced by triangulation. And once the Earth's

James Cook. (Wikimedia Commons/ National Maritime Museum)

orbital radius had been measured, those of all the other planets would also be known, from Kepler's Third Law.

Having read Jeremiah Horrocks' paper, Halley realised that transits of Venus would be far more suitable for the purpose than those of Mercury, as it moves more slowly and presents a much larger disc. He knew that no transit of Venus would occur within his own lifetime, but wrote a paper in which he laid down the challenge for astronomers of a future generation.

In 1761, several expeditions were mounted by the British and French governments, to observe the transit from diverse locations. The results were of limited value; accurate timings were hampered by the now famous "black drop" effect. But in 1769, both countries resolved to try again; now that the black drop was known, methods could be devised to compensate for it when reducing the data.

The British Royal Society, with the financial support of King George III, mounted three expeditions, to the North Cape at the Arctic tip of Norway, Hudson Bay in Canada, and "a suitable island" in the South Pacific. The chosen island was Tahiti, which was one of the few Pacific islands whose latitude and longitude had been measured precisely. That voyage was undertaken by HM Bark *Endeavour*, commanded by Lt. James Cook RN (1728–1779); it was in fact the primary purpose of Cook's famous round the world voyage, during which he discovered Australia. While the Royal Society chartered the ship for the purpose of the observation, the Admiralty decided that it would be followed by an expedition to find the "southern continent", which was by then rumoured to exist.

The *Endeavour* carried several telescopes and three scientists – astronomer Charles Green (1734–1771), the eminent naturalist Sir Joseph Banks (1743–1820) and the Swedish naturalist Daniel Solander (1733–1782). All three, and Cook himself, would carry out observations of the transit.

On arrival in Tahiti, the crew built a settlement and observatory named Fort Venus. On the day, they successfully observed the transit, though there were discrepancies between the timings recorded by the individual observers.

Cook, of course, went on to complete his circumnavigation of the globe, and succeeded in discovering the southern continent. Green sadly died during the return voyage, but his observations contributed to the overall effort.

Using the observations from all the British and French expeditions, the value obtained for the AU was within 0.8% of the correct value – quite remarkable, given the limited accuracy of the observations!

Edward Charles Pickering

3 February marks the centenary of the death of Edward Charles Pickering (1846–1919), who served as Director of Harvard College Observatory for 42 years, from 1877 until his death, and established it as an observatory of international renown. In 1882, he invented a method of photographing multiple stellar spectra simultaneously.

Edward Charles Pickering. (Wikimedia Commons)

Pickering is perhaps best known for his employment of women as "computers", at a time when few women achieved careers in science. In those days, a "computer" was an assistant who carried out mathematical calculations by hand, to reduce the data from scientific observations.

In the early years of his directorship, Pickering was dissatisfied with the work of his then all-male team of "computers", and is said to have remarked, "My Scottish maid could do better!" The maid in question was Williamina Paton Stevens Fleming (1857–1911), a Scottish immigrant and former teacher, who had taken work in his household after being abandoned by her husband. In 1881, Pickering hired her as an assistant, and taught her to analyse stellar spectra. She proved more than capable, and became a respected astronomer in her own right; among other things, she discovered the Horsehead Nebula.

Over the next few years, Pickering recruited over 80 women, and established an exclusively female team of "computers", which became known, in politically incorrect times, as "Pickering's Harem"! Several of them became eminent astronomers, including Annie Jump Cannon (1863–1941), Antonia Maury (1866–1952) and Henrietta Swan Leavitt (1868–1921). Cannon, together with

Fleming and Pickering himself, was primarily responsible for developing the Harvard Stellar Classification system, which is still used today, and which formed the basis of the Henry Draper Catalogue. Leavitt, of course, is famous for her discovery of the period-luminosity relationship for Cepheid variable stars, which enabled the first measurements of the distances of other galaxies.

It is also worth noting that, unusually for the time, Pickering recognised the potential of people with disabilities; both Annie Jump Cannon and Henrietta Swan Leavitt were deaf.

Robert Goddard: *A Method of Reaching Extreme Altitudes*

Another centenary this year is that of the publication of a paper which laid the foundations of many of the principles of spaceflight.

Robert Hutchings Goddard (1882–1945), Professor of Physics at Princeton, was one of the most important pioneers of the Space Age. He was a prolific inventor, who filed a total of 214 patents. But more importantly, he carried out many early experiments in rocketry – initially self-financed, but from 1916 with the sponsorship of the Smithsonian Institution and the National Geographic Society. His publicly stated goal was to develop sounding rockets to investigate the upper atmosphere; his private one was to develop rockets to travel into space. He refrained from publicising the latter for fear of ridicule, as few scientists at that time took the idea seriously.

Robert Hutchings Goddard with rocket. (NASA)

Small solid-fuel rockets had been in use for centuries, as fireworks and even as military weapons – though the latter with little success. Goddard was the first to propose two concepts which proved vital for spaceflight – liquid-fuelled rockets and multi-stage rockets. He filed patents for both as early as 1914.

During the First World War, he designed a portable rocket launcher for the US military – the forerunner of the bazooka – but the prototype was not built until shortly before the Armistice, so the weapon was never used.

In 1926, Goddard built and launched the first liquid-fuelled rocket. By 1941, his rockets had achieved altitudes of 2.6 km and a speed of 885 km/h. He also successfully controlled their flight, by means of three-axis stabilisation, gyroscopes and steerable thrust.

It was in late 1919, however, that the Smithsonian published a paper which Goddard had written three years earlier, entitled *A Method of Reaching Extreme Altitudes*. Together with the earlier work of Konstantin Tsiolkovsky, which was then little known outside Russia, the paper is regarded as one of the pioneering works of the science of rocketry.

Most of the paper was concerned with sounding rockets, and detailed experimental relations between propellant, rocket mass, thrust and velocity. However, a final section proposed the concept of launching rockets into space, and suggested that they could be made sufficiently powerful to achieve escape velocity. He proposed, as a thought experiment, sending a small rocket to the Moon, with a payload of flash powder, which would ignite on impact to provide proof that the vehicle had reached its target. He even proposed the now familiar concept of the ablative heat shield, to enable spacecraft to return to Earth.

On 12 January 1920, the *New York Times* published a story on Goddard's paper, which sensationalised his proposals for spaceflight. There was a common misconception at the time, even among some scientists, that rockets could not operate in a vacuum, since the exhaust gases had to "push against" something, namely the atmosphere. (The principle behind the rocket is, of course, simply Newton's Third Law; the expulsion of the exhaust gases at high speed is the "action", and the "reaction" causes the vehicle to accelerate in the opposite direction). The following day, an unknown author, who apparently shared that misconception, wrote an editorial in the same paper, entitled *A Severe Strain on Credulity*. While he acknowledged that Goddard's proposals for sounding rockets were feasible, the author ridiculed his ideas about spaceflight, and even claimed that Goddard did not understand elementary physics! He wrote:

> "That Professor Goddard, with his 'chair' in Clark College and the countenancing of the Smithsonian Institution, does not know the

> relation of action and reaction, and of the need to have something better than a vacuum against which to react – to say that would be absurd. Of course he only seems to lack the knowledge ladled out daily in high schools."

Ironically, it was precisely because Goddard *did* understand "the relation of action and reaction", that he knew that rockets could operate in space!

There was a humorous sequel to this episode. On 17 July 1969, the day after the launch of Apollo 11, the paper published a somewhat belated, and presumably tongue in cheek, apology. In a short piece headed *A Correction*, it referred to the 1920 editorial, and concluded:

> "Further investigation and experimentation have confirmed the findings of Isaac Newton in the 17th Century, and it is now definitely established that a rocket can function in a vacuum as well as in an atmosphere. The *Times* regrets the error."

Apollo 11

This year's most obvious anniversary is, of course, the 50th of one of the greatest human technological achievements in history – the first landing of humans on the Moon by the Apollo 11 mission. It is hardly necessary to describe the mission here in great detail!

Following four "workup" missions, which tested the spacecraft and techniques in stages, Apollo 11 was launched on 16 July 1969, crewed by Commander Neil Armstrong, Command Module Pilot Michael Collins and Lunar Module Pilot Edwin "Buzz" Aldrin. Significantly, while most of NASA's astronauts at that time were serving military officers, Armstrong, who would become the first man to set foot on the Moon, was a civilian – a fact which served to emphasise that the mission's purpose was peaceful.

Apollo 11 was a "bare minimum" mission, intended simply to land two men on the Moon and return them safely to Earth, to prove that it could be done, and to fulfil President John F. Kennedy's goal of doing so by the end of the decade. Significant exploration and scientific research would follow on the later missions.

Apollo 11 Crew. (NASA)

On 20 July, Armstrong and Aldrin descended to the lunar surface in the patriotically named Lunar Module *Eagle*, while Collins remained in orbit in the Command Module *Columbia*. They touched down in the Mare Tranquillitatis at 20h 17m GMT; Armstrong named their landing site Tranquility Base.

Eagle remained on the Moon for only 21½ hours, and the astronauts made a single EVA, or "moonwalk". At 02h 56m GMT on 21 July – still the evening of 20th in the US – Armstrong became the first human to set foot on another world. He remained on the surface for 2 hours 31 minutes, and Aldrin about half an hour less. Apart from deploying a couple of scientific experiments and collecting some rock samples, their duties were mainly ceremonial; they planted the American flag, and spoke to President Richard Nixon in the White House.

Apollo 11 returned to Earth on 24 July, after a flight time of a little over 8 days, splashing down in the Pacific Ocean.

Two other centenaries are described in detail elsewhere in this book. One is that of the founding of the International Astronomical Union, described in Susan Stubbs' article *100 Years of the International Astronomical Union*. The other is that of Sir Arthur Eddington's solar eclipse expedition, which he used to verify the bending of light by the Sun's gravity, as predicted by Einstein's General Theory of Relativity; this is described in this author's article *In Total Support of Einstein: Eddington's Eclipse, 1919*.

The Cassini-Huygens Mission to the Saturn System

Carl Murray

At 4:55 a.m. Pacific time, on 15 September 2017, after more than 13 years in orbit, one of the most successful planetary missions of all time came to a fiery end as the Cassini spacecraft plunged into the atmosphere of Saturn and was vaporised. A spacecraft that was first planned in the early 1980s, launched in 1997 and reached Saturn in 2004 had finally exhausted the propellant it needed

A spectacular mosaic showing Saturn and its ring system in October 2013 produced from 36 Cassini images taken through three different filters. The hexagonal feature surrounding the north pole is clearly visible as is a bright, narrow wave of clouds at a latitude of 42 degrees north. (NASA/JPL-Caltech/SSI/Cornell)

to make course corrections. If it was left marooned in orbit there was always the possibility that it could one day collide with one of Saturn's biologically interesting moons and risk contaminating it. Rather than take that chance the mission planners made the decision to destroy the spacecraft and allow its vaporised components to become part of the planet it had faithfully studied for more than a decade. This is the story of this phenomenally successful mission, its main discoveries and its final hours.

The Saturn System

Following a brief glimpse of the Saturn system by the Pioneer 11 spacecraft during its 1979 flyby, our first detailed look at this jewel of the solar system came in November 1980 and August 1981 when the two stalwarts of planetary exploration, the Voyagers 1 and 2 spacecraft, returned data on the planet, its moons and rings that would be unsurpassed for more than 20 years. What the Voyagers revealed helped to formulate the questions that the next generation of spacecraft would attempt to answer. One particularly intriguing object was Saturn's largest moon, Titan. Larger than the planet Mercury, the nature of Titan's surface was essentially unknown because of the thick atmosphere and haze layer that surrounded it. There was active speculation that the conditions were right for lakes or even a global ocean of liquid methane on the surface. It was clear that Titan was an obvious target for a future mission. Other puzzles revealed by the Voyager results included the nature of bright and dark hemispheres on Iapetus, the origin of the smooth regions on the surface of Enceladus, the cause of the bizarre structure in Saturn's narrow F ring as well as bigger questions such as the mass, age and ultimate fate of Saturn's impressive ring system.

The Mission

The Cassini-Huygens mission was a collaboration between NASA, ESA and ASI, the Italian Space Agency. NASA would lead the development of the Cassini spacecraft which would remain in orbit around Saturn for at least four years; it would also provide the launch vehicle. ESA would be responsible for the Huygens probe which would be delivered by Cassini into Titan's atmosphere early on in the mission and descend to its surface. It was decided that scientists from US institutions could apply to participate in the six instruments on Huygens while their European counterparts could apply to be involved with the twelve instruments on Cassini. The result was a truly international mission from the start and a role model for trans-Atlantic scientific collaboration. Instrument selection for Cassini and Huygens was announced in 1990 with a planned launch date in late 1997 and arrival at Saturn in mid-2004. Following the successful completion of its four-year nominal mission, the Cassini project received funding for an Equinox Mission (2008–2010) and a subsequent Solstice Mission (2010–2017). In the end Cassini travelled almost 8 billion kilometres and sent back 635 GB of data to participating scientists from 27 nations.

Saturn

Orbiting Saturn for 13 years gave scientists the opportunity to study the gas giant in unprecedented detail as it completed almost half of an orbit around the Sun. It was already known from Voyager images that Saturn had a hexagonal, jet stream feature encircling its north pole and Cassini had the opportunity to study it in detail as the northern hemisphere moved from winter into spring and started to warm up. Cassini also detected vortices at Saturn's north and south poles, each resembling hurricanes on Earth, only 50 times larger. After patient searching the cameras on Cassini also observed lightning associated with storms on Saturn but with strikes that were 10 times more powerful than their terrestrial counterparts.

Although it was known that Saturn generates giant storm systems in its northern hemisphere approximately once every 30 years – these "Great White Spots" had been observed since the 19th century – the next one was not due until roughly 2020, long after the end of the Cassini mission. However, Saturn kindly obliged by producing one of these spectacular storms in December 2010 and Cassini's instruments were able to study it for more than a year as it spread right around the northern hemisphere.

If a planet has a solid surface, then it is fairly straightforward to measure its rotational period – all that is needed are tracking observations of any obvious features on its surface. However, in the case of the gas giants this becomes more problematic because the only visible features are clouds moving at different speeds at different latitudes and there is no solid surface! The proven alternative is to use instruments such as the magnetometer to characterise the

A false-colour image of the spinning vortex at Saturn's north pole. The feature is 2,000 km across with wind speeds as high as 540 km/h. The colour red indicates low cloud while green indicates high cloud. (NASA/JPL-Caltech/Space Science Institute)

planet's magnetic field with the assumption that the rotation of the field is an intrinsic measurement of the planet's rotation. But the Cassini data showed that Saturn's "day" can vary between 10.6 and 10.8 hours and can even depend on the hemisphere being measured. Observations made during the Grand Finale stage of the mission may finally help to understand exactly what is happening in Saturn's interior to produce such strange behaviour.

Titan

On 14 January 2005 the Huygens probe spent over two hours parachuting through the thick atmosphere of Titan relaying priceless, in situ data back to the Cassini spacecraft. While its batteries lasted the probe continued transmitting from the surface for more than an hour after landing. Descent images showed the characteristic dendritic pattern of flowing liquid from uplands to an apparent

A stereographic projection of the surface of Titan around the Huygens landing site. The image was obtained from an altitude of 5 km above the surface on 14 January 2005 by the Descent Imager/Spectral Radiometer (DISR) instrument on the Huygens probe. (ESA/NASA/JPL/University of Arizona)

shoreline. Images from the surface revealed what appeared to be a dried-up lake bed with boulders and pebbles embedded in a surface showing the tell-tale signs of a previously flowing liquid. There had always been speculation that lakes or even oceans of liquid methane and ethane could exist on the surface, but it would take several more years before definitive evidence of this finally emerged thanks to the instruments on Cassini. Although the atmosphere of Titan is predominantly composed of nitrogen (as is the case on Earth) there is also a significant methane component. We now know that, on Titan, it periodically rains methane giving rise to lakes near both the north and south poles, with more lakes in the north which, during the Cassini mission, had yet to receive the full heat of a Titan summer. In fact, there is believed to be a "methane cycle" on Titan operating in much the same way as the "water cycle" does on Earth. Both objects have noticeable seasonal changes and weather systems. This provides yet another

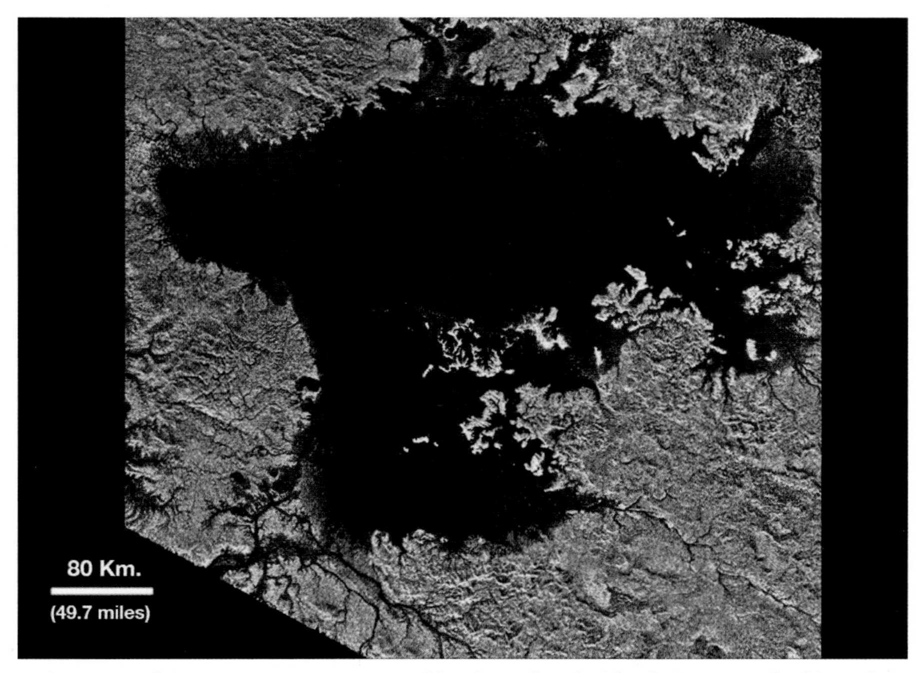

Radar image of the Ligeia Mare region near Titan's north pole. The dark areas in the false-colour image are lakes of liquid hydrocarbons such as methane and ethane. (NASA/JPL-Caltech/ASI/Cornell)

remarkably similar parallel between the processes on Titan and those on Earth. It is important to note that the average surface temperature on Titan is −179°C and so it is too cold for liquid water to exist there. However, the latest models for Titan's interior suggest that there is a global ocean of liquid water and ammonia at a depth of 55 to 80 km below the surface.

Enceladus

Perhaps Cassini's most remarkable and far-reaching discovery came from an unexpected source. Since the time of Voyager, Saturn's moon Enceladus was something of an enigma with large regions devoid of craters (implying a young surface) and the fact that it was embedded in the middle of an extensive, faint, diffuse ring of ice particles, Saturn's E ring, with no obvious signs of a source. Cassini's magnetometer detected distortions in Saturn's magnetic field near Enceladus' south pole leading to speculation that the moon was an active source of particles. Spectacular confirmation came when Cassini images revealed plumes of material spewing from parallel grooves in the south polar region of Enceladus. The analysis showed that, even though Enceladus gets

A cutaway view of Enceladus illustrating possible hydrothermal activity giving rise to the plumes observed by Cassini. This is a natural occurrence in Earth's oceans. (NASA/JPL-Caltech)

only 1% of the sunlight that the Earth receives, it somehow manages to have a source of liquid water beneath its surface, probably due to tidal heating, and it is these ice particles that continually supply the E ring. Further flybys and a subsequent analysis of Enceladus' shape and gravity field showed that there is likely to be a global ocean of liquid water underneath its surface. The suite of instruments on Cassini was employed to analyse the composition of the plumes. They revealed that as well as particles of water ice there were traces of salts, again indicative of a global ocean, and even molecular hydrogen described by one Cassini scientist as like "a candy store for microbes" drawing parallels with the processes that occur near thermal vents on the ocean floor back on Earth. Cassini did not find any evidence for life beneath Enceladus but what it did was demonstrate that the conditions for life to exist can occur in some unexpected places and this will have a profound effect on strategies for the search for life elsewhere in the universe.

Other Moons

In the course of its 294 orbits around Saturn the Cassini spacecraft had 162 targeted (i.e. close) flybys of its moons. The close approach to Iapetus on New Year's Eve 2004 revealed a ridge extending almost three quarters of the way around the moon's equator giving it a distinct "walnut" appearance. While there was speculation that this could be due to a primordial ring around Iapetus, a more likely explanation is that the ridge is a by-product of a cooling process early in the moon's history. Cassini scientists were finally able to analyse remotely the surface coating that gave Iapetus its dark, leading hemisphere and the consensus is that it was due to dust from the outer, retrograde moon, Phoebe, spiralling inwards and being swept up by Iapetus. Observations from NASA's Infrared Spitzer Space Telescope in 2009 revealed that Phoebe possessed a tenuous ring, probably as a result of meteoritic impacts onto its surface. The Phoebe ring was subsequently detected by Cassini and its physical properties seem to match those of the material coating Iapetus.

The moon Tethys, in common with Iapetus and most of Saturn's main moons (and indeed our own Moon) keeps one face towards the parent planet – this is a consequence of tides acting over the age of the solar system. This means that there is a well-defined leading and trailing hemisphere as the moon orbits the planet. In a peculiar parallel with the surface of Iapetus, there are colour

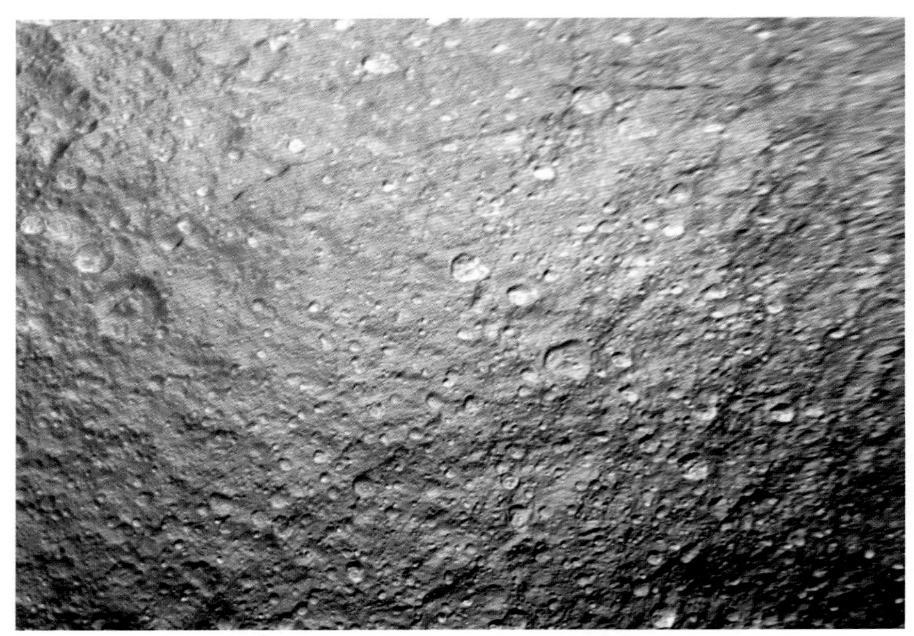

An enhanced-colour mosaic of the surface of the moon Tethys covering an area 490 km by 415 km. The curved red streaks are of unknown origin but could be exposed ice with chemical impurities or perhaps even material from an out-gassing event from within Tethys. (NASA/JPL-Caltech/Space Science Institute)

differences between the leading and trailing hemispheres of Tethys. The moon also has strange, several hundred kilometres-long, red streaks on its surface, with an unknown origin. What is known is that Tethys – along with several other moons – orbits within the E ring and so is exposed to bombardment from its icy particles. There are also thought to be collisions with energetic electrons orbiting within Saturn's magnetosphere.

Cassini's cameras also discovered six new moons of Saturn during the mission. Methone, Pallene and Anthe were all found orbiting between Mimas and Enceladus. Polydeuces was discovered orbiting near the trailing Lagrange point, L5, of the main moon, Dione, while the long-suspected Daphnis was found orbiting in the narrow Keeler Gap in Saturn's A ring. Finally, Aegaeon was discovered embedded in the G ring between the Janus-Epimetheus pair and Mimas. Observations showed that Methone, Pallene, Anthe and Aegaeon were

The moon Daphnis (at the right) orbiting in the 42 km-wide Keeler Gap in Saturn's A ring. Ring particles passed by the moon on the outside are perturbed and form wave structures which eventually damp due to collisions. (NASA/JPL-Caltech/Space Science Institute)

all associated with their own tenuous rings. This clear association between rings and embedded moons is common to most of the small satellites orbiting near the planet. It is likely that – as with the Phoebe ring – continual meteoritic impacts on the moons provide the source material for the associated rings. High resolution images of the moons Atlas, Daphnis and Pan showed prominent equatorial bulges which are thought to

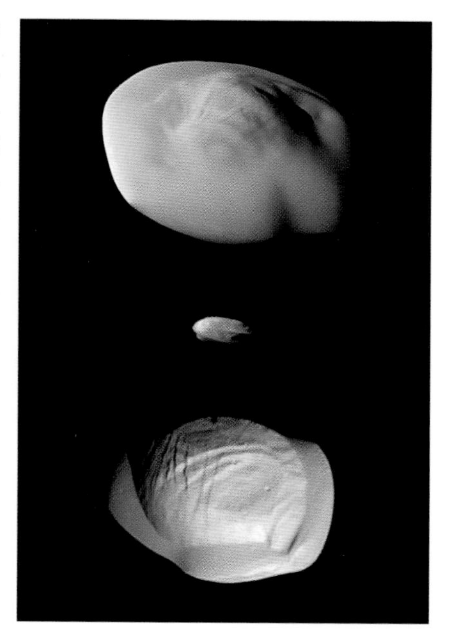

Cassini images of the small moons Atlas, Daphnis and Pan at the same scale showing their unusual shapes and surface features. All images were taken in 2017 towards the end of the mission. (NASA/JPL-Caltech/Space Science Institute)

be created by the re-accretion of nearby ring material. Pan, the moon orbiting in the Encke Gap in the A ring, had the most unusual shape with more than a passing resemblance to a giant piece of ravioli.

Rings

Although Jupiter, Uranus and Neptune all have rings, Saturn's ring system is by far the largest and most massive that the solar system has to offer. It has been known for some time that the rings (mostly composed of icy particles of varying sizes from dust to mountains) and small moons are closely linked. A detailed analysis of Saturn's A ring confirmed that all of its structure can be explained by the existence, at specific locations, of gravitational resonances between the moons and the particles that make up the rings. However, the massive B ring is much more of a problem. Cassini images confirmed that the Encke and Keeler Gaps in the A ring each has an associated moon (Pan and Daphnis, respectively) that probably cleared these locations of ring particles. The difficulty is that there are other gaps in the rings with no obvious orbiting satellite, despite numerous searches. Just to complicate things, some of these gaps contain narrow rings leaving open the possibility that rings themselves

A natural colour composite of two Cassini images of Saturn's B ring between 98,600 km and 105,500 km from the centre of the planet. The different ringlets are part of the B ring's "irregular structure" which is still unexplained. (NASA/JPL-Caltech/Space Science Institute)

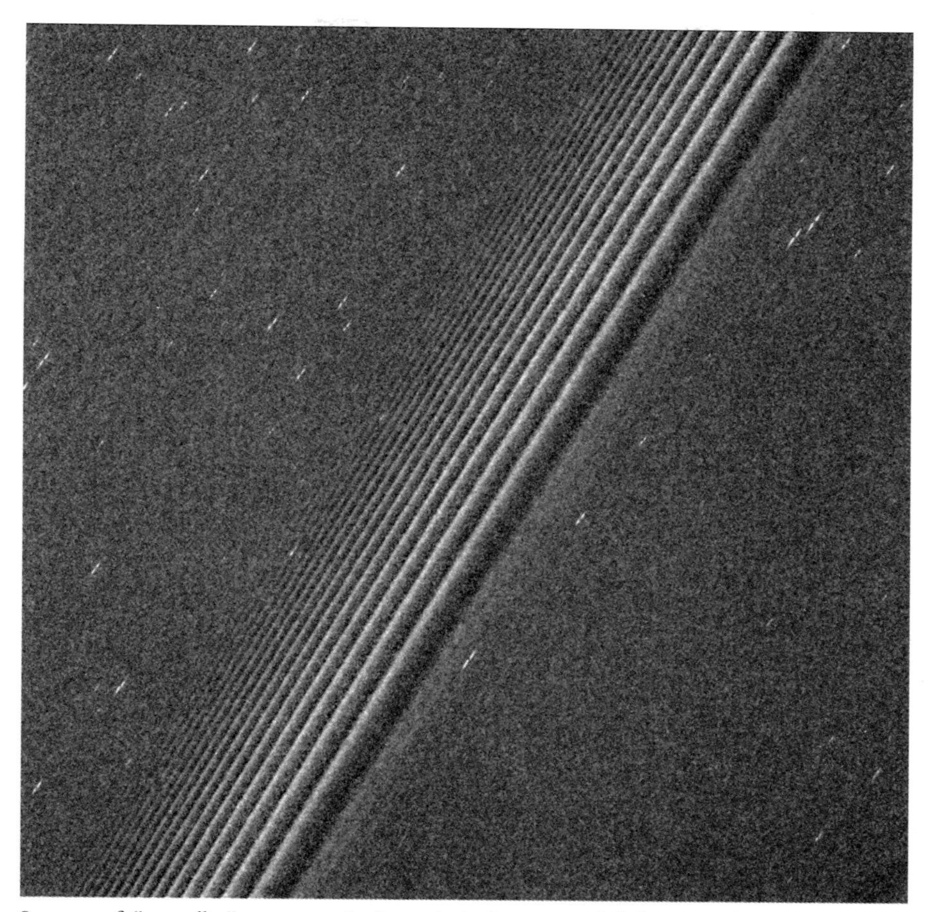

Swarms of "propeller" structures in Saturn's A ring at a radial distance of 129,000 km. The main, diagonal, parallel features at the centre of the image are spiral waves due to a gravitational resonance with a moon orbiting outside the rings. Each of the short, bright, parallel streaks is a "propeller" which has a small object at its centre perturbing adjacent ring particles. The resolution is 385 m/pixel. (NASA/JPL-Caltech/Space Science Institute)

can act to clear a gap. At least Cassini was able to answer the question of what happens when a small moon (or large ring particle – there is really no distinction) is not massive enough to clear a gap. Computer simulations predicted that such an object would still affect the orbits of nearby particles and create a characteristic pattern called a "propeller" to accompany it around

Saturn. Hundreds of such structures have now been observed and tracked by Cassini's cameras and some have even been seen to be migrating in an apparently random manner. High resolution images of the bizarre, twisted F ring showed how collisions with small, nearby moonlets – themselves perturbed by the persistent gravitational tugs of the moon Prometheus – produced the rings ever-changing, unusual structure.

A spiral density wave structure in Saturn's B Ring caused by a gravitational resonance with the moon Janus. This location, near a radial distance of 96,000 km from Saturn is one of the few places where such structures are seen. The resolution is 530 m/pixel. (NASA/JPL-Caltech/Space Science Institute)

Some of the "big" questions Cassini was designed to answer concerned the rings' origin and age. Both of these are related to the mass of the rings, a key quantity Cassini was finally able to estimate near the end of the mission thanks to dedicated tracking of its radio signal. The preliminary answer suggests a mass ~50% that of the moon Mimas with a corresponding age of perhaps only 200 million years, an uncomfortably small fraction of the 4.6 billion-year age of the solar system. Indeed, the rings could have originated from a relatively recent break-up of an icy moon or a passing giant comet that was tidally disrupted by Saturn. Perhaps we are lucky enough to see just the latest version of a ring system which comes and goes?

The Grand Finale

After studying the Saturn system from orbit for 13 years, Cassini had a final close flyby of Titan on 22 April 2017 and the resulting gravity assist meant that it would undergo a succession of 22 passages through the narrow gap between the innermost D ring and the upper atmosphere of the planet before finally burning up on 15 September 2017. These final orbits were invaluable for making the most detailed measurements yet of Saturn's gravitational and magnetic fields, thereby placing constraints on models of the planet's interior. On the final orbit several Cassini instruments operated right up until the very end, some sampling the molecules of the upper atmosphere, others

An illustration showing the Cassini spacecraft breaking up as it enters the atmosphere of Saturn on 15 September 2017. (NASA/JPL-Caltech)

measuring impacts on the spacecraft. The vital bits of data were received until the spacecraft's firing thrusters were no longer able to maintain radio lock with NASA's Deep Space Network dish at Canberra, Australia. First the X-band signal was lost and then a few seconds later the S-band. Cassini ended its life the same way Huygens had done more than 12 years earlier – taking invaluable scientific data right to the very end.

The Cassini-Huygens mission has not just re-written the textbooks – it has produced a revolutionary store of digital knowledge about the Saturn system. This is a legacy which will not be surpassed until the next generation of robotic spacecraft reach Saturn in decades to come.

Science Fiction and the Future of Astronomy

Mike Brotherton

One of the primary jobs of science fiction is to try to forecast the future in an entertaining and perhaps adventurous way. Usually this task involves the extrapolation of current trends and represents a reflection of aspects of our own society – for better or worse. When evaluated by future generations, these science fictional experiments tend to miss the mark more than they hit it. Still, it can be fun and interesting to do. One area for which science fiction has tended to hit the mark a bit more often concerns technology, forecasting the development of space travel, robots, and ever smaller and more powerful computers … eventually – 1950s stories featured planet-sized computers and astronauts using slide rules.

Astronomy is a science very much driven by technology, increasingly via developments in computation and robotics, and as such is particularly amenable to an adventure in science fictional extrapolation. First, let's set the stage, since the science in recent years has overtaken even some fairly recent science fiction, and both continuously push each other to (literally) higher and higher levels.

Anyone who looks at the sky and thinks seriously about what is up there is an astronomer, professional or not. Before telescopes, careful naked eye observations of celestial positions revealed the planets as distinct from stars, and transitory objects like comets and supernovas showed that the heavens were not perfect and unchanging, but evolving. Telescopes revealed more – the phases of Venus, the moons of Jupiter, and the existence of spiral nebulae that are in fact entire galaxies of stars.

Photography gave us the ability to record images and to share them with a degree of fidelity beyond that of human drawing. Spectroscopy let us know what stars, and other objects in the sky, are actually made of, without being able to take samples and analyze them in a lab. And beyond chemical composition, astronomers discovered the Big Bang, dark matter, neutron stars, black holes, dark energy, neutrinos, cosmic rays, gravitational waves, exoplanets, and more …

Fantastic progress has been made during the past few decades, despite the fact that there remain persistent and stubborn mysteries. From Lost in Space and Star Trek in the 1960s when black holes were theoretical objects, to the centre of the Milky Way where we can now plot stars orbiting a supermassive black hole that exists there weighing between four and five million solar masses, and we have gone from no known exoplanets in the early 1990s to the point now of knowing how common planets are around other stars in our galaxy – they are rather common!

Technology, more than anything else, has been the driving force behind this remarkable advancement. We have developed space-based telescopes to observe the ultraviolet, X-ray, and Gamma-ray parts of the spectrum inaccessible from the ground (e.g., the Hubble Space Telescope also known as HST, the Chandra X-ray Observatory, and the Fermi Gamma-Ray Telescope), and to obtain the time-series photometry necessary to find exoplanets eclipsing their central stars (Kepler). We have developed ground-based telescopes using millimetre-wave interferometry providing the angular resolution to image proto-planetary disks and see forming star systems (e.g., the Atacama Large Millimeter Array, ALMA), to use adaptive optics (AO) to provide space-based level imaging quality from the ground, and to detect gravitational waves from cosmic events like black hole mergers across the universe (e.g., Laser Interferometer Gravitational-Wave Observatory, LIGO).

Technology does not magically appear, like Athena from the head of Zeus. Advances in astronomy over the next decade are already clearly mapped out, as major telescopes both on the ground in space now take routinely many years to move from conception to deployment. Dozens of ambitious new facilities are now in various stages of development, several of which promise major scientific revolutions.

The first game changer coming in the very near future is the James Webb Space Telescope (JWST), scheduled for a 2021 launch, and sporting a 6.5 metre diameter primary mirror operating with instruments that will work over a wavelength range from 0.6 to nearly 30 microns. Robotics will allow the telescope to unfold like a flower at the L2 Sun-Earth Lagrangian point. Sometimes thought of as Hubble's successor, the emphasis on the infrared part of the spectrum will make it a rather different animal in practice, one that will permit the exploration of the high-redshift universe and lead to a fundamental

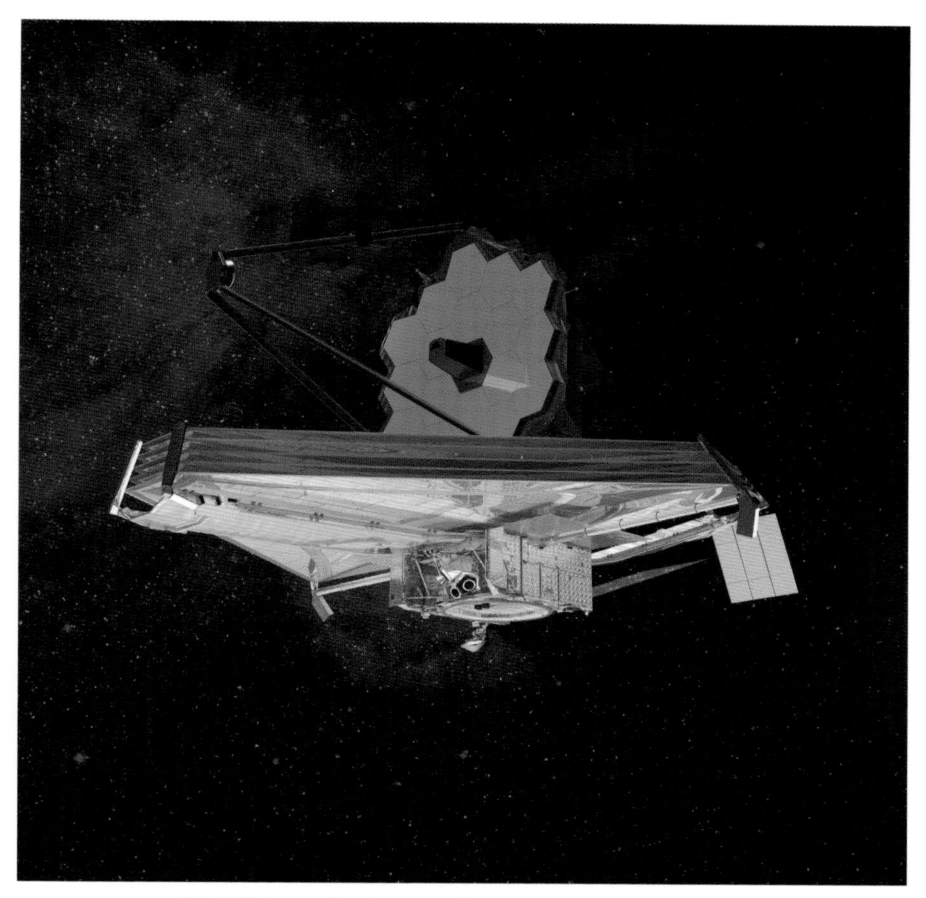

Artist's concept of NASA's James Webb Space Telescope. (NASA)

new understanding of the formation of the first galaxies, among a number of science goals.

The 2020s will bring a new generation of larger telescopes online: the Giant Magellan Telescope (GMT, 24.5 metres diameter, Chile), the Thirty Metre Telescope (TMT, 30 metres diameter, Hawaii), and the Extremely Large Telescope (ELT, 39.3 metres diameter, Chile). Compared to the current crop of large telescopes with primary mirrors in the 8–10 metre class, these new telescopes will have approximately an order of magnitude more light-collecting area. They will also cost around an order of magnitude more – a 30-metre class telescope comes with a price tag in the $1 billion USD range, less than that of JWST (around $10 billion USD). While these telescopes will all feature cutting-edge technology ranging from AO to the latest and greatest detectors, in some

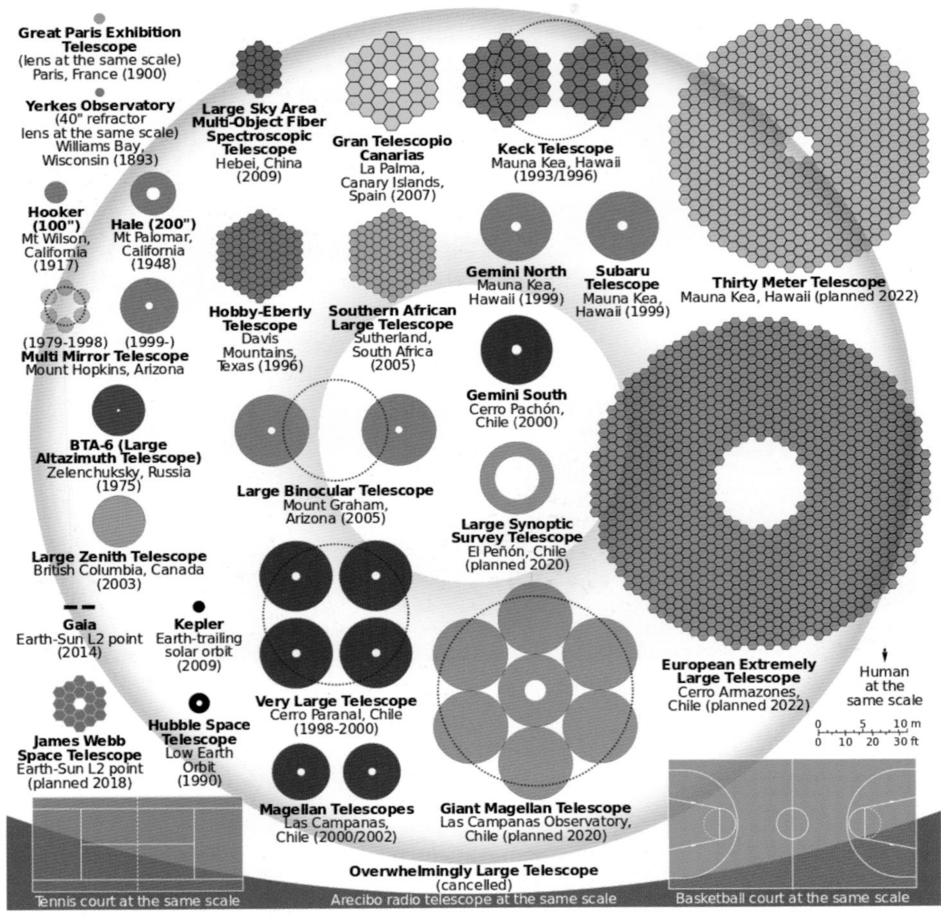

Comparison of nominal sizes of primary mirrors of notable optical telescopes. Dotted lines show mirrors with equivalent light-gathering ability. Of particular interest here for the near future are the James Webb Space Telescope, the Large Synoptic Survey Telescope, and the new generation of large ground-based telescopes including the Thirty Metre Telescope, the European Extremely Large Telescope, and the Giant Magellan Telescope. (CMG Lee)

ways they may represent a last hurrah for conventional ground-based telescope technology. While one can imagine an even more advanced generation of football-field sized ground-based optical telescopes, the physical scales are likely starting to approach some practical limits.

Finally there is the Large Synoptic Survey Telescope (LSST), a half a billion USD facility that will open up the astronomical time domain like never before. A ground-based telescope in Chile with a 8.4 metre primary mirror, the unique aspect of the LSST is its record-breaking etendue, the product of its collecting area and field of view (9.6 square degrees, about the size of 40 full moons), which will enable it to image the sky at unprecedented depth and speed. By the early 2020s, the LSST will be providing photometry every few days for some tens of billions of objects and ultimately detecting and cataloguing many trillions of objects in very deep, combined images. Estimates indicate that the LSST will issue some ten million alerts nightly,

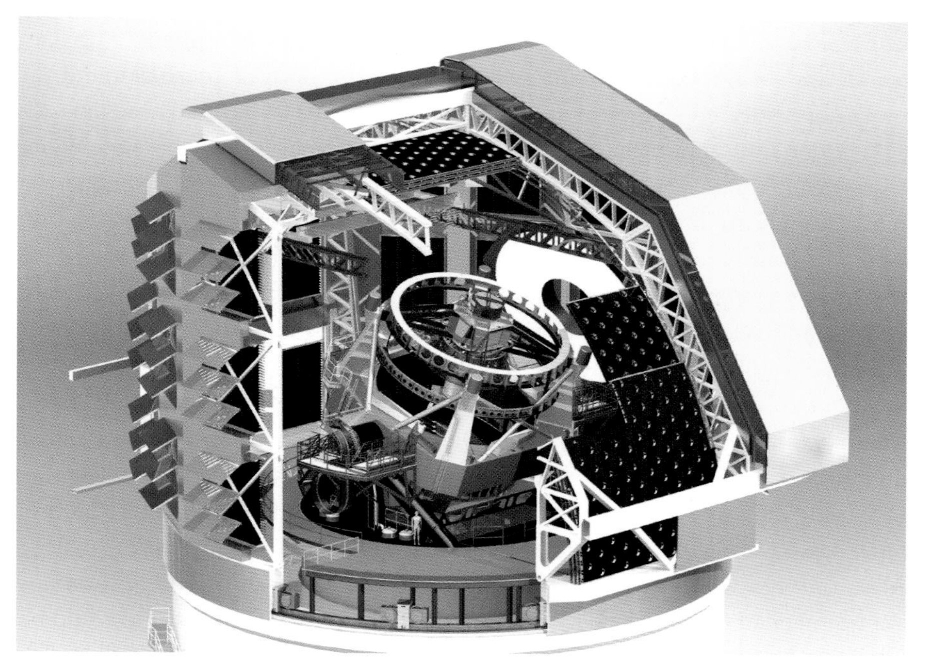

A three-dimensional rendering of the baseline design of the dome with a cutaway to show LSST within. (LSST Project Office)

indicating the detection of source variability or position change. We already know that variability can be used to detect exoplanets orbiting stars, to identify black holes ripping apart stars, and to spot supernovas that can be used for cosmological applications. The images will also identify moving objects and will be invaluable to discovering new faint objects in our solar system (e.g., small asteroids, Kuiper Belt Objects). What are probably even more exciting are the variable objects that LSST will discover that are currently unknown, opening up new scientific frontiers. And in keeping with the science fiction theme, if there's a telescope so far imagined that could spot alien spacecraft moving around in the solar system, it's this one.

The technology and trends are easy to predict, but the discoveries are not. The time-domain is very likely to show us new phenomena. The exact nature of dark matter and dark energy are unknown, but will be much better characterized in the near future and limit the possibilities – or be unravelled and understood entirely. We should be able to see directly how the first stars, galaxies, and black holes formed. More space-based telescopes exploring the universe at X-ray and infrared wavelengths are under development. The data trends are clear, too. More data. Faster computers. Both of these demand yet more sophisticated statistical analyses. Various forms of artificial intelligence (AI) under several names (e.g., machine learning, neural nets) are now being employed to deal with the fire hose of numbers. As the LSST shows, the requirements are literally astronomical, and will only get bigger in the future. The nature of big data itself has largely driven out the romantic image of the lone scientist making a breakthrough, and more and more large team collaborations lead the way. Within the same trends of computing and robotics driving big collaborations, however, also lie the seeds of their demise.

Robotics is already removing the physical labour involved with some multiplexing instruments (e.g., the Dark Energy Spectroscopic Instrument on the Kitt Peak National Observatory 4 metre Mayall Telescope). Remote observing and robotic telescopes are becoming common. Data reduction pipelines have been standard for years now in astronomy, and adding a little AI to the mix will only accelerate that trend. More and more often, humans are not needed for the standard operations of current telescopes, and could be eliminated entirely in the future. The exact timescale is unclear. Some predict human-level artificial intelligence within the next two decades. Even if it's

When humans return to the Moon, their visits will likely extend beyond the week-long trips of Apollo. One possible mission is to establish a lunar observatory with a radio telescope built into the lunar surface. (NASA)

slower than that, it is not unimaginable that machine minds could reduce and analyse data in the near-future. And likely much more.

Despite the advances of AO and the development of the 30-metre class telescopes on the ground, the future is space. As long as humans are still investing in space, astronomy will advance in step. Clouds and atmosphere are bad – both for purposes of resolution and the absorption of wavelengths of light, local pollution, limited baselines and apertures. There are a number of smart choices involving the next steps of astronomy in space, although they are limited by economics and politics.

We may get radio astronomy on the far-side of the moon (not the same thing as the dark-side), shielded from Earth's cell phones, at least for a while. Radio silence is as precious to radio astronomers as dark skies are to optical astronomers. We should also get bigger and better space telescopes over all

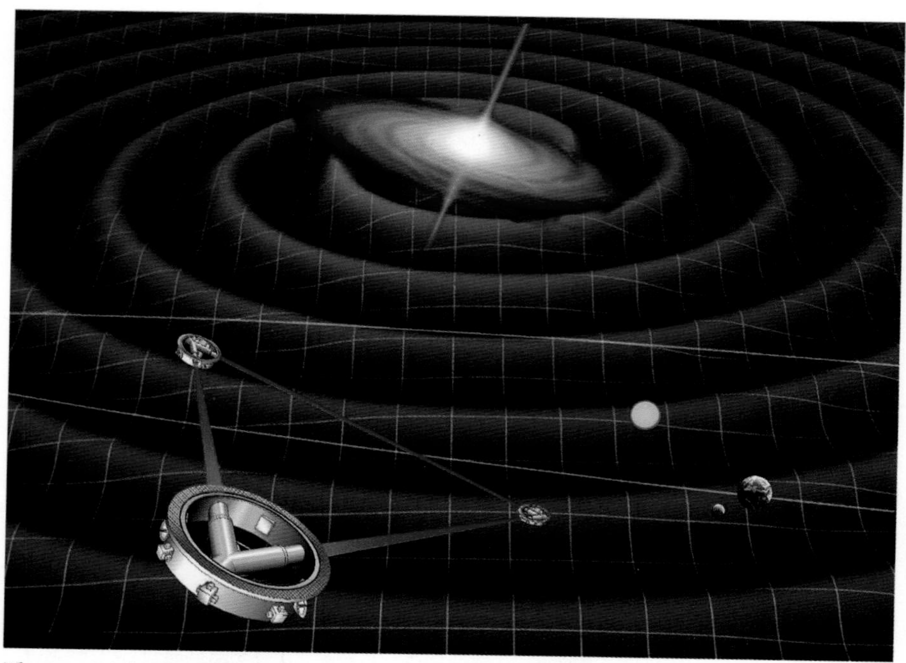

The proposed LISA mission will detect gravitational waves in space using a trio of satellites, separated by millions of kilometres. Lasers will be employed to measure the minute changes in their relative distance induced by impinging gravitational waves. (ESA)

wavelengths, as well as other facilities that can detect gravitational waves (e.g., the Laser Interferometer Space Antenna, or LISA) and cosmic rays – charged particles accelerated to relativistic speeds by supernovas, active galactic nuclei, and perhaps some other types of violent events in the universe.

One of the more innovative ideas on the drawing boards is for a telescope that can image terrestrial worlds around other stars using a special flower-shaped occulting plate flying thousands of kilometres in front of a space telescope. This "Starshade" will nearly perfectly cancel out the light of the star revealing the orbiting planets. Clever optical designs are stuff of the future, but the real science fiction is in the breakneck advancements of robots and computers.

Asteroid mining companies have already formed. Choice targets are already being picked out. Humans in space are expensive. Robots will be sent whenever possible, guided by AI for routine operations. Such commercial activities,

The Starshade deployed in space to cancel the light of background stars, enabling the imaging of their exoplanets. (NASA/JPL/Caltech)

dealing with processing materials in space and building infrastructure will enable an entire new generation of autonomous telescopes. These will be built and controlled in space by tireless digital minds using equally tireless robots with a range of sizes, down to micro-robots or perhaps even true nanotech.

Imagine a solar system sporting hundreds if not thousands of giant telescopes operating across the electromagnetic spectrum, combining as interferometers for the benefits of angular resolution, watching the whole sky all the time, sharing data, sharing analysis, sharing insights, and hopefully sharing all this with the humans who initiated their development.

Where can astronomy go from there? Micro-robots building giant distributed space telescopes is already the stuff of science fiction. There is a clear destination, however, when space-based capabilities and machine intelligences become commonplace, and that destination is some 550 astronomical units (AUs) away or more.

A key prediction of Einstein's general relativity, extensively tested and verified, is that matter bends light. Concentrations of matter therefore can literally act as lenses, focusing background light. In essence, any mass concentration is a

A deep-space telescope designed to take advantage of the gravitational lensing of the sun, which produces hugely magnified imaging of background objects, as envisioned by Claudio Maccone in his 2009 book Deep Space Flight and Communications. (Claudio Maccone)

potential telescope. In the solar system the greatest mass concentration is the sun, and its gravitational focus ranges from at least 550 AUs to a few thousand AUs, depending upon the detailed geometry of the background targets, the telescope being used, and issues involving the solar corona.

The ultimate goal of observational astronomy is to collect as many photons as possible of every possible wavelength from every direction at the highest angular resolution achievable. Moreover, astronomy has been expanding beyond just light to include cosmic rays, gravitational waves, neutrinos, and likely dark matter and dark energy in the future. There is perhaps yet one additional step to take in our exploration of the astronomy of the future.

Science fiction writers and futurists have posited that the next stage of an advanced civilization is to capture all the energy emitted by its home star. This might be accomplished by creating a surrounding sphere, called a Dyson

Artists' concept of a Dyson sphere. Notice the little moon or planet on the left side, being ravaged for raw materials. All the light from the enclosed sun can be used by an advanced civilisation, while the outer surface could itself be covered with advanced detectors turning the entire structure itself into a telescope. This image – called Shield World Construction – is by Adam Burn. (www.flickr.com/photos/djandywdotcom/31437348556)

sphere after the theoretical physicist and mathematician Freeman John Dyson who popularized the idea. The solar radiation would be absorbed on the inner surface and available to power whatever projects such an advanced civilization might choose to undertake. From the perspective of astronomy, however, there would now be a giant surface with very dark skies, not even zodiacal light, that could be covered in multi-wavelength detectors. No need for telescopes to focus light – the entire outer surface of the Dyson sphere would itself be a giant telescope, recording the times, directions, and energies of every photon striking it. Our entire world would be a telescope, looking up at the universe, watching for the next celestial mystery to unravel. In this proposed science fiction adventure, the solar system itself would evolve into a self-aware entity, wondering about the universe. An astronomer.

Asaph Hall – Man of Mars

Neil Norman

Asaph Hall was born in Goshen, a small town in Connecticut, U.S.A, on 15 October 1829. His father, also called Asaph Hall, was a manufacturer of fine wooden clocks (a popular line of trade in America in the early 19th century) and had an extensive library to which the young Asaph had full access. This was to stimulate his young mind and spark in him a lifelong thirst for knowledge.

Asaph Hall. (USNO/Wikimedia Commons)

Tragedy was to befall the family when Asaph Hall senior passed away in 1842 during a business trip to Georgia. This left the remaining Hall family members facing difficult financial circumstances. His mother Hannah then came upon the idea of operating a cheese factory, with the intention of paying off the mortgage on the family farm, although this venture was doomed to last for just three years before failing. The 16-year-old Asaph pulled out of the local district school. Taking up an apprenticeship as a carpenter in 1845 it was three years later that he began work as a journeyman carpenter through which he earned his living for six years. It was during this period that he was able to perfect skills that would later prove invaluable for when his supervisory eye was to be cast over the construction of observing shelters on numerous astronomical expeditions.

With a desire to extend his higher education, he enrolled in the recently-formed New York Central College, McGrawville, New York in 1854, where he mixed with a group of students who were less into classical studies and more into the adventurous ways of life. It was here that he was to meet a mathematics teacher by the name of Chloe Angeline Stickney, who was completing her final year at the college. She shared many of Asaph's principles and the chemistry between them was instantaneous. A whirlwind romance ensued culminating in the couple being married in March 1856 and shortly afterwards moving to the newly-established observatory at the University of Michigan. It was there that Asaph studied under the German astronomer Franz Friedrich Ernst Brünnow (1821–1891), then director of the observatory.

Angeline Stickney Hall. (U.S. Naval Observatory)

Their stay at the university lasted just three months, after which a lack of funds led to Asaph and his wife taking up teaching jobs at Shalersville Institute in Ohio where they stayed for a year. Asaph had by this time decided that his future lay in astronomy and in order to pursue this goal the couple moved to Cambridge, Massachusetts, where he met the astronomer George Phillips Bond (1825–1865). In spite of Bond informing Asaph, quite bluntly, that he should abandon astronomy or he would starve (a reference to the low paid nature of the work) Asaph took a low paid job at the Harvard College Observatory. Once there, he became an observer and computer of orbits and impressed those around him by his willingness to work. At this point, his wife had taught him German and with this new skill he read Brunnow's *Lehrbuch der Sphärischen Astronomie* (Handbook of Spherical Astronomy). In 1858 he published the first of his numerous mathematical and astronomical articles in scientific journals. During this period he also generated extra income by computing almanacs and by observing Moon culminations at a dollar per observation.

In 1862 Asaph moved to Washington to take the post of assistant astronomer at the U.S. Naval Observatory (USNO). However, the onset of the American Civil War (fought between 1861 and 1865) was to take both a physical and mental toll upon him. He became ill with jaundice and was subject to the exhausting process of seeing several friends be either wounded or killed during the war.

It was to take him a full two years to recover, after which his wife was to play a large part in his next career move. She put his name forward for the role of full professor of mathematics at the USNO, the upshot of which was that Asaph was promoted to the post.

This Lunar Orbiter 4 image shows the 39 km diameter disintegrated lunar crater Hall, located on the south eastern shores of Lacus Somniorum, which was named in honour of Asaph Hall. (NASA/James Stuby)

In 1869 and 1870 Asaph intended to travel on two expeditions to view solar eclipses, the first from the east coast of Siberia (7 August 1869) and the second from Sicily (22 December 1870). He also planned to lead a party to Vladivostok to observe the 9 December 1874 transit of Venus. Although inclement weather and lack of adequate photographic apparatus prevented these expeditions from fulfilling their expectations, he was more successful as leader of the expeditions to Colorado to observe the eclipse of 29 July 1878 and to Texas to observe the transit of Venus on 6 December 1882.

In 1875, Hall was put in charge of the impressive 26-inch refracting telescope, constructed by Alvan Clark & Sons and acquired in 1873 and which at the time was the largest telescope of its type in the world. His first discovery with this telescope was made in December 1876 when he noticed a white spot on Saturn. His observations of this spot, made through more than sixty rotations of the planet, enabled him to establish the first definitive rotation period for Saturn since William Herschel attempted the same in 1794.

In 1877 Mars came to a particularly favourable opposition, approaching to within 56 million km (35 million miles) of the Earth on 5 September of that year. Asaph decided to undertake a systematic search for possible moons of the planet. He was aware that theoretical proposals indicated that any moons of Mars must orbit quite close to the planet, as otherwise the gravitational effects of the Sun would overpower the attraction of Mars.

"The chance of finding a satellite appeared to be very small ..." Hall wrote, "... so that I might have abandoned the search had it not been for the encouragement of my wife." Angeline Hall was an enthusiast and Angelo, the third of their four sons, later claimed that she "… insisted upon her husband's discovering the satellites of Mars."

Hall first glimpsed the small object that was eventually named Deimos on 11 August 1877, noting it in his records as "… a faint star near Mars." Fog from the nearby Potomac River put a halt to the search, following which the next few nights were frustratingly cloudy. However, after a few days Asaph managed to resume the search, and located Phobos on 17 August.

He disclosed his observations to Simon Newcomb (1835–1909), professor of astronomy at the U.S Naval Observatory. Newcomb erroneously believed that Hall, in his modest conversation, was reluctant to recognise the 'Mars Stars' as satellites, and so took for himself an undeserved credit for their discovery in the wide press coverage that followed. For many years Hall quietly harboured a grudge against Newcomb, who eventually offered his apologies. In recognition of his discoveries, Asaph was awarded the Lalande Prize of the French Academy of Sciences (1877) and the Gold Medal of the Royal Astronomical Society (1879).

Following his discovery of the Martian moons, Hall commenced a programme of observation (with Newcomb) aimed at improving the positional measurements and precise determination of the orbits of planetary satellites, which included not only those of Mars, but those of Saturn, Uranus and Neptune. Amongst other revelations, this resulted in the realization by Hall in 1884 that the position of the elliptical orbit of Saturn's moon, Hyperion, was retrograding by around 20° per year.

The prominent feature seen at lower right of this image is the 9 km diameter impact crater Stickney, the largest crater on the Martian moon Phobos. It was named in honour of Angeline Stickney Hall who had been a driving force behind her husband's efforts to discover the moons of Mars. (NASA/JPL-Caltech/University of Arizona)

Hall's memoir relating to the orbit of Iapetus, another of Saturn's moons, was described by the mathematical astronomer George William Hill (widely considered as the greatest master of celestial mechanics of his time) as one of '… the most admirable pieces of astronomical literature …', even comparing its clarity and precision to the work of the highly respected German astronomer Friedrich Wilhelm Bessel (1784–1846).

In addition to his work on planetary satellites (for which he was presented with the Arago Medal of the French Academy of Sciences in 1893), Hall was a diligent observer of double-stars, carrying out numerous investigations of binary star orbits. In 1892 he showed that the two components of 61 Cygni were physically related. He also investigated stellar parallaxes and the positions of the stars in the Pleiades star cluster.

A prolific writer of papers, his published works include a Catalogue of 151 Stars in Praesepe (1870); Observations and Orbits of the Satellites of Mars (1878); Observations of Double Stars Made at the United States Naval Observatory (1881 and 1892); Orbits of Oberon and Titania (1885); The Orbit of Iapetus (1885); The Six Inner Satellites of Saturn (1886); Observations for Stellar Parallax (1886); and Saturn and its Rings 1875–1889 (1889).

As we can see, Asaph Hall was never one to be idle and, as well as acting as a consulting astronomer and non-resident Director to the Washburn Observatory in Madison, Wisconsin from 1887, he was elected to the National Academy of Sciences in 1875 (where he served as Home Secretary for twelve years and Vice-President for six). He also served as president of the American Association for the Advancement of Science in 1902, and was associate editor of the Astronomical Journal from 1897 to 1907. Asaph was served his mandatory retirement from the US Naval Observatory in 1891, when aged 62, but continued to work as a volunteer observer using the 26-inch telescope.

A year later, on 3 July 1892, his beloved wife Angeline died and in 1894 Hall left Washington for Connecticut. In 1896 he went to Harvard University as a lecturer in celestial mechanics and taught there until 1901. After five years of teaching he finally retired to his rural home in Goshen, Connecticut and married Mary Bertha Gauthier, continuing to live in Goshen until his death on 22 November 1907 while visiting his son Angelo in Annapolis, Maryland.

Getting the Measure of Double Stars

John McCue

Many stargazers are finding the observation of double stars a rewarding pursuit in these days of light pollution in towns and cities. Although the scourge of artificial light in the urban sky drowns out the delicate views of distant galaxies

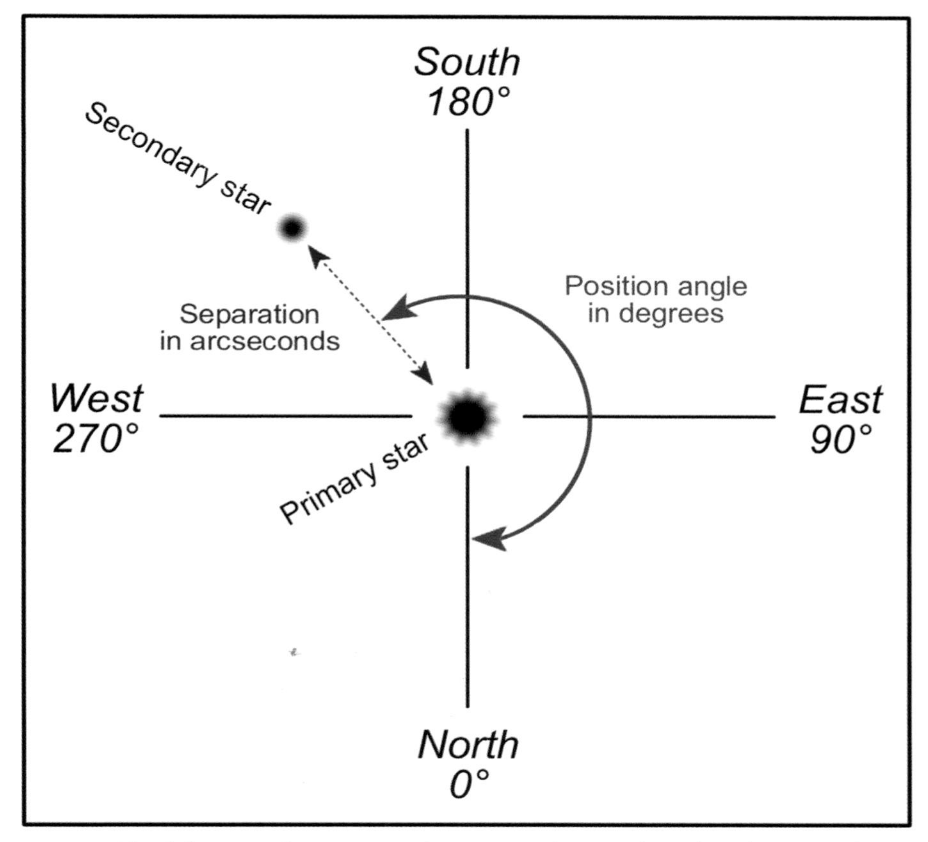

Figure 1. The definitions of separation and position angle. Seen through a telescope without a star diagonal, the NS compass points are inverted, and the EW points are reversed. (Brian Jones/ Garfield Blackmore)

and Milky Way nebulae, town observers can still enjoy the many varied and multi-coloured double stars. However, to turn your views into quantitative knowledge requires a measurement of the configuration of the two stars, as seen through your eyepiece. The two parameters that pin down the relative positions of the components of a double are known as the separation and the position angle (PA), as shown in Figure 1.

An accumulation of these measures, over time (often many years) for any particular binary star, enables their orbits around each other to be worked out. This is the only direct way of determining a star's mass. A binary star is a gravitationally-bound system, as opposed to an apparent double seen because of a lining-up effect, just as one would look up and see a nearby bird in the line-of-sight of a distant aeroplane. Many thousands of double stars are known in our own locality of the Milky Way galaxy. In fact, as a single star, our sun is in a minority. Yet there are fewer binary orbits known than one might imagine. In these modern times, here are three methods used to measure the position angle and separation of a double star.

Astrometric Eyepiece Method

Firstly, we can use the astrometric eyepiece, as shown in Figure 2. Telescope manufacturers Meade and Celestron produce eyepieces with inscribed angular

Figure 2. The Meade astrometric eyepiece. (Meade US and Opticstar Ltd., UK)

and linear scales that are internally illuminated (from the attached barrel) and can thus be seen superimposed on the view of the double star. The scale is quite complex, but the position angle and separation can be measured quite quickly and easily with a little practice.

The double star separation is measured using the scale across the middle of the reticle as a ruler, as seen on the Meade version in Figure 3, but first the scale of that middle ruler, as judged against the real starlit night sky, must be determined. This is done by allowing a star to drift across the 50 divisions of this middle linear scale, and timing its passage. This should be done several times and a mean taken. As high a magnification as possible should be used in order that the timing error is reduced, which in practice means using a telescope of long focal length (at least 1,000mm) with the 12mm astrometric eyepiece, though a Barlow lens is sometimes used. Converting from a time in seconds to an arc angle in seconds, requires a multiplication by 15.0411, since a sidereal day of 23h 56m 4s is equivalent to 360°. But this is only true at the celestial equator, so multiply also by cos δ, where δ is the declination of the star being observed. A star at around 50° to 70° declination is usually suitable,

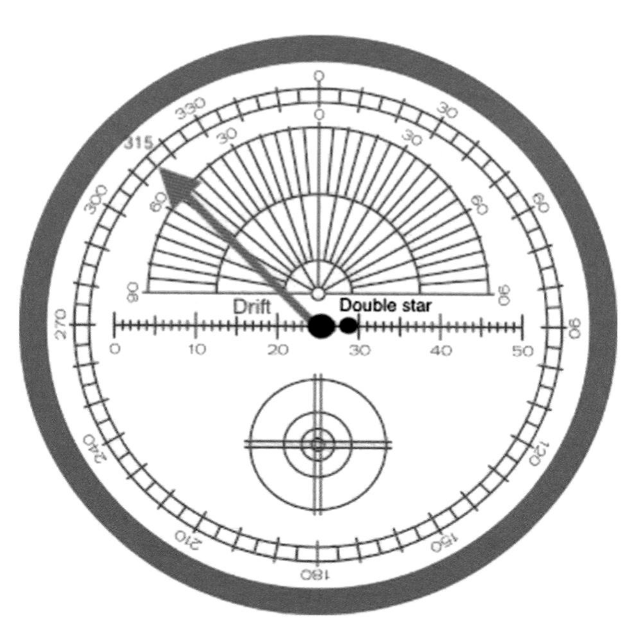

Figure 3. The view through the Meade astrometric eyepiece showing the drift method of measuring the position angle. (Meade US and Opticstar Ltd., UK)

and this will make timing easier anyway as the passage of the star across the same middle scale will take longer away from the equator. In fact, an exact 60° declination will make the calculation easier as cos 60° is 0.5. Now divide by 50 to get the scale constant in arc seconds per division. If you keep using the same optical arrangement this constant need not be determined again, and it can be used for any double star at any position in the sky. The chosen target double for the night can be lined up anywhere on the middle scale, with the telescope drive running, so the number of divisions between the primary and secondary stars can be read from the scale, then converted to arc seconds.

Next, there are many and varied ways of measuring the position angle (PA) of your target double, though the basic idea is to allow the double star to drift across the reticle to reach the protractor scale fully around the side. Knowing N, S, E and W in the eyepiece view of your double star, you can tell in which quadrant the PA of the double will be, as in Figure 1, and hence get a very rough idea of the value of the PA. Keeping the orientation of the reticle as it was after measuring the separation, move either component (usually the primary) to the centre of the linear scale (it may already be there). Switch off the telescope drive, let the star drift until it reaches the outer protractor scale, then start the drive, thus stopping the star on the protractor scale. Read off the angle. If it doesn't fall near your rough estimate, add 180°. This correction will give the correct quadrant. If this answer comes above 360°, just subtract that 360° to give you the PA of your double star – repeat several times to obtain a mean. In the example of Figure 3, the protractor reads 315°. If that falls within your estimated quadrant, all is well and 315° is your observed PA. If not, add 180°. This gives 495°, so subtract 360°, giving 135° as your observed value.

Crosshair Method

The astrometric eyepiece described above is quite expensive so, secondly, there is a very cheap modus operandi – you only need a simple crosshair eyepiece and a stopwatch.

When you have the double star in your crosshair eyepiece, turn off the telescope drive and turn the crosshairs until one arm is aligned with the way the stars are drifting; that is E–W direction, of course. Adjust your telescope view (with slow motion controls, or the motor drive keys if available) until the double is 'upstream' of the wires, and time the difference between the

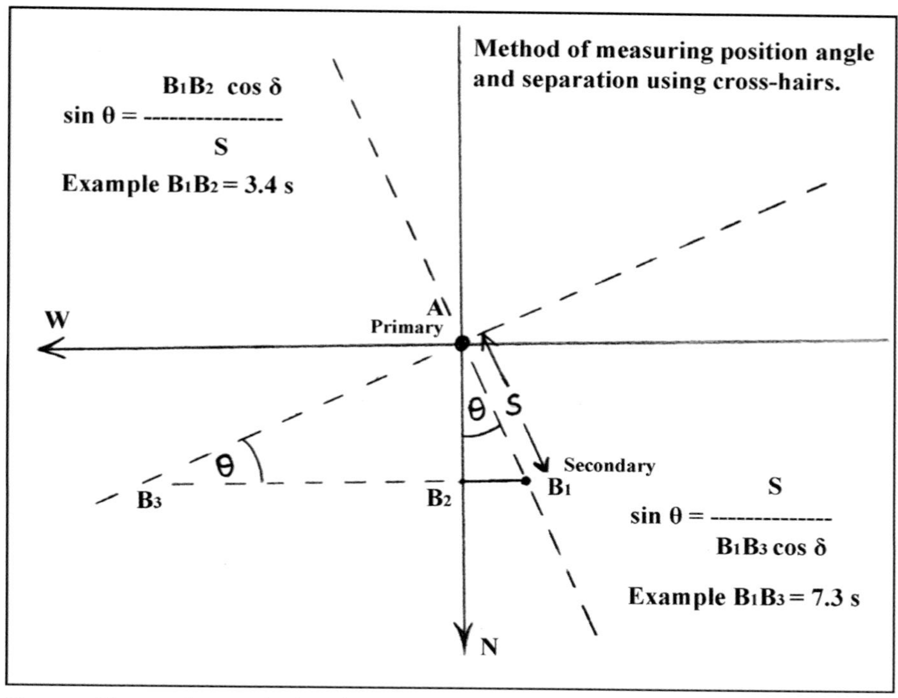

Figure 4. The crosshair method. (John McCue)

primary crossing the N–S wire and the secondary doing so. In Figure 4, the primary is shown at the central wire crossing but that doesn't have to be the case; both stars just have to cross the N–S wire. Make a note of the time on your stopwatch. This will be time B_1B_2.

Next, turn the crosshairs until one arm is aligned with the two stars themselves (the dotted lines in Figure 4), then time the difference (again the drive will be turned off) between when the secondary crosses that arm (with the primary crossing the central wire crossing point) and when it crosses the other arm. Again, the primary doesn't have to go through the centre, but if not, time the difference between when the two stars cross the non-aligned arm, no matter whether on the top side or lower side of the middle. This will be the time B_1B_3.

The diagram in Figure 4 also shows the two equations that are solved to give the values of s (separation) and θ (position angle). Note also that you substitute for the value of δ, the declination of the double star. When you have calculated s, turn this seconds of time into seconds of angle by multiplying by 15.0411 (or just 15 will suffice for this level of measurement accuracy). It also gives two example times, B_1B_2 = 3.4 seconds, and B_1B_3 = 7.3 seconds. The equation should give separation, s = 37.5" (37.4" when multiplying by 15) and position angle, θ = 43.0°, if the declination δ is 60°.

Notice that if the position angle is more than 90°, but less than 180°, the primary will still be leading the way across the wires, and the same equation holds good, but the calculated PA will be 180° – θ. If the PA is between the 180° and 270°, the secondary will now lead the way, the same method and equation can be used, and the calculated PA will be 180° + θ.

Finally, if the PA is between the 270° and 360°, the secondary will still lead the way, the same method and equation can yet again be used, but the calculated PA will be 360° – θ. If the PA is close to 0° or 180°, then the timing becomes difficult, unless the double is quite widely separated. Some experience with a spreadsheet such as Excel, will allow a speeding up of the calculation.

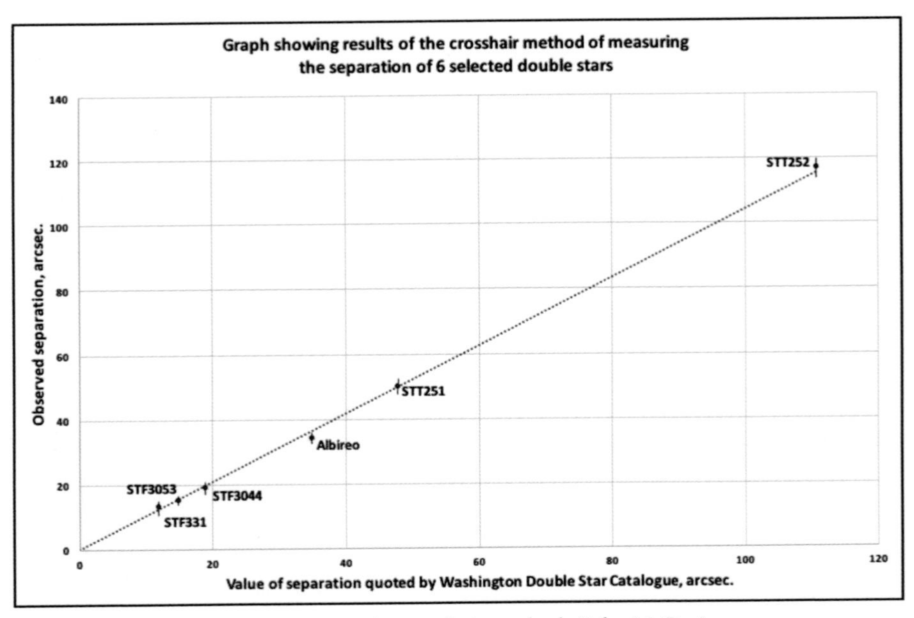

Figure 5. Observed separations using the crosshair method. (John McCue)

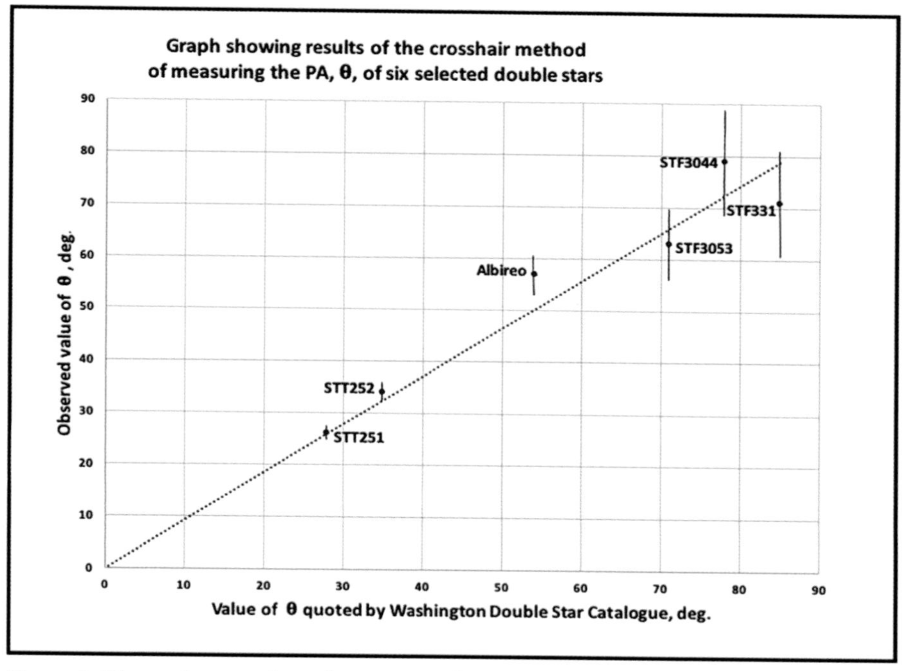

Figure 6. Observed acute values of position angles using the crosshair method. (John McCue)

The author used this method on six selected double stars, and the results are shown in Figures 5 and 6, which show the accuracy of the observations (how close they are to the accepted value on the sloping line), and their precision (the size of the error bars). It can be seen that the method works well on the double stars' separation, but on their PA (acute angle) the precision is low beyond about 50°.

British Astronomical Association double star observers Philip Rourke, Neil Webster and the author compared the aforementioned two methods for the following eleven double stars, and obtained the results listed in Figure 7, which are shown in tabular and graphical form.

Star	Separation " (crosshair)	PA ° (crosshair)	Separation " (eyepiece)	PA ° (eyepiece)	Separation " (WDSC)	PA ° (WDSC)	Observer
μ Boo	102	170	109	171	109	170	JM/NW
α CVn	18	223	19	227	20	229	JM/NW
η Cas	14.6	325	13	325	13	326	PR
ν Dra	63	310	62	316	63	310	PR
ζ Psc	20.6	60	23.6	65	23	63	PR
1 Cam	9.7	308	11.8	312	10.3	307	PR
η Per	27	300	29.5	295	28.3	301	PR
ε Peg	137	318	147.5	320	145	318	PR
β Cyg	35.6	58.7	35.4	55	34.4	54	PR
γ And	9.4	71	10.6	65	9.8	63	PR
56 And	205.6	296.5	200	297	203	299	PR

Figure 7. Tabulated values of comparison measurements (WDSC refers to those published in the Washington Double Star Catalogue)

The graphs in Figures 8 and 9 are encouraging because the dotted lines are where the results should fall if the two methods agree, and so they do. However, this does not mean that the methods are giving values close to the accepted catalogue values, but the tabulated values, and Figures 5 and 6, show that such correspondence is good. Note that the separation values for 1 Cam and γ And

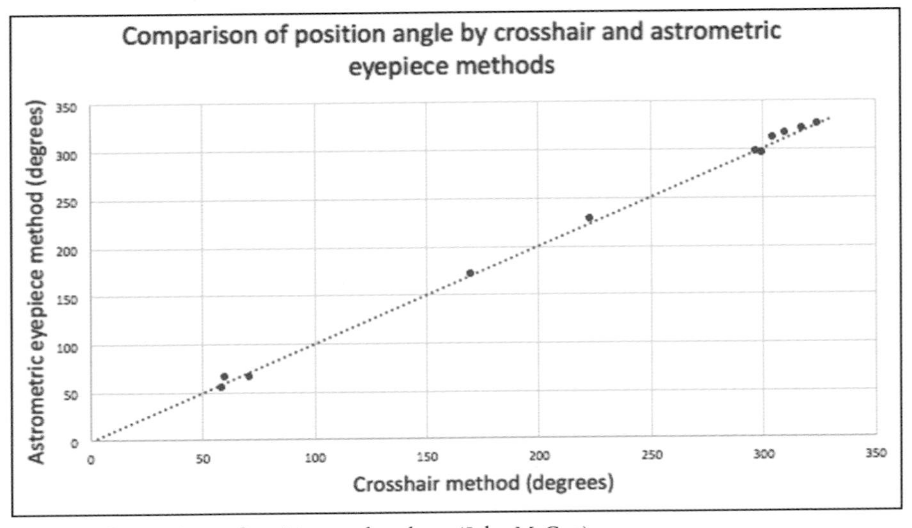

Figure 8. Comparison of position angle values. (John McCue)

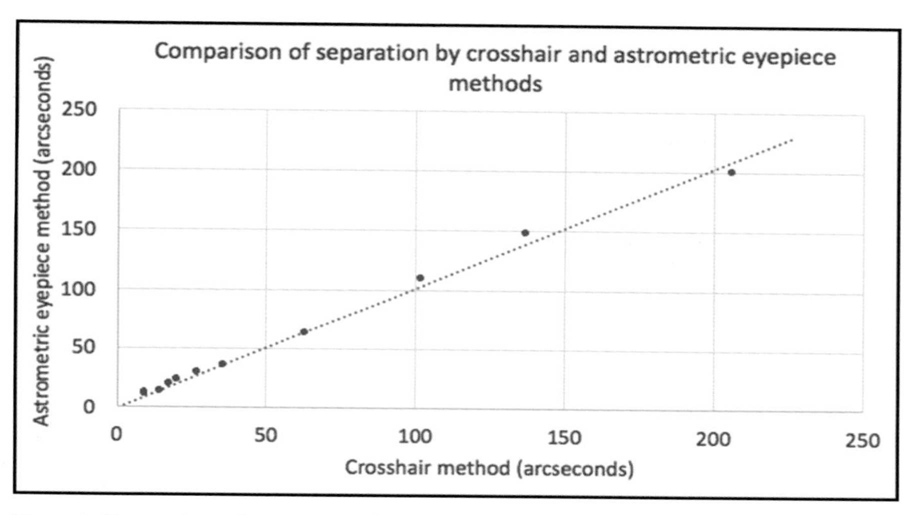

Figure 9. Comparison of separation values. (John McCue)

are so close in value that, with the chosen scale, their points almost merge into one.

Observing double stars requires a telescope with a focal length as long as possible (at least 1000mm) so giving the widest split between the components. Some observers use a Barlow lens, as mentioned above, to enhance the focal length further, but when measuring the separation on the middle scale of the astrometric eyepiece the same optical configuration must be used as that which determined the scale constant. The same is not true for the crosshair method as the time of drift is independent of the magnification employed.

Dedicated Software Method

Thirdly, a method well worth exploring for the astrophotographer and computer user is to measure the double star parameters from a FITS (Flexible Image Transport System) image of the system; it provides highly accurate results, and a brief summary will be given here. The FITS format is the most commonly used way of storing and transporting astronomical images; it carries with it a file of data about the image, such as telescope details and exposure. The modern CCD camera for dedicated use with a telescope captures images in this format, and DSLRs give excellent images in other formats, which can

Figure 10. An image of Omicron (o) Cygni by Graham Darke of the Sunderland Astronomical Society. (Graham Darke)

be converted to the FITS format. Figure 10 is an image of Omicron (o) Cygni taken by Graham Darke with a 140mm triplet refractor (980mm focal length) and a Canon 600D camera. The final image was produced from 12 exposures of 10 seconds each (ISO 800) and stacked in Deep Sky Stacker software.

The FITS image of the double star can then be digitally imprinted with a right ascension and declination grid (WCS – World Coordinate System) of that area of the sky by matching it, for example, with the UCAC (US Naval Observatory CCD Astrograph Catalogue), which is available online.

With such an invisible grid on the double star image, professional-standard software such as Aladin, with its many analytical functions, can give an immediate, and highly accurate, separation and position angle of the double star. Aladin is an interactive sky atlas allowing the user to measure astronomical images of double stars and other stellar targets. Accessed at **aladin.u-strasbg.fr** it also links with other astronomical databases.

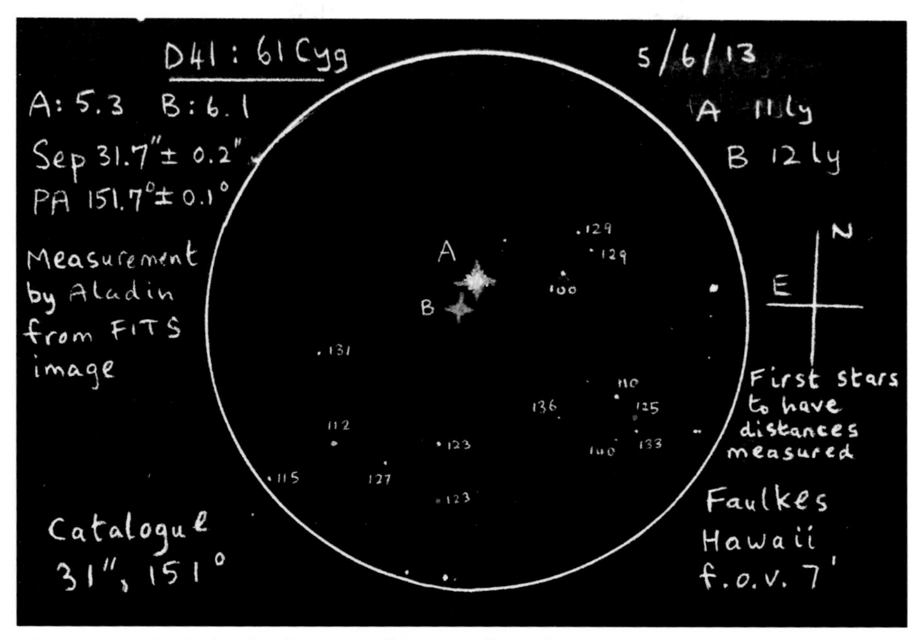

Figure 11. Author's sketch of 61 Cygni from a Faulkes Telescope image. (John McCue)

Figure 11 shows a sketch made from an image obtained remotely with the Faulkes North Telescope (2m Ritchey–Chrétien system). The downloaded FITS image was analysed with Aladin software. The Faulkes Telescopes are a global network of robotic telescopes for education and research, **www.faulkes-telescope.com**

The reader will encounter other methods of measuring separation and PA, some from an earlier time of mechanical measurements (such as a micrometer), but the methods presented here are straightforward and accessible to everyone.

100 Years of the International Astronomical Union

Susan Stubbs

2019 marks the centenary of the founding of the International Astronomical Union (IAU) which came into being on 28 July 1919 in Brussels, Belgium, at the Constitutive Assembly of the International Research Council. Its mission has always been to promote and safeguard the science of astronomy through international collaboration. As an organisation for professional astronomers, its work and achievements may be little known to the amateur astronomy world; it is perhaps best known as the organisation that demoted Pluto from a planet to a dwarf planet in 2006 during their 26th General Assembly in Prague (still a contentious decision for some amateurs!) It continues as the body

Planet Definition Voting at the IAU 26th General Assembly in Prague in 2006, with what appears to be a somewhat dejected looking Pluto gazing out from the table. (IAU)

which maintains the standardisation and conventions for the naming of all astronomical bodies and any of their surface features.

The inaugural meeting in 1919 included the seven initial member states of Belgium, Canada, France, United Kingdom, Greece, Japan and the United States of America. These were followed by Mexico and Italy within a short period, and by the time of the first General Assembly in Rome in 1922 they had been joined by Australia, Brazil, Czechoslovakia, Denmark, the Netherlands, Norway, Poland, Romania, South Africa and Spain, bringing the total membership to 19 countries.

The birth of the IAU was preceded by a long history of cooperation in astronomical research between countries and institutions in solar physics and astrophysics. The creation of the IAU in 1919 incorporated into the new body pre-existing organisations, such as the International Union for Cooperation in Solar Research (formed in 1905) and the International Carte du Ciel (initiated in 1887), and immediately became the leading astronomical organisation in the world at that time. The concept of more effective research when many can contribute was well understood. The political ethos of the time prevented some countries, such as Germany and Austria, joining until decades later, but all continents and cultures are now represented in membership of the IAU.

In addition to more than 12,000 individual members in 101 countries around the globe, the IAU has 79 professional astronomical communities and societies affiliated to it, including the Royal Astronomical Society in the United Kingdom, the Swedish Academy of Sciences, the Australian Academy of Science and the Indian National Science Academy. The governing body is the General Assembly, which includes all members, and the Union is administered from its head office at the Institut d'Astrophysique de Paris in France. The current President is Dr Silvia Torres-Peimbert, an astronomer from the Institute of Astronomy at the Universidad Nacional Autonoma de Mexico, and the General Secretary is Dr Piero Benvenuti of the Department of Physics and Astronomy at the University of Padua, Italy.

The IAU Founding President in 1919 was Benjamin Baillaud and the General Secretary the Yorkshire-born Alfred Fowler (1868-1940). Since then the ranks of elected officials have included many famous names from the history of astronomy, including Arthur Stanley Eddington, Jan Hendrik Oort, Bart Bok, Bernard Lovell, Frank Watson Dyson and Willem de Sitter.

Individual membership of the IAU is currently limited to those who are professional scientists involved in astronomical research. There is currently an intention to open up membership to PhD students, but not to amateur astronomers. However, interested organisations, such as astronomy societies and schools, can be affiliated. Membership has grown from 207 individual members and seven founding countries in 1919 to the current individual membership of over 12,000 and 79 countries.

Benjamin Baillaud (1848–1934), a specialist in mathematics and celestial mechanics from the Observatoire de Paris, was the Founding President of the IAU. (Observatoire de Paris)

During the lifetime of the IAU the population of the world has grown from 2 billion to 7 billion people; the corresponding growth of IAU membership is therefore faster than might be expected purely due to relative expansion in numbers of astronomers. There has also been

Jean-Michel Jarre, master of ceremony, is seen with the distinguished panel of speakers at the official opening of the International Year of Astronomy 2009 at UNESCO. (IAU/José Francisco Salgado)

a change in gender balance with increasing numbers of female members, although still overall a majority of men.

Following on from the very successful International Year of Astronomy (IYA) in 2009 the IAU formulated a strategic plan for the next decade. Sponsored by the IAU and supported by the United Nations and UNESCO, the IYA celebrated 400 years since Galileo had first used a telescope to look at the heavens. The IAU document *Astronomy for the Developing World- Building from IYA 2009* was adopted by the General Assembly in August 2009 and laid out the vision, specific goals and implementation plans for the Union's activities going forwards.

Traditionally much of its budget had gone to supporting professional activities, such as the sharing of research and results at annual symposia, albeit with significant educational activities such as the International School for Young Astronomers. There was now a greater realisation of the need for public outreach and education and the IAU now considers dissemination of astronomical knowledge, particularly to developing countries, as one of its most important roles. The Strategic Plan for 2010-2020 has three main areas of focus; Technology and Skills, Science and Research, and Culture and Society. This acknowledges the importance of the use of technology used in astronomy and its overlap into everyday life, and the enormous value of international collaboration. Examples include the international contribution to the construction and operation of the Atacama Large Millimetre Array, European Space Agency missions such as Rosetta, Hubble and ExoMars, and international cooperation in public education.

Much of the IAU's income is used for educational projects, such as their annual International School for Young Astronomers, which is open to undergraduate and postgraduate students. These were established in 1967 and are held regionally throughout the world often, but not exclusively, in areas where students may have less opportunity to be exposed to cutting edge astronomy research and teaching. The lecturers are experts from centres of excellence across the world and the school runs for three weeks of lectures with additional practical and observational experience. The IAU also sponsors education for school teachers of astronomy through the Network for Astronomy School Education (NASE), with courses held worldwide. NASE began in 2010 and links IAU members and teachers in each country to facilitate courses for both primary and secondary school teachers teaching aspects of astronomy.

The programme of scientific meetings that the IAU organises is an important part of their promotion of international collaboration in astronomy. These are a key activity for them and around one third of their income goes towards grants for participation in the annual symposia and assemblies. The General Assembly has met every 3 years since 1922 (with the exception of the period 1938 to 1948, due to the Second World War) for discussion of scientific matters and those related to statute and process. An Extraordinary General Assembly was held in February 1973 in Warsaw to commemorate the quincentenary of the birth of the astronomer Nicolaus Copernicus - this at the request of the Polish section of the IAU.

In addition to the General Assemblies, many scientific meetings are held around the world, often as output from one of the many groups within the IAU. The Union functions through a number of divisions, commissions and working groups covering all aspects of astronomy, education and public outreach. There are nine divisions, including Fundamental Astronomy, Education Outreach and Heritage, Galaxies and Cosmology, and Sun and Heliosphere. There are 35 commissions which are coordinated through the nine divisions, which include areas of interest such as astrometry, radio astronomy, history of astronomy, and gravitational wave astrophysics. There are also 54 working groups which cover aspects such as stellar spectroscopy, astronomy for equity and inclusion, public outreach, and the publication of the public access newsletters and journals. The working groups cover defined tasks for a set period and may be established by the Executive Committee, divisions or commissions.

Amongst the roles of the IAU are the definition of fundamental physical and astronomical constants, promotion of educational activities in astronomy and international collaboration in education and research, and astronomical nomenclature for celestial bodies and their surface features. It is for the latter that the IAU is perhaps best known outside of professional astronomical circles, and the decision about Pluto in 2006 remains widely controversial over a decade later. The Minor Planet Centre at the Smithsonian Astrophysical Observatory is responsible for the designation of minor bodies in the Solar System under the auspices of Division F of the IAU. The Working Group for Planetary System Nomenclature is responsible for naming of surface features, and following the flyby of Pluto by the New Horizons spacecraft in July 2015 named several newly-seen surface features. The Working Group on Star Names (Division C) was formed in May 2016 to formalise the names and spellings of commonly

Remote and enigmatic, Pluto resides on the edge of the Solar System in a region known as the Kuiper Belt. In July 2015, NASA's New Horizons space probe flew past Pluto, offering the first detailed look at this small, distant world, and its largest satellite Charon, transforming these mysterious bodies into worlds with distinct features. Pluto's first official surface-feature names are marked on this map, compiled from images and data gathered by New Horizons during its historic flight through the Pluto system, (NASA/JHUAPL/SwRI/Ross Beyer)

used colloquial names for stars. In February 2017, 17 newly discovered exoplanets were named by the Working Group on Small Bodies Nomenclature (Division F) – for the first time suggestions were invited from the public and decisions made by public vote.

The Union is committed to education and public outreach, and has an excellent and informative website at **www.iau.org** which has information regarding IAU activities and meetings; news features; newsletters and journals; and an accessible image and video section. It also has links and signposts to other websites with information about careers in astronomy, constellations, naming of astronomical bodies and so on as well as a useful Frequently Asked Questions section. There is also much educational content. The AstroEDU section of the website **astroedu.iau.org/en** has activities for 6-14 year olds including planetary maps, making a sundial and investigating constellations. The Union is also active on Twitter via **@IAU_org** and Facebook (search for

An artist's impression of an exoplanet seen from the surface of its moon. Thousands of exoplanets have been found to date and many more will undoubtedly be discovered in the months and years to come. The IAU dissociates itself entirely from the commercial practice of selling names of planets or stars. These practices will not be recognised by the IAU and their alternative naming schemes cannot be adopted, (IAU/L. Calçada)

The International Astronomical Union), with regular communication of news items of interest to the general public.

In summary, the International Astronomical Union has seen significant growth since its birth in 1919. Whilst it remains an organisation of membership for professional astronomers only, there has been a welcome shift in emphasis in recent years towards public outreach and education, building on the initial stated aim to promote and safeguard the science of astronomy. There is, perhaps, more that could be done in making astronomy more accessible to the general public by improving the educational content of their website, particularly for young people, and better links with enthusiastic amateur astronomers. Collaboration with, and the provision of representatives to, other scientific bodies and international organisations such as the United Nations and UNESCO (as with the International Year of Astronomy in 2009 and International Year of Light in 2015) will be ongoing. As they proceed into the next 100 years, change and development will undoubtedly continue.

In Total Support of Einstein: Eddington's Eclipse, 1919

Neil Haggath

This year sees the centenary of one of the most important experiments in the history of science – Sir Arthur Stanley Eddington's eclipse expedition, which verified the bending of light by gravity, and provided the first evidence in support of Einstein's General Theory of Relativity.

Albert Einstein (1879–1955) first published his Special Theory of Relativity in 1905, which dealt with the rather restricted special case of unaccelerated motion, and proposed such concepts as the constant velocity of light in vacuum and time dilation. This was followed ten years later by his General Theory of Relativity, which proposed that gravity is not a "force", as stated by Newton, but rather a "bending" or curvature of space – or more accurately, spacetime – by the presence of matter, which then affects the motion of moving objects, which follow the straightest possible path through curved space. We have all seen the analogy of placing a heavy object on a trampoline, and rolling a marble around it – the presence of the mass causes the normally two-dimensional surface to curve in the third dimension.

Half a century later, the physicist John Archibald Wheeler (1911–2008) summed it up very succinctly: "Matter tells spacetime how to curve, and spacetime tells matter how to move."

As General Relativity was published in German during the First World War, it initially did not become widely known outside the German-speaking world, though the Dutch physicist Willem de Sitter (1872–1934) attempted to publicise it further afield. One man was largely responsible for introducing the theory to the English-speaking world – Arthur Stanley Eddington (1882–1944), who as well as being an outstanding astronomer, became one of the finest popularisers of science who have ever lived.

Eddington, at that time, was Professor of Astronomy at Cambridge, and Director of the university observatory. (Incidentally, his predecessor in that chair had been another great science populariser, Sir Robert Stawell Ball). He was also Secretary of the Royal Astronomical Society, which meant he was the first person in Britain to receive de Sitter's letters and papers. Furthermore, he was one of the very few people to instantly understand General Relativity; he immediately became one of Einstein's earliest supporters, and set out to publicise his work.

Arthur Stanley Eddington. (Wikimedia Commons)

Rather stupidly, due to the War, many British scientists flatly refused to pay any attention to a theory proposed by a German – despite the fact that Einstein had left Germany years earlier and taken Swiss citizenship, and had refused all invitations to return to his homeland and help with the war effort. The eminent physicist Sir Oliver Joseph Lodge, whose son had been killed serving as an army officer, dismissed the theory out of hand, proclaiming to anyone who would listen that "German science" had killed his son! Eddington, being a Quaker, was a pacifist and vehemently opposed to the War, so he ignored such prejudice – though he received much personal criticism for his support of Einstein.

One of the most important predictions of General Relativity was that the path of light would be bent by gravity. The effect would only be noticeable if the bending was due to an extremely large mass, such as a star. Eddington and the then Astronomer Royal, Sir Frank Watson Dyson (1868–1939), realised that if a star could be observed almost in line with the Sun, then it would appear to shift position with respect to other stars further away, due to the bending of the path of its light by the Sun's gravity. Even then, the effect would be tiny – Einstein's equations predicted that the deviation would be a mere 1.75 seconds of arc – but it would, at least in theory, be measurable.

Of course, the only time that stars can be observed close to the Sun is during a total solar eclipse! Eddington and Dyson realised that that part of the theory could be tested, by photographing the star field around the Sun through a telescope during totality, and comparing the image with one taken of the same star field in

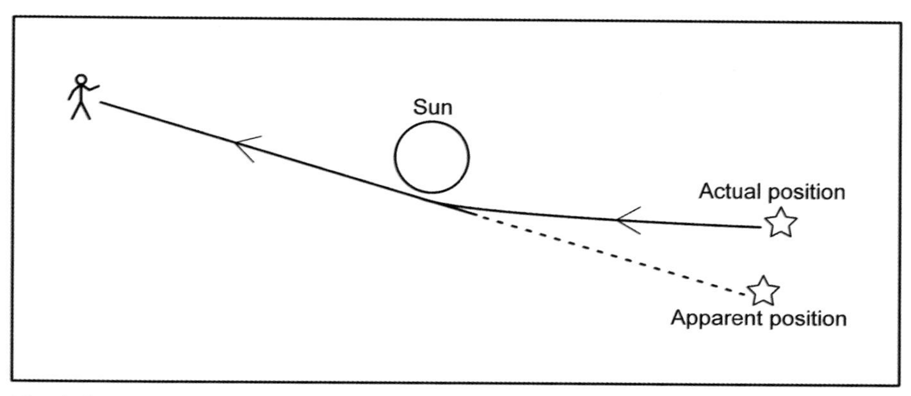

The deflection of the path of a star's light by the Sun's gravity (greatly exaggerated). (Neil Haggath / Garfield Blackmore)

the night sky a few months earlier. Unfortunately, not just any total eclipse would do; due to the slow speed of photographic emulsions in those days, a couple of minutes of totality would be barely enough time to expose even a couple of plates! They needed an eclipse with as long a duration of totality as possible.

By incredible luck, just such an eclipse was due to happen in the very near future! On 29 May 1919, one would occur, which would cross South America, the Atlantic and Africa; its maximum duration of totality, in the Atlantic off the west coast of Africa, would be no less than 6 minutes 51 seconds – only 40 seconds less than the maximum theoretically possible! Furthermore, it would occur among the stars of the Hyades cluster, providing an excellent star field for the purpose.

This eclipse, in fact, belonged to Saros Series 136, the series of exceptionally long eclipses which has become well known to present day astronomers, and which includes the 1991 eclipse in Hawai'i and Mexico and that of 2009 in China. The central eclipse of the series, two saroses after Eddington's in 1955, had, at 7 minutes 8 seconds, the longest totality in recorded history! I myself was defeated by the weather in both 1991 and 2009, and so achieved the unenviable record of not seeing two successive eclipses of this remarkable series!

Eddington and Dyson persuaded the British Government of the importance of this experiment, and obtained funding; somewhat optimistically, as the War was still raging, they planned to mount two expeditions to observe the eclipse on both sides of the Atlantic – one to Sobral in Brazil, and one to the island of

Príncipe off the west coast of Africa. The duration of totality in Príncipe would be about six and a half minutes.

It was in fact the importance of this work which led to Eddington being exempted from compulsory military service. When conscription was introduced in 1916, he applied for "conscientious objector" status on religious grounds, but was denied – somewhat bizarrely, as opposition to war is a mainstay of Quaker beliefs – despite stating a willingness to serve in the Red Cross, or as a harvest labourer. Dyson, however, testified as to the vital scientific importance of the planned eclipse expeditions, and Eddington was granted an exemption on those grounds.

The end of the War in 1918 meant that the expeditions were able to go ahead safely. Eddington and Dyson themselves went to Príncipe, while Andrew Crommelin led the expedition to Sobral, which was intended as a backup in case of bad weather in Príncipe.

Eddington and Dyson almost *were* defeated by the weather; most of the day was cloudy, but the sky partially cleared in time for totality, and they observed the eclipse through drifting clouds. They exposed sixteen plates, but the clouds ruined all but two. Those two images, however, proved to be of vital importance.

Back in Britain, Eddington compared the plates with images of the same star field taken in the night sky the previous January. Sure enough, the stars close to the Sun's limb were apparently shifted; their measured deviation was within two standard deviations of that predicted by General Relativity. Crommelin's plates from Brazil, however, did *not* show the expected deviation – but the images were of inferior quality.

Eddington triumphantly presented his results to a meeting of the Royal Society on 6 November 1919. While there was some controversy – Lodge, for example, walked out in disgust and refused to acknowledge the results – he was generally acclaimed. The following day, the story made headlines around the world, and made Einstein an instant celebrity.

Eddington became a household name in his own country, as the leading populariser of science of his time. He became an accomplished radio broadcaster, renowned for his "fireside chat" manner of speaking, and his ability to explain complex concepts in layman's as well as scientific terms. He lectured extensively on Relativity, and was mainly responsible for popularising it in the English-speaking world. In 1923, he collected a series of his lectures into *Mathematical Theory of Relativity*, which Einstein himself described as "… the finest presentation of the subject in any language." He was knighted in 1930.

He was not without his critics, however. Many have claimed that his observations in Príncipe were not sufficiently accurate to verify the bending of light – though many more modern experiments have verified the phenomenon beyond all reasonable doubt. Some even accused him – and a few still do to this day – of falsifying his data to show his desired conclusion, and rejecting Crommelin's results because they did not agree with his own. However, it was soon established that Crommelin's images were unreliable due to a defect in his telescope – an explanation widely accepted by contemporary astronomers – while a re-analysis of Eddington's data in 1979, using modern techniques, confirmed his conclusions.

One of Eddington's images of the 1919 eclipse, with the stars used for the measurements highlighted by horizontal tick marks. (Wikimedia Commons/F. W. Dyson, A. S. Eddington, and C. Davidson)

It is often stated that Eddington "proved Relativity", or "proved Einstein right". This is not strictly true. By the strictest definition, a scientific theory can never be absolutely "proven", in the way that a mathematical theorem can, as that would require testing it to infinity! All we can say is that a theory is strongly supported by evidence, and so is *almost certainly* correct; the more evidence is found to support it, and the more stringent the tests it passes, the more confident we can be that it is correct.

Eddington did, however, obtain the first direct experimental evidence in favour of General Relativity, which led to it becoming widely accepted by the scientific community. In the century since, the theory has passed every experimental test applied to it, with observed results matching predictions to an accuracy of better than one part in a billion; it is now supported by such overwhelming amounts of evidence that it's as close to being established fact as any theory *can* be.

The First Micro-Quasar

David M. Harland

This article celebrates the 40th anniversary of a momentous discovery in 1978–1980 of an entirely new class of astronomical object, now known as a micro-quasar.

In 1972 David H. Clark, a New Zealand radio engineer starting a fellowship at the University of Sydney, Australia, teamed up with James Caswell, a Brit working at a nearby research establishment, to undertake a survey of the radio supernova remnants in the southern Milky Way. They used the 64-metre radio dish of the Parkes Observatory in New South Wales and the Mills Cross telescope at Molonglo, near Canberra.

By 1974 they had discovered 30 new remnants, rejected 12 objects which had been proposed as possible supernova, and gained improved data on many of those that were confirmed.

One of the objects investigated was W50 in the constellation Aquila. This designation meant it was the 50th item in a catalogue of radio sources issued by Dutch astronomer Gart Westerhout in 1958. It was not apparent at visible wavelengths.

The Clark and Caswell study established that point-like radio sources are often either within the bounds of, or very close to radio supernova remnants. Statistical evidence suggested these were unlikely to be simple line of sight coincidences.

Michael Large and John Sutton at the Mills Observatory checked out the Clark-Caswell list, but did not detect any radio pulses to indicate that any of the point sources were pulsars.

Moving to England to the Mullard Space Science Laboratory, Clark, writing up the Parkes-Molonglo results, made contact with Paul G. Murdin, an optical astronomer at the Royal Greenwich Observatory at Herstmonceux in Sussex, who was about to start a 3-year tour at the Anglo-Australian Observatory that was based in Sydney and operated the 3.9-metre Anglo-Australian Telescope

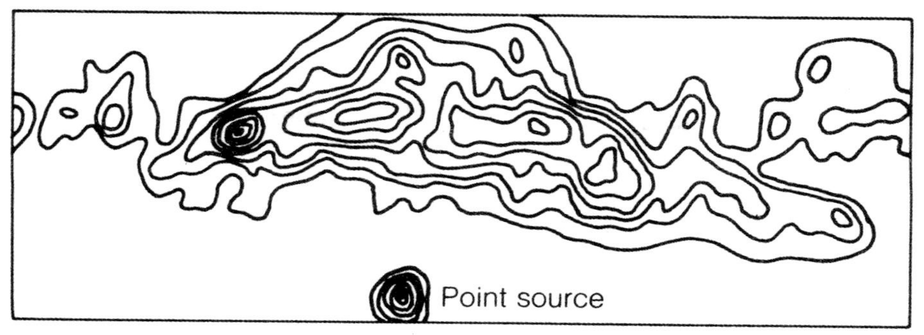

Figure 1. Clark and Caswell's radio map of W50.

on Siding Springs Mountain near the town of Coonabarabran, several hundred miles inland. Amongst several telescopes at Siding Springs was the 1.2-metre UK Schmidt.

The 'radio map' of W50 produced from the Molonglo observations had a resolution of about 10 arcsec. The nebula was a banana-shaped feature, rather than a near-symmetrical form, and the point source was close by, on the inner side of the arc.

On graduating from the Tata Institute of Fundamental Research in Bombay in India, Thanasamy Velusamy went to the University of Maryland in 1969 for his doctorate. This involved making radio observations of supernova remnants in the northern skies using the 300-foot and 140-foot antennas of the National Radio Astronomy Observatory at Green Bank, West Virginia. He published a map of W50 in 1974.

Because these observations were made at a higher radio frequency than those of Clark and Caswell, Velusamy was better able to lift it out of the diffuse background of radio emission from the Milky Way and thus capture the full extent of the remnant. The arc observed by Clark and Caswell was on the northern rim.

The point source in the Molonglo map was listed on Velusamy's map with the identifier 4C 04.66. This implied that radio astronomers at Cambridge had spotted it in the mid-1960s. Clark and Caswell suspected the object found by that survey was really the bright 'knot' in the arc, rather than the point source close by, but they couldn't prove it.

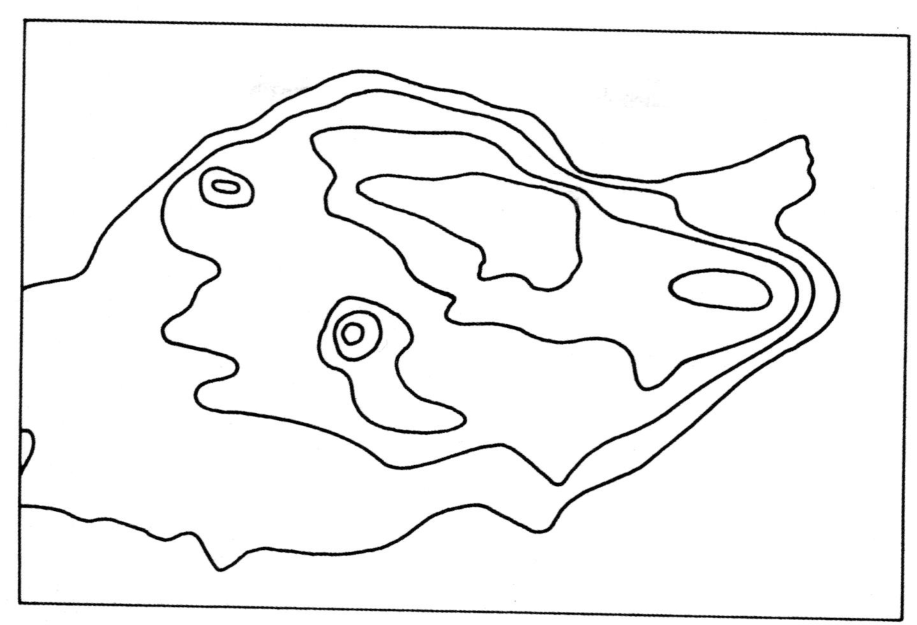

Figure 2. Velusamy's radio map of W50.

In 1975 Caswell briefly returned to England, to Cambridge, and drew the point source in W50 to the attention of Sir Martin Ryle, who was in charge of radio astronomy at the university, suggesting further observations be made by the 5-km interferometer there.

In compiling the first all-sky survey at X-ray frequencies in the early 1970s the Uhuru satellite didn't detect any emissions coincident with the radio point source at the centre of W50.

In 1976 Frederick Seward, an X-ray astronomer from California, visited the University of Leicester in England to study X-ray data provided by the Ariel 5 satellite launched in October 1974. He discovered a source that was too weak to have been seen by Uhuru and in any case was variable. It was catalogued as A1909+04. Although the position was not precisely defined owing to the way the instrument operated, Seward suggested the object might be associated with the W50 supernova remnant.

At a Royal Astronomical Society meeting in 1976 to discuss results from Arial 5, John Shakeshaft, one of the Cambridge radio astronomers, alerted by

Caswell to the presence of the radio point source in W50, reported it to flare erratically. This was noteworthy because most point radio sources maintain a constant brightness. This observation established a link between a supernova remnant, a flaring point-like radio source, and a highly variable X-ray source.

In June 1978, Clark joined Murdin at the AAT to see if they could find an optical counterpart for the point source in W50. Clark used a position that he and David Crawford at Sydney had determined when compiling a list of point-like radio sources close to the galactic plane. It became the 439th item in this Clark-Crawford list. It was a selection of these sources that Clark and Caswell had examined in the Parkes-Molonglo survey.

When Clark and Murdin turned the AAT onto the target on 28 June, they ignored the bright star that was lying in the middle of the field of view, and examined the spectrum of each of the faint stars without finding anything to attract their interest. Finally, they took a look at the bright star. Upon seeing bright emission in its spectrum, Murdin exclaimed, "Bloody hell! We've got the bastard!"

The data comprised a pair of spectra obtained by the AAT. One was of the 'red' end of the spectrum taken on 28/29 June and the other was of the 'blue' end taken the following night. The 'red' spectrum exhibited strong emission lines from hydrogen (particularly H-alpha) and number of fainter lines (some of which were clearly from helium). In addition, the 'red' spectrum exhibited some emission at wavelengths that were not so readily interpreted.

Stars possessing emission features in their spectra are interesting because they indicate that something is 'going on'.

That same night Murdin checked with a catalogue of over 5,000 such stars compiled by Lloyd R. Wackerling of the Northwestern University campus in Evanston, Illinois, and published by the Royal Astronomical Society in 1970. This was a valuable listing, because it cross-matched with earlier surveys. It had no entry for the star in W50.

The position of the star on the sky could be measured with great accuracy, but uncertainties in the radio and X-ray observations made it difficult to state with certainty that the three sources were the same object. The most accurate positional measurements of the radio source would have been obtained by the 5-km interferometer at Cambridge, so Clark asked Shakeshaft to examine his recent data of the point source and provide his best estimate of its position. In

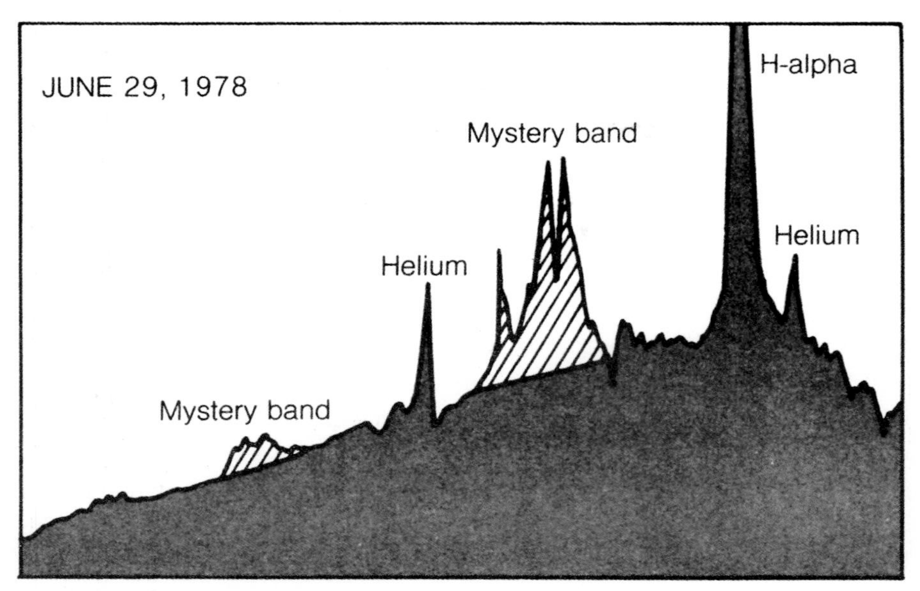

Figure 3. Clark and Murdin's first 'red' spectrum.

principle, the interferometer ought to be able to determine a position accurate to 0.1 arcsec, which was comparable to (if not slightly better than) the optical image.

Clark had relocated from MSSL to RGO in September 1977. On returning from Australia in July 1978 he received a visit from Bruce Margon, an optical astronomer from the University of California, Los Angeles. In chatting, Clark told Margon about the emission line star in W50. Margon's interest was stars associated with X-ray emission. Because this star was coincident with both a radio point source and an X-ray source, Margon promised to examine the star in California.

As a favour Murdin asked a colleague, David Allen, to take further spectra of the star during Allen's allocation of time on the AAT. These were taken on the nights of 13, 14, 15 and 17 July 1978. The bright emission lines were still present, but the wider features were absent and there was now a 'wing' on the H-alpha line. It was now evident that the emission features were variable over time.

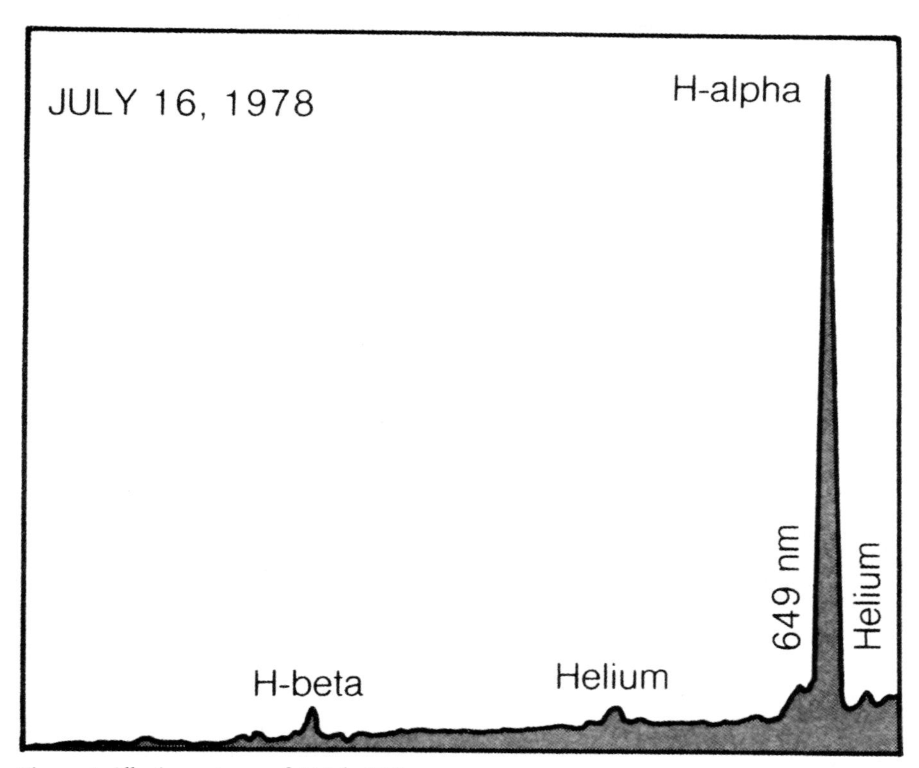

Figure 4. Allen's spectrum of 16 July 1978.

In August, Murdin also asked Ian Glass at the South African Astrophysical Observatory to obtain the first infrared observations of the star. These showed an 'infrared excess'.

To further broaden the coverage, Murdin asked a friend at RGO involved in the International Ultraviolet Explorer satellite to check the star; it had not been detected in that part of the spectrum.

In September 1978, Clark visited Seward in Boston, Massachusetts, where the latter was planning X-ray targets for the Einstein satellite that was soon to be launched. As a co-discoverer of the A1909+04 X-ray source, Seward was delighted to hear that it appeared to be coincident with an emission line star, and assured Clark it would be inspected by the new satellite.

At a conference in Boulder, Colorado, a few days later, Clark approached William Liller of the Harvard Observatory and asked him to investigate that establishment's archive of images, to find out if the star displayed long-term variability.

In the final week of September, an IAU circular reported the detection of highly variable radio from the 433rd item in a catalogue of H-alpha emission objects published in the *Astrophysical Journal* in 1977 by Bruce Stephenson and Nicholas Sanduleak of the Warner and Swasey Observatory of the Case Western Reserve University in Cleveland, Ohio, where a program had been initiated in the 1960s to survey the central portion of the Milky Way to find objects which displayed emission lines, particularly H-alpha. By placing an objective prism on their telescope, they could efficiently survey a lot of sky. Although the resulting spectra were of low resolution, this was sufficient to detect relevant star-like objects, many of which proved to be quasars in the extragalactic realm.

The radio observations of this star (reported by the IAU) were by Ernest R. Seaquist of the David Dunlap Observatory near Toronto, Philip C. Gregory of the University of British Columbia, and P. C. Crane of the NRAO between 15 August 1977 and 30 June 1978. They had selected it for study because it was listed as having strong emission lines.

The surprise for Clark and Murdin was that this was the *same* star that they had found to have strong emission lines. When Murdin had checked whether it was a known emission line star, he was unaware of the Stephenson-Sanduleak catalogue. Why wasn't it in the Wackerling list? The reason probably lies in a comment by Stephenson and Sanduleak that many of the 455 items they listed were variable, and hence likely to have been missed by previous surveys. By sheer coincidence, the last night of Seaquist's radio observations was also the one on which Clark and Murdin made their first observations of the star using the AAT, and as yet they possessed the only high resolution spectra of it.

A follow-up circular from the IAU coined the designation 'SS433' and the name stuck.

In choosing to study SS433 for its strong emission lines, Seaquist and his colleagues were unaware that it had been examined by Clark and Caswell at radio wavelengths for its likely association with the W50 supernova remnant. Nor did they know that its radio variability had been noted by the Cambridge astronomers and discussed by Shakeshaft at the RAS meeting. And they did not know of the link to the A1909+04 X-ray source. And conversely, the fact that the optical counterpart was already known to display emission lines was not known to Clark and Murdin.

All of this illustrates the difficulty that astronomers face in keeping up to date in a rapidly advancing field, with objects being listed in many different, specialised catalogues. A lot of data was also being passed around privately, well ahead of formal publication.

In early October 1978 Clark received a letter from Liller in Harvard, who, when delving into the archive of old images, realised he had checked out the same star in response to a request by Seaquist three months earlier. This had not been immediately evident to Liller because Seaquist had listed the 433rd star in the Stephenson-Sanduleak catalogue, whereas Clark had specified the optical counterpart of the A1909+04 X-ray source. It was another illustration of the pitfalls of having so many catalogues.

On 29/30 September 1978 Margon obtained his first spectrum of the star. He used the 3-metre telescope of the Lick Observatory on Mount Hamilton, near San Jose, California. In addition to the emission lines of hydrogen and helium, there were the strange emission features. He set himself the task of accounting for the latter.

Margon asked a colleague, Remington Stone, to take more spectra, which Stone did on nights between 23 and 26 October with a 0.6-metre telescope at Lick. These spectra revealed that the strange emissions occurred at different wavelengths at different nights. It was also realised that there were emission features on either side of the H-alpha line, and that these 'drifted' in opposite directions away from the line. Nothing like this had ever been observed in a stellar spectrum!

Margon made the bold assumption that the strange emission features, which were outshone only by the H-alpha line, were a hydrogen emission. Somehow, this was not only displaced in the spectrum, it was also drifting away from the H-alpha line.

After Seaquist's announcement of radio variability associated with SS433, Augusto Mannano and colleagues used the 1.8-metre telescope of the Asiago Astrophysical Observatory operated by the University of Padua in Italy to take 23 spectra between 10 October and 10 December 1978. In mid-November they reported via the IAU the presence of unusual emission features in the spectrum that drifted in wavelength from night to night. They were unaware that spectra had been taken months previously by Clark and Murdin at the AAT and more recently at Lick by Margon and Stone. They therefore unknowingly beat their rivals into print.

When Clark and Murdin came to write their first paper about the star, they decided to publish one of the spectra obtained by Allen because it proved the presence of hydrogen and helium emission lines without drawing attention to the mysterious features. Although the paper was not published in *Nature* until 2 November 1978, the authors circulated it in preprint form to alert interested parties.

Meanwhile, Caswell and the Cambridge astronomers published a paper in *Nature* on 7 December 1978 reporting observations of a number of point-like radio sources they suspected were associated with supernova remnants.[1] This paper gave a detailed account of the radio variability of the W50 point source that Shakeshaft had discussed at a conference in 1976.

So now there were formal publications reporting that there was a variable point-like radio source in W50 that seemed to be coincident with an emission line star exhibiting strange features that drifted in wavelength. The correlation with the A1909+04 X-ray source was suspected, but had not yet been proved.

The fact that Aquila would be lost in daylight from early December 1978 to the end of February 1979 imposed a pause in optical observations. This hiatus provided an opportunity to consolidate results.

Margon laid down the challenge to theorists at a conference on Relativistic Astrophysics in Munich, West Germany, in mid-December when he gave an informal account of the way in which the emissions were drifting in terms of wavelength.

The obvious cause of displacement in a spectrum was a Doppler shift. But what could cause the H-alpha to be split into three components: one primary line that was seemingly stationary, and two displaced lines, one to each side, which, in the observations available to date, were seen to drift away from the central line.

The displacements in Margon's spectra implied speeds of 40,000 km/sec. How could something associated with a *star* display such a velocity?

---

1. In fact, future observations would show only two of the eight sources in the Cambridge list to be 'local', the others were extragalactic.

In January 1979 theoreticians began to offer suggestions in the form of a blizzard of preprints.

In Cambridge, Andrew Fabian and Martin Rees adapted a model that had been developed for radio galaxies (and quasars in particular) to a star with a pair of 'jets' aimed in opposite directions. They said that the jets accelerated matter from the central source outward at relativistic speeds, and the drifting emission lines derived from material that had been carried down the jets and cooled sufficiently to radiate H-alpha. The jet that was aimed partially in our direction would provide an emission with a blue shift and the jet aimed away from us would provide a feature with a red shift. The relatively small amount of drift in the displaced lines would be caused by differences in speed of the glowing 'blobs'. This appeared to explain the spectra presented by Margon at the conference in December.

Armed with the additional spectra from Asiago, Mordehei Milgrom of the Weizman Institute of Science in Israel developed a rather more refined model and circulated preprints at much the same time as Fabian and Rees publicised their interpretation. The longer timespan of the Italian spectra demonstrated a key fact that was not evident in the spectra obtained by Clark and Murdin and by Margon and Stone. The emission features to either side of the H-alpha line were not simply drifting slightly about a fixed displacement wavelength (as if material was varying in speed as it travelled down jets at fixed orientations to our line of sight), the two emission features were moving symmetrically about a fixed mean wavelength that was slightly longward of the seemingly static H-alpha line. Furthermore, the migration of the blue-shifted line was a *mirror image* of the red-shifted one.

Milgrom interpreted this symmetrical drift in terms of the axis of the jets rotating in space on a timescale of months. When the jets were closest to our line of sight, the Doppler would be maximum and when they were side on it would be zero. A significant point in favour of Milgrom's model was that he could explain why the two emission features did not merge with the H-alpha line when there was no Doppler shift. The offset of the crossover wavelength to longward of the H-alpha line was due to the relativistic speed of material accelerated in the jet. In Special Relativity, a clock that moves at a very high speed with respect to an observer will appear to run slowly – the greater the speed the slower the clock. This is called time dilation. Since the process by

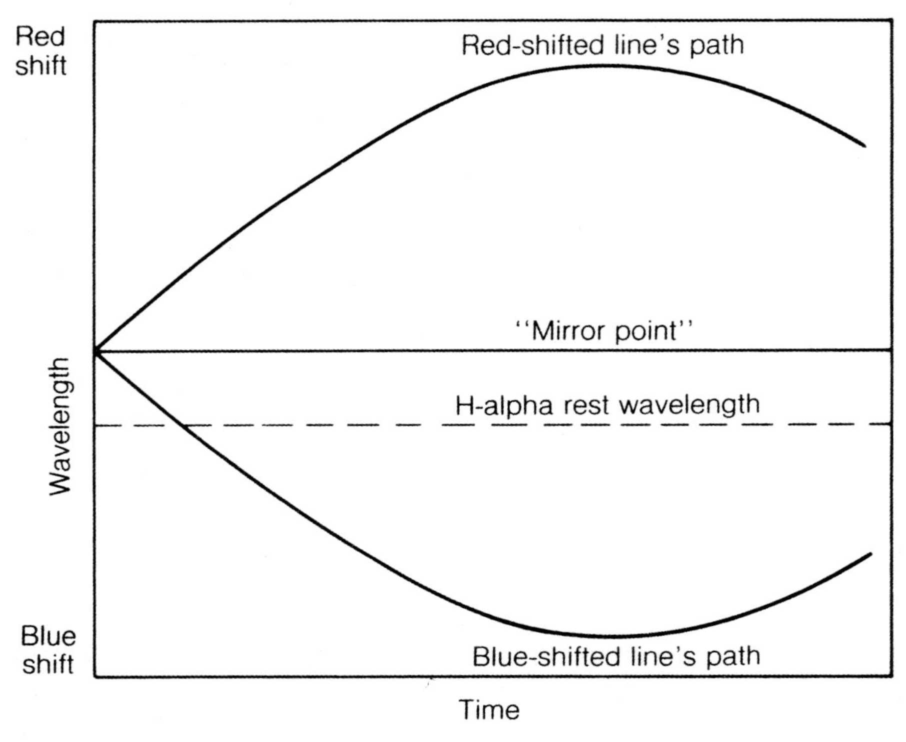

Figure 5. Milgrom's mirror image deduction.

which an atom radiates light would seem to slow, we would perceive that as shifted to longer wavelengths. Of course from the point of view of the atom, everything is normal. This relativistic effect is known as Transverse Doppler shift. It only becomes noticeable at speeds exceeding about 0.05 that of light. The shifts in Margon's spectra had certainly implied speeds that would cause such a spectral displacement.

Milgrom was able to perform this analysis because he was the first to note that the *mean* wavelength of the two drifting emission features remained at a fixed wavelength and that this was displaced longward of the H-alpha line. It was a major step forward.

When SS433 reappeared in the sky at the end of February 1979 it became a popular target.

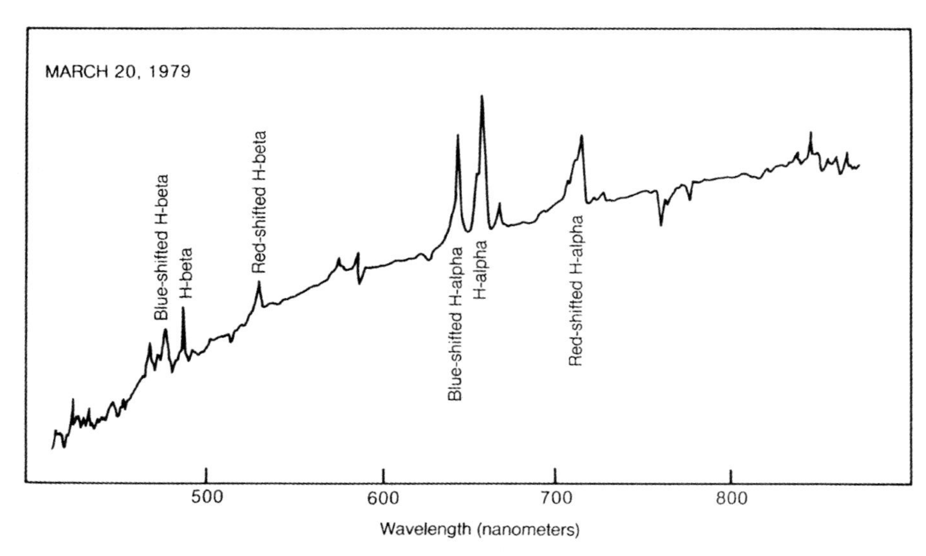

Figure 6. Margon found each line to be tripled.

Margon's initial spectra of the new observing season clearly demonstrated that *all* of the emission lines had three components, not just the H-alpha line. Furthermore all of the redshifted lines for both hydrogen and helium implied identical speeds of recession and all of the lines shifted in the other direction implied identical speeds of approach. In each case, the crossover wavelength was displaced by the same amount from the 'stationary' line.

An analysis by James Liebert and colleagues at the Steward Observatory in Tucson, Arizona, of all the data available confirmed Milgrom's prediction that the spectral shifts were cyclic.

Peter G. Martin of the University of Toronto, a member of Liebert's team who was visiting Cambridge, joined with Murdin to reassess all of the AAT spectra obtained in June–July 1978. This early observation assisted in fixing the periodicity of the cycle at about 160 days.

Next Liebert, Murdin and Martin applied Milgrom's physical model to all of the spectroscopic data obtained from all observers. The transverse Doppler at the crossover wavelength implied the material in the jets was travelling at 80,000 km/sec, which was 0.25 the speed of light. And the jets were not only rotating, they were precessing. The model indicated them to be inclined 21.5°

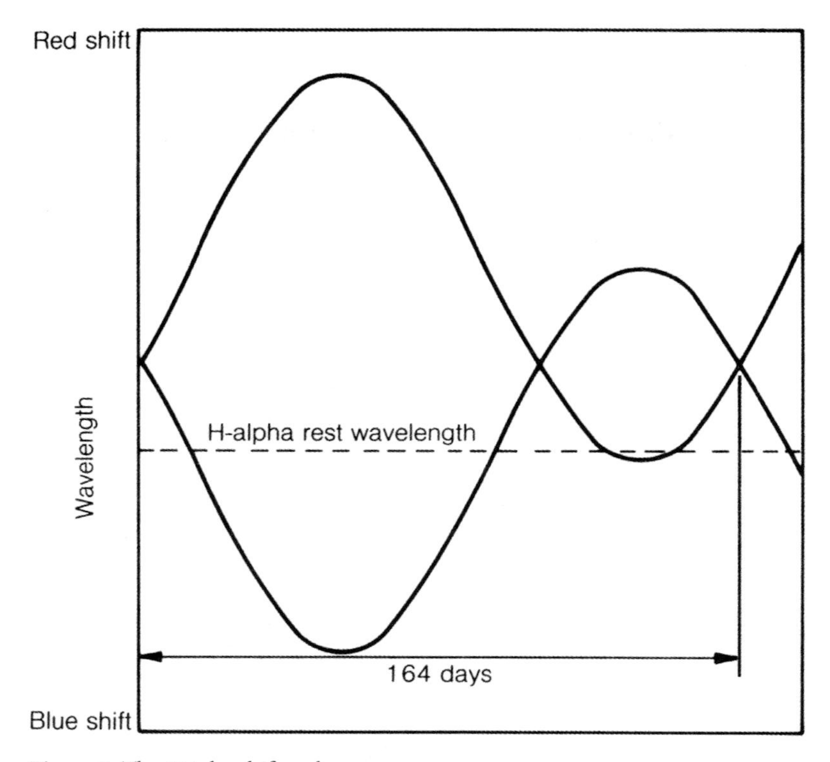

Figure 7. The 164-day drift cycle.

to an axis that was tilted 79° to the line of sight. Their study was reported via the IAU on 11 May 1979. At a meeting of the American Physical Society the previous month, Margon had reported a similar analysis made by himself and George Abell in Los Angeles. They had calculated a period of 164 days, a 78° angle to the line of sight, a 17° offset to that axis, and a speed of 0.27 of light for the material in the jets.

So spectroscopic observations that had been baffling when they were made in 1978 were now being modelled in remarkable detail.

Meanwhile, other observers were busy.

In 1978–79 Barry Geldzahler of the University of Pennsylvania was at the Max Planck Institute for Radio Astronomy in Bonn, West Germany. He was studying whether point radio sources were physically linked with supernova

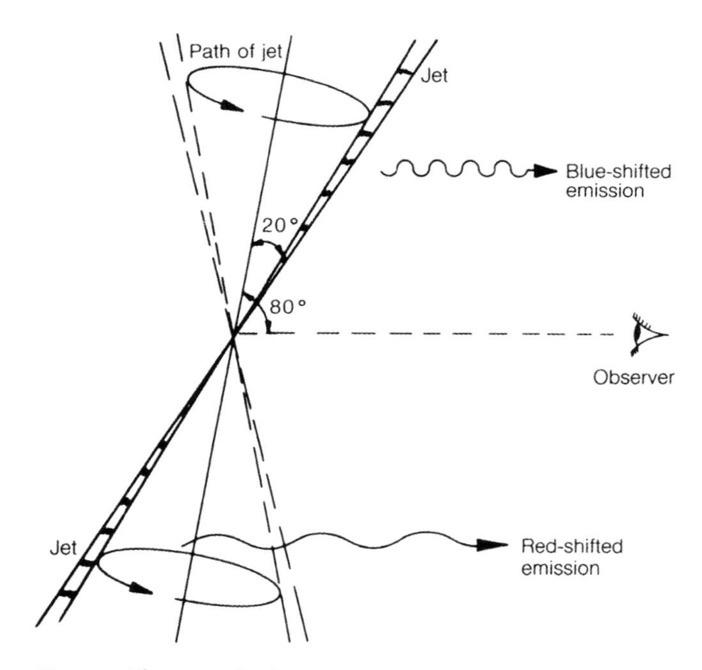

Figure 8. The precessing jets.

remnants. On 3/4 August 1978 he observed W50 with the 100-metre dish at Bonn, having heard of its point-like source a few months earlier from Clark. His high-resolution map of W50 revealed the existence of 'ears' sitting east and west of the remnant. He circulated this map as a preprint in April 1979.

On 3 June 1979 William Zealey obtained a 'red' image of W50 using the AAO's Schmidt telescope and recorded an arc of filamentary nebulosity on each of the eastern and western edges of the radio remnant.

These optical filaments were independently discovered by Sidney van den Bergh of the Dominion Astrophysical Observatory in Canada, when he took pictures of W50 using the Schmidt of the Palomar Observatory in California. He announced this via the IAU on 17 August 1979.

Meanwhile, Murdin, who knew of Zealey's images, and was visiting the South African Astrophysical Observatory, obtained a spectrum of the eastern filament on 23 August 1979. Most 'old' supernova remnants display spectra dominated by hydrogen emission. Although the spectrum of the filament of W50 contained

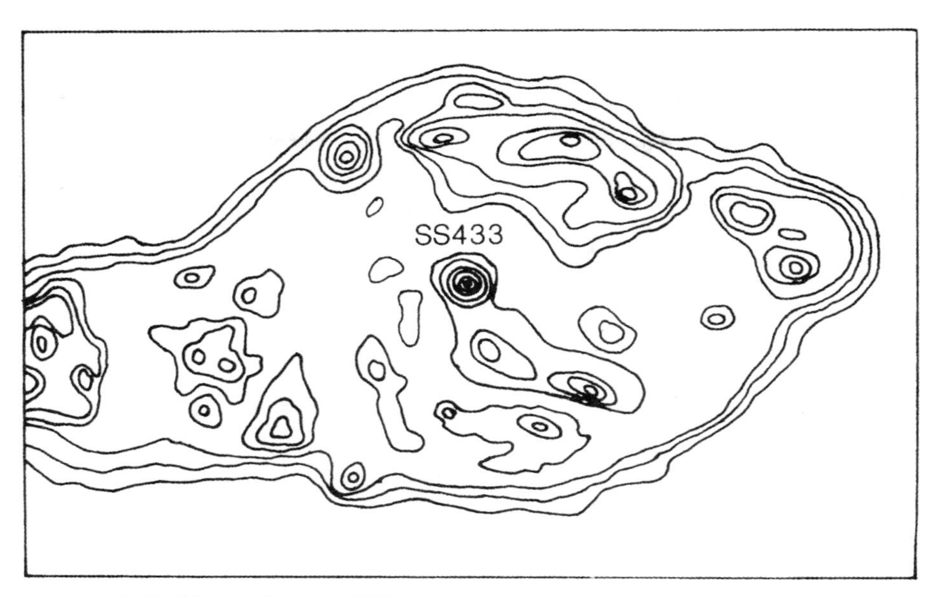

Figure 9. Geldzahler's radio map of W50.

hydrogen emission, this was outshone (by a factor of five in intensity) by nitrogen emission. This was reported in an IAU circular on 10 September.

Was the interstellar medium in that region enriched with nitrogen prior to the supernova? Or had the supernova that created the W50 nebula released a lot of nitrogen?

Van den Berg suggested that the progenitor was a Wolf-Rayet star. Such stars were first studied by the French astronomers C. J. E. Wolf and G. A. P. Rayet in 1867. These rather rare, very hot stars are thought to be so massive that they shed large amounts of material to space in the form of strong stellar winds. Although they show emission features, they characteristically do not have hydrogen emission lines. However, as they shed the hydrogen of their outer atmospheres they reveal the nucleosynthesis that produces elements of heavier atomic mass. In particular, some Wolf-Rayet stars display emissions from carbon and others show emissions from nitrogen. Hence Van den Berg suggested that a 'WN' star had exploded and produced a nebula enriched in nitrogen.

Further radio studies confirmed the existence of the hypothesised jets on either side of the point source. What is more, these radio jets were aligned in such a manner as to suggest they were responsible for the 'ears' in the radio map of W50 and the matching optical filaments. These high-resolution radio observations were made by Ralph Spencer of the Jodrell Bank Observatory, using the initial MERLIN interferometer in the UK centred on the 250-foot Lovell telescope. The discovery of the radio jets was revealed informally in May 1979.

Their existence was confirmed by Geldzahler, using a network of radio telescopes that spanned the longitude range from Spain to California – the width of Earth – in order to achieve a resolution of a few thousandths of an arcsec; and also by Richard Schilizzi at Westerbork using radio telescopes distributed across Europe.

In April 1979 Seaquist had obtained X-ray observations by the Einstein satellite that showed a point source precisely coincident with SS433. This confirmed the correlation based on the less accurately defined position of A1909+04. Clark became aware of this informally. And in October of that year, Seaquist obtained Einstein data at a higher level of sensitivity which revealed the existence of X-ray jets co-aligned with the radio jets. The radio emission fades with increasing distance from the point source but the X-ray intensity increases. The X-ray jets extend farther out and broaden to create 'lobes' coincident with the 'ears' in the radio map and the optical filaments. The Einstein results were published in *Vistas in Astronomy* in 1981.

Seaquist observed SS433 in October 1980 using the newly commissioned Very Large Array interferometer in New Mexico. This data revealed that the

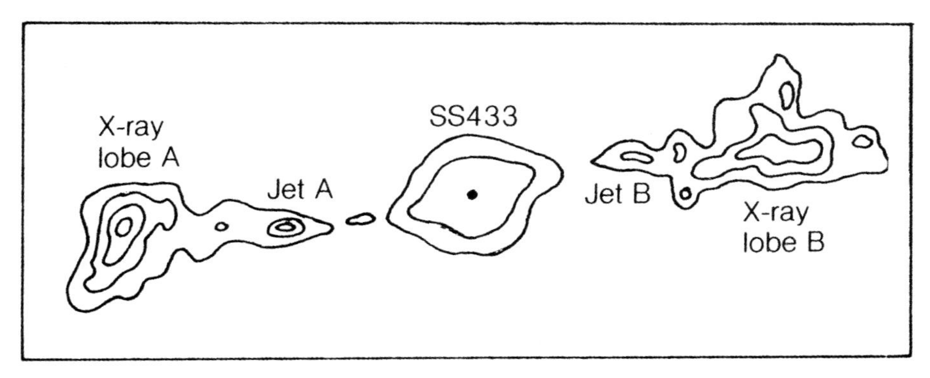

Figure 10. Seaquist discovered that the X-ray jets extend far beyond the radio jets and broaden to create lobes.

precession of the radio jets was causing them to hose out material in a helical pattern, rather like a lawn sprinkler.

Even as the drifting emission features in the spectrum of SS433 attracted attention for their novelty, David Compton and colleagues at the Dominion Astrophysical Observatory had obtained 48 high-resolution spectra between April and July 1979 to study the 'stationary' lines. They discovered that the lines were cycling in wavelength with a period of 13.1 days. The conclusion that SS433 is a binary star system was announced via the IAU on 6 August 1979.[2]

The 13.1-day spectroscopic periodicity was confirmed by an optical light curve built from observations in 1979 and 1980 by Anatol Cherepashchuk, a Russian astronomer at the AAO. This displayed both primary and secondary minima as components were eclipsed on the line of sight.

Interestingly, Margon also realised that the intensities of the 'stationary' emission lines was varying on a 13.1-day cycle.

The news that SS433 was a binary system enabled the theorists to update their models to include the transfer of material from one star to the other and the likely role of an accretion disk.

Searches for radio pulses on timescales ranging from several milliseconds to hundreds of seconds were negative, so if one of the components was a neutron star it was not showing any evidence of its rotation period.

Also, there was no indication of 'flickering' to suggest that the majority of the light in the continuum spectrum was being produced by an accretion disk.

It was well established that any star that evolves into a white dwarf cannot exceed 1.4 solar masses. A star with a mass of up to about 3 solar masses can leave behind a neutron star by undergoing a supernova explosion. A star that exceeds 3 solar masses will either completely destroy itself or create a black hole.

Since the progenitor of the supernova that produced the W50 remnant left behind a star in a binary system, the issue was whether this was a neutron star or a black hole. Neither candidate would directly contribute much light to the continuum. That left the primary star and an accretion

2. Although this detection was a classic case of a spectroscopic binary, it must be appreciated that the magnitude of the wavelength shift was less than one thousandth that displayed by the drifting emission features, which is why the fact that SS433 was a binary had not previously been realised.

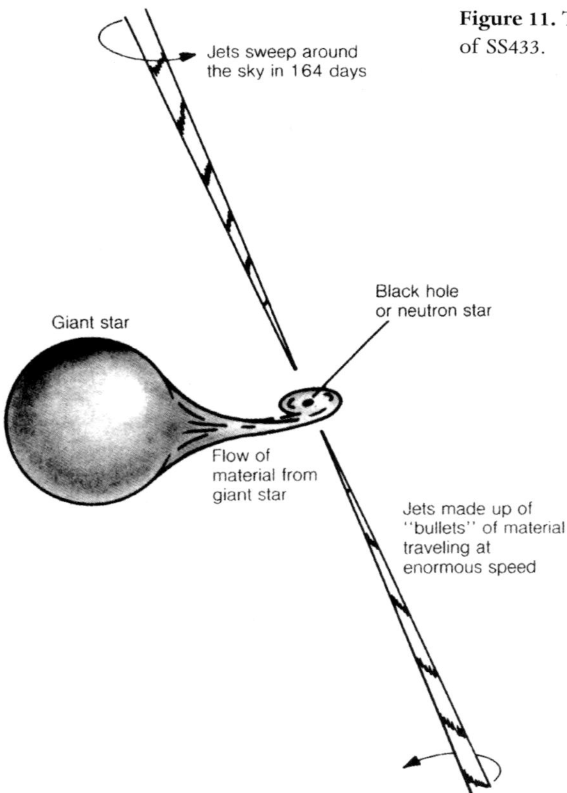

Figure 11. The inferred general characteristics of SS433.

disk. The absence of flickering argued against an accretion disk being the main continuum source.

In early 1980, Clark and Murdin modelled the visible and infrared regions of the spectrum on the basis that the main continuum source was the primary star, and reasoned that it must be a very hot star. This indicated SS433 was a high-mass binary system.

Although the evidence was not definitive, it implied that the binary was a 'hot' star of up to 20 solar masses with an intense stellar wind, around which was orbiting a neutron star (possibly a black hole) with a period of 13.1 days. Material from the primary was accumulating in an accretion disk around the secondary. The compact component was generating relativistic jets that were precessing with a period of 164 days. This explained the emission features in the spectrum that drifted in a cyclic manner. The fact that the jets created the 'ears' in the W50 nebula confirmed the physical association between the star system and the supernova remnant.

In this scenario, the 'stationary' H-alpha emission was produced by a 'hot spot' where matter from the primary met the accretion disk. Although it was intrinsically bright, the disk was outshone by the primary star. Nevertheless, the disk proved to be a source of 'hard' X-rays, and the fact that these were absent for roughly 2 days during each orbit meant the primary was eclipsing the disk. The velocity dispersion of the strong stellar wind from the primary contributed the 'wings' that broadened the 'stationary' emission lines.

The progenitor of the supernova may well have been a Wolf-Rayet star whose great mass meant it was short-lived. Although the current primary is itself rather massive, it would have been the 'junior partner' in the original system.

In later years a number of other systems were found which display similar characteristics to SS433, including Cygnus X-1 (the first proven instance of a black hole) and Cygnus X-3 (an object whose eruptive behaviour made it the centre of attraction in the early 1970s and is still somewhat mysterious).

Although the details differ from one case to another, all such systems were classified as 'micro-quasars' because the relativistic jets issued by the compact component bear a similarity to those formed by the super-massive black holes that lurk at the centre of active galaxies.[3]

This account of the events of 1978–1980 is based on *The Quest for SS433: The Discovery of the Astronomical Phenomenon of the Century*, by David H. Clark, which was published in 1985 by Viking Penguin in the USA and in 1986 by Adam Hilger in the UK. Obviously, with 200 pages available, David explains the process of discovery in much greater detail than was possible in this article, so I strongly urge readers to track down a copy of the book.

I am grateful to David for reviewing this article and for allowing me to use some of his diagrams.

3. I discussed super-massive black holes in the Yearbook of Astronomy 2018.

Father Lucian Kemble and the Kemble Asterisms

Steve Brown

In addition to the 88 officially recognised constellations, there are many unofficial patterns in the night sky, known as 'asterisms'. One well-known asterism is the Summer Triangle, formed by the bright stars Deneb (in Cygnus), Vega (in Lyra) and Altair (in Aquila). Other examples include the Plough in Ursa Major, the Keystone in Hercules and the Coathanger in Vulpecula. To these can be added the three particularly attractive asterisms identified by Franciscan Friar and amateur astronomer, Lucian Kemble.

Lucian Kemble is seen here on the right with John Dobson (telescope maker and creator of the Dobsonian telescope) in October 1990 at Pigeon Lake, Alberta, Canada. (Reproduced courtesy of the Royal Astronomical Society of Canada / RASC Archives)

Born Joseph Bertille Kemble on 5 November 1922 on a farm near Pincher Creek in Alberta, Canada to Charles and Ida Mae Kemble, he gained an appreciation for the night sky from an early age with the encouragement of his parents. This interest continued through his service as a radio operator for the Canadian Air Force in the Second World War and developed into a passion after the War when he entered the Franciscan Friars. Kemble was ordained as a priest in 1953 and took the name Lucian. He spent most of his vocation preaching from retreats at Mount St. Francis on the outskirts of Cochrane, Alberta and at St. Michael's, near Lumsden, Saskatchewan.

Kemble was interested in astronomy throughout his life and inspired all those he met with his passion for the night sky. Like many amateur astronomers he began with naked eye observing and using a pair of binoculars to learn his way around the night sky. Graduating to telescopic observing, he could often be seen with his Celestron 5 in the car park of the retreat centre outside Cochrane with fellow astronomers and enthusiasts. One early notable observation was a nova in the constellation of Cygnus in August 1975 (nova V1500 Cygni). He also observed comets West in 1976, Hyakutake in 1996 and Hale-Bopp in 1997.

Lucian Kemble became an accomplished visual astronomer and kept detailed records and sketches of his observations, including many of faint galaxies listed

Likened to a stellar waterfall cascading into a celestial pool, the meandering line of colourful stars forming Kemble's Cascade extend to the misty patch of light emanating from the open star cluster NGC 1502. (Steve Brown)

in the New General Catalogue. Having joined the Royal Astronomical Society of Canada (RASC) in 1971, he was awarded the RASC Messier Certificate in 1980; the Astronomical League of America Herschel 400 Certificate in 1981 (Kemble helped to create the Astronomical League's Herschel 400 list); the RASC Chilton Prize in 1989; and the Webb Society award of excellence in 1997. He also had numerous observations published in Sky & Telescope and Astronomy magazines.

The first of Kemble's asterisms was made known to the world in 1980 after one of his sketches was featured in Sky & Telescope magazine. In Walter Scott Houston's Deep-Sky Wonders column, Kemble described "...a beautiful cascade of faint stars tumbling from the northeast down to the open cluster NGC 1502." This distinctive pattern of stars, discovered by Kemble while sweeping the sky with a pair of 7×35 binoculars, was given the descriptive name Kemble's Cascade by Houston. The name has stuck, and this is without doubt the best known of the three asterisms associated with Lucian Kemble. Located in the constellation Camelopardalis, Kemble's Cascade is a linear grouping of more than 20 colourful stars ranging from fifth to tenth magnitude and extending for over two and a half degrees. This beautiful chain of stars seems to flow into the seventh magnitude open cluster NGC 1502 which is located at one end of the asterism, the overall effect being that of a celestial waterfall cascading into a pool.

The second of his asterisms is simply known as Kemble 2, although it is often referred to as the "Mini-Cassiopeia". Located in the constellation of Draco, Kemble 2 is a group of half a dozen seventh and eighth magnitude stars forming a distinctive 'W' shape, which is indeed similar to the 'W' of Cassiopeia, albeit considerably smaller at only a third of a degree across. Kemble had written about this asterism in an article that was never published.

The third, and perhaps least well-known, of Kemble's asterisms is Kemble's Kite, which can be found in the constellation of Cassiopeia, near the fifth magnitude star Gamma (γ) Camelopardalis in the neighbouring constellation Camelopardalis. Resembling a diamond-shaped kite with a tail, the head of Kemble's Kite is made up of seven stars ranging in magnitude from six to eight with the tail of the kite formed from a curving line of magnitude seven to eleven stars flowing away from the head. The whole asterism is about one and a half degrees or so in length.

Best observed through large binoculars or small telescopes with low magnification is the crooked line of stars forming the "W" shape of Kemble 2, which is clearly reminiscent of a mini-Cassiopeia. (Steve Brown)

Another aspect of Kemble's love for astronomy was that he enjoyed inspiring and encouraging others to take up the hobby. He often led observing sessions that lasted well into the early hours of the morning, frequently with Mozart, Bach or Rachmaninoff playing in the background and hot apple cider on hand to keep everyone warm. He was one of that rare breed of person who see the natural wonder in everything around them and seem to infect others with their enthusiasm for life. One of his sayings was "We may be made from dust, but it's star dust!" When asked if observing the cosmos made him feel insignificant, he would often reply "Au contraire, I am as big as that which I contemplate." This is a surely the feeling that many astronomers have experienced when gazing up at the night sky.

Father Lucian Kemble died on 21 February 1999 in Regina, Saskatchewan, aged 76. He had an enduring love of astronomy, and at the time of his death

Located in the northern reaches of Cassiopeia near its border with neighbouring Camelopardalis are the five colourful stars forming the distinctive diamond-shaped head of Kemble's Kite, below which we see a curving line of fainter stars forming the tail. (Steve Brown)

he was planning to photograph the forthcoming conjunction of Jupiter and Venus. His legacy includes not only three beautiful asterisms for us to enjoy but also a love for astronomy in the many people he inspired in Canada and around the world. His astronomical sketches, observation notes, books and correspondence were donated to the Royal Astronomical Society of Canada's Regina Centre after his death, leaving a lasting record to inspire generations to come. To honour his contribution to astronomy, the asteroid 78431 Kemble, discovered by Andrew Lowe on 16 August 2002 and located in the inner regions of the asteroid belt, was named in his honour.

My thanks are due to Jack Estes (The Astronomical League), Vance Petriew (Royal Astronomical Society of Canada, Regina Centre), Barry Thorson, Janine Myszka (Sky & Telescope), Bob Mokry (Order of Friars Minor Franciscans of Canada) and Kevin Lynch (Order of Friars Minor Franciscans of Canada) for their help and assistance with my research for this article.

Mira 'The Wonderful'

Roger Pickard

Introduction

Located in the constellation of Cetus (the Whale), the famous star Mira, or Omicron (o) Ceti, is the prototype of a class of variable star termed Long Period Variable Stars (LPVs). This is, for all intents and purposes, a single star which we see varying due to pulsations taking place within the star itself, or more strictly within the outer layers of the star. The period of Mira is around 332 days, with a typical maximum brightness of around magnitude 3 and a typical minimum of around 9.5 or thereabouts. So although you can see the star without optical aid when it is at or near its brightest, you will need either a good pair of binoculars or a small telescope, and a dark observing site, to see it at minimum.

Early History

The Dutch clergyman and amateur astronomer David Fabricius (1564–1617) is credited with the discovery of this famous star in 1596. He had used it as a reference star for marking the position of Jupiter, but when he next looked some weeks later the star had brightened, and later still it had faded from view and was believed to have probably been a nova.

Fabricius recorded the star again in 1609, by which time other astronomers had begun to make observations themselves. In 1603 the German astronomer Johann Bayer (1572–1625) labelled the star Omicron (o) Ceti, but little further attention was paid to it until Dutch astronomer Johannes Phocylides Holwarda (1618–1651) rightly surmised that the observed variability was a recurring process and deduced that the variations in brightness took place over a period of around 11 months. In his *Historiola Mirae Stellae* (a Short History of the Wonderful Star) published in 1662, the Polish astronomer Johannes Hevelius (1611–1687) gave this object its more common name of Mira (Latin for 'Wonderful') because no other stars known at that time behaved in such a manner.

In this image, measuring some 3 degrees square and showing stars down to magnitude 13, Mira is the bright red object to the upper right. Many of the stars seen here can be identified on the 5 degree comparison chart for Mira, which can be found in the article *Some Interesting Variable Stars to Observe in 2019* elsewhere in this volume. (Digitized Sky Survey 2. Acknowledgment: Davide De Martin)

More Detailed Information

Mira is very easy to identify with the naked eye when at or near its brightest, its location being at right ascension 02h 19m 20.8s and declination −02° 58' 39" (2000.0). This places it roughly half way between the two stars Gamma (γ) Ceti, the southernmost of the circlet of stars forming the head of Cetus, and Baten Kaitos (ζ Ceti), as shown on the chart.

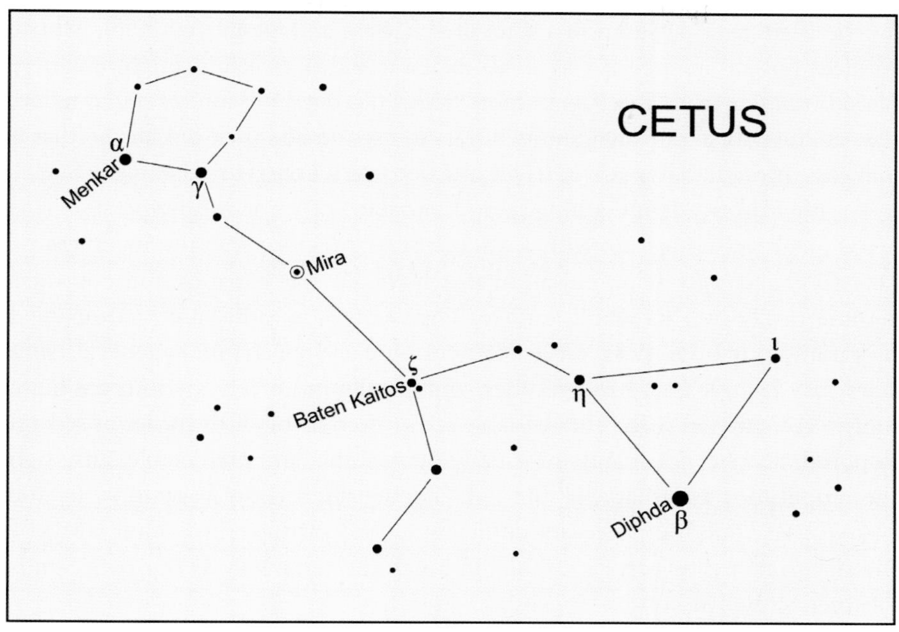

Chart of Cetus (the Whale) showing the location of Mira. (Brian Jones / Garfield Blackmore)

Mira is the prototype 'Mira type variable', a red giant star whose outer layers pulsate as noted above. As well as causing the star to expand and contract slightly, these pulsations also produce changes in its surface temperature. During the cooling phases very simple molecules form, which then dissociate when the surface warms up again. However, while they exist, the molecules absorb some of the light being emitted by the star, causing 'irregularities' in the rise and fall, and the brightness of the star not varying at a constant rate.

Mira stars were once much like our own Sun, but are now in the very late stages of stellar evolution. As cool red giant stars, they are found in the high luminosity portion of the Asymptotic Giant Branch (AGB) of the Hertzsprung–Russell diagram. These stars generally have larger radii, higher luminosities, lower temperatures and lower surface gravities than our Sun. As a result of the low surface gravity, the outer atmosphere is tenuous and loosely bound and forms an envelope around the star. Pulsations of this cool, weak outer atmosphere give rise to the brightness variations seen in Mira-type stars.

There are over 6,000 known stars of this class and all are red giants whose surfaces pulsate in such a way as to increase and decrease in brightness over periods ranging from about 80 to more than 1,000 days. It is interesting to reflect that in due course our own Sun will start to expand and then pulsate, so that it will pass through that phase of its life and become a Long Period Variable Star.

So, how do we define a long-period variable (LPV) or Mira-type star? It is a red-giant pulsating star with a well-defined period, generally in excess of 100 days, and a form of pulsation in which the star expands and contracts symmetrically over its whole surface. The changes in radius are accompanied by variations in brightness, surface temperature and spectrum. In general, Mira stars vary by at least 2.5 magnitudes, and usually more. The periods are quite long in comparison with other variable stars (often being of the order of several hundred days or more) and are usually quite stable and predictable, although the amplitudes and shape of the light curves often show variations. In due course, these pulsations become sufficient to puff off the entire outer layers of the star to form a planetary nebula, in much the same way as our own Sun is destined to evolve.

The spectra of Mira stars are usually of the M, S and C types, Mira itself being of spectral type M5e-M9e. In spite of its diameter being around 650 million km (compared with the Sun's diameter of just less than 1.4 million km), Mira is only slightly more massive than our Sun at just 1.2 solar masses.

Mira and its Companion

Interestingly, Mira is also a double star with a white dwarf companion which is also a variable star, designated Mira B, or VZ Ceti. It had been suspected that Mira may have a companion, and in the 1920s American astronomer Alfred Harrison Joy (1882–1973) of Mount Wilson Observatory noted an anomaly in the spectrum of Mira when it was at minimum. At Joy's request, Robert Grant Aitken of Lick Observatory observed the star with the 36-inch refractor for visual confirmation of a companion. The two stars complete an orbit around each other every 500 years or so and are separated by about 0.5 seconds of arc. This equates to around 100 astronomical units (AU) at the distance of Mira from us of between 350 and 400 light years. According to the General Catalogue of Variable Stars (GCVS), VZ Ceti varies with a range of between magnitudes 9.5 and 12 and has a possible periodicity of about 13 years.

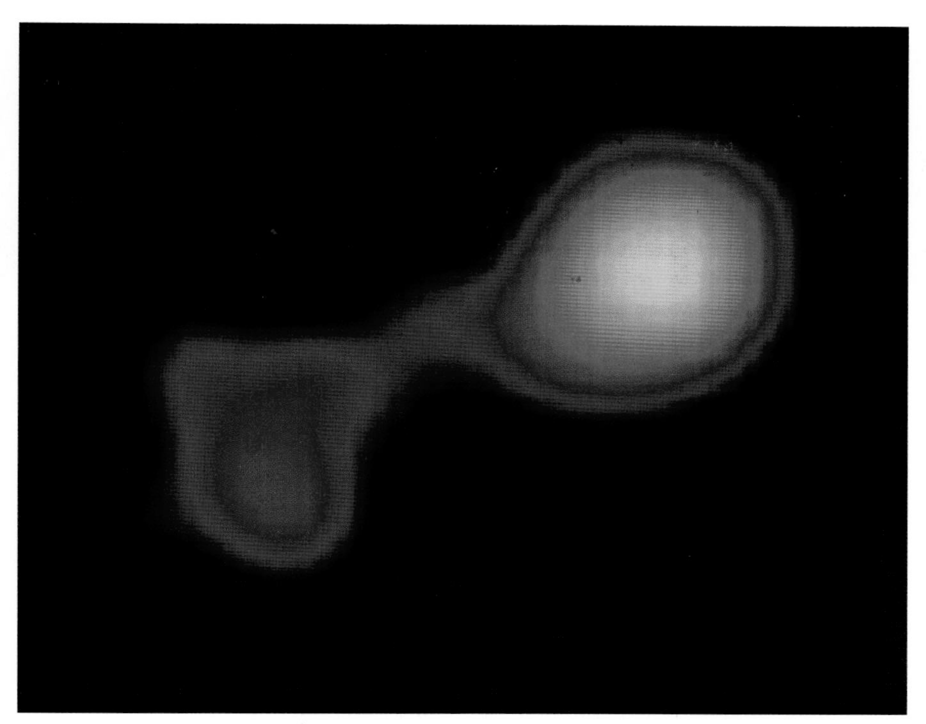

This image, taken by NASA's Chandra X-ray Observatory, measures a little over 1 arc second per side and shows the highly evolved red giant star Mira A (right) and its white dwarf companion Mira B (left). The gas that Mira A is losing from its upper atmosphere is pulled towards Mira B, forming a gaseous bridge between the two stars and the accumulation of an accretion disk around Mira B. (X-ray: NASA/CXC/SAO/M. Karovska et al)

Binary systems such as this are also known as "symbiotic star" systems, Mira and VZ Ceti being the nearest variable star of this type to the Sun. In 2007, the Galaxy Evolution Explorer (GALEX) orbiting space telescope (designed to observe galaxies at ultraviolet wavelengths) discovered that Mira was speeding through space at 130km per second, leaving in its wake a tail of material some 13 light years long and revealing a history of Mira's mass loss over a period of some 30,000 years. Infrared observations of Mira have also shown that deviations in the shell of gas surrounding it in the direction of the companion star indicate that it will continue to affect the outer shell of material as it evolves towards the planetary nebula stage.

In this GALEX image, Mira's comet-like tail stretches an amazing 13 light-years across the sky (equivalent to over three times the distance from our Sun to Proxima Centauri, the closest star to our solar system). The tail extends from Mira itself (located on the right of the image but lost in nebulosity) to another star on the far left. (NASA/JPL-Caltech)

Observing Mira

The BAA Variable Star Section (VSS) produces three charts to help with observing Mira, which one you use depending on how bright (or faint) it is at the time. These are a 60 degree chart to assist in its location and for when it is near maximum brightness; an 18 degree chart for those occasions in between maximum and minimum; and a 5 degree chart for when Mira is fainter and needing a pair of binoculars or a small telescope to observe it.

The brightness changes do not repeat exactly from one 11 month cycle to the next. Some maxima are brighter than others and the date of maximum can only be roughly predicted in advance (see below). In addition, Mira often brightens very rapidly as it approaches maximum, sometimes by several magnitudes in just a few weeks.

Several years ago the maxima were very difficult to observe as the star was situated behind the Sun when this happened. However, from 2017 on, Mira has become more conveniently placed in the evening sky to allow observations to be made.

The Next Maximum and Minimum of Mira

The next maximum of Mira is likely to be around October/November 2019 with a minimum occurring around June/July 2019. Charts for Mira are available to download from the Section's website at **www.britastro.org/vss/xchartcat/omi-cet_.html** although the 5 degree chart for this star is included in the article *Some Interesting Variable Stars to Observe in 2019* elsewhere in this volume.

In 2017 Mira attained a magnitude of 10.25 at minimum. This was one of the faintest minima ever recorded for Mira, the last time it was as faint being as far back as 1930!

Mira is at maximum for at least a week before any fading is seen. With a period of 332 days, one observation per week is more than enough to monitor the maximum and decline, the star not varying quickly enough to warrant more frequent observing. Over-observation of slowly varying stars can lead to bias whereby you are suspecting the star to be of a certain magnitude the next time you observe it, this often being precisely what you see!

More Observations Always Wanted

Although the BAAVSS already has over 17,000 observations of Mira dating back to 1885, there is no need to think observations can stop there! With professional astronomers always keen to observe even well-known stars with ever more sophisticated equipment, they also want to know how their observations compare with the latest visual observations. We often receive requests for observations of certain stars (even "standard" long period variables) as professionals probe ever deeper into the mysteries of these and other stars.

So, please consider observing Mira (and any other variable stars should the "bug" bite you) and submitting your results to the VSS database. There is plenty of information on the Section's website detailing how to make and submit observations of variable stars, and plenty of general help and guidance is available via our "mentoring" scheme.

When you are observing Mira "The Wonderful" it is worth bearing in mind that you are gazing at an enormous pulsating red giant star, slowly decreasing in size as it nears maximum and then expanding again as it fades. Add to this the mass lost to its tiny companion star, and the amazing 13-light year long tail left behind as it speeds through space, and you soon realise just why this is such an aptly named star.

Some Interesting Variable Stars to Observe in 2019

Roger Pickard

I have listed the stars below, firstly some suitable for observing with binoculars and then some which go fainter and so will require a telescope. Comparison charts for ALL of these stars can be found on the BAA VSS website at **www.britastro.org/vss**

Each star is followed by its type (for example, SRb indicates that it is a semi-regular star of type b – for a further explanation of type see the American Association of Variable Star Observers Variable Star Index at **www.aavso.org/vsx**). Next we have the typical range in magnitude of the star, then its period and finally a recommendation of how frequently the star should be observed. Below each table are additional notes that I feel may be helpful. An asterisk entered after the star name denotes that a comparison chart has been provided for that star, and a double asterisk indicates that a light curve diagram is given. Comparison charts and light curve diagrams can be found at the end of the article.

Binocular Variables

STAR	TYPE	RANGE	PERIOD	OBSERVING FREQUENCY
U Cam *	SRb	7.0 / 9.4	2,800 days	No more than about 5 days
V CVn	SRa	6.5 / 8.6	192 days	5–7 days
SS Cep	SRb	6.7 / 7.7	90 days	5 days
AF Cyg **	SRb	6.4 / 7.7	92.5 days	5 days
TX Dra	SRb	6.8 / 8.2	78 days?	5 days
Y Lyn *	SRc	6.6 / 8.3	110 days	7 days
Z Psc	SRb	6.4 / 7.5	156 days	7 days
R Sct	RVa	4.2 / 8.6	146 days	7 days

- Be careful when observing **U Cam** when it is faint as there is a field star close by.
- **V CVn** is a nice star to follow.
- There was a deep fade of **SS Cep** in 2016, but will it be repeated in future years?

- There is quite a lot of scatter in the light curve of **AF Cyg**, but nonetheless this is a nice star to follow.
- There is plenty of variation with **TX Dra**, so this is a good star to follow.
- When observing **Y Lyn**, do not get confused with a nearby star when the variable is at or near its faintest.
- **Z Psc** has quite a small amplitude (range in brightness from maximum to minimum), so particular care must be taken to ensure that you have identified it correctly and made the correct estimate.
- **R Sct** is another good star to follow, although it does have quieter periods.

Telescopic Variables

STAR	TYPE	RANGE	PERIOD	OBSERVING FREQUENCY
Z And	ZAND	7.7 / 11.3	NA	Nightly
X Cam	UGZ	7.4 / 14.2	144 days	5–7 days
T Cas **	Mira	6.9 / 13.0	445 days	10–14 days
R CrB *	RCB	5.7 / 15.0	NA	Nightly
o Cet (Mira) *	Mira	2.0 / 10.1	332 days	7–10 days
Chi Cyg *	Mira	3.3 / 14.2	408 days	7–10 days
SS Cyg	UGSS	7.7 / 12.4	~50 days	Nightly
RU Peg **	UGSS + ZZ	9.0 / 13.0	74 days	Nightly

- **Z And** is a great star to follow.
- **X Cam** is another good star to follow.
- **T Cas** is another nice star to follow although the range has shown a reduction in more recent years. This star used to show a 'hump' on the rising branch of its light curve as well, but this seems to have changed more to a 'double' peak at maximum.
- **R CrB** can go very faint, as it did in 2008-2011, 2013, 2014 and 2016. Even at the time of writing (early 2018) this star has still not retained its 'normal' magnitude of around 5.7 / 6.0 and is hovering around magnitude 7.
- **o Cet (Mira)** has a close companion when faint, so great care must be taken around minimum. This is a great star to follow even though it does drop behind the Sun for at least a couple of months of the year.
- Although **Chi Cyg** is often cited as a good star for beginners (which it is when at or near maximum magnitude) it can become lost in the surrounding rich field of faint stars when approaching minimum. This is still a good star

to follow, especially as it can reach magnitude 4 at maximum or even a tad brighter on occasions.

- **SS Cyg** is always worth watching.
- **RU Peg** has a close companion, but you quickly get used to that. Nice star to follow with its regular outbursts.

Eclipsing Binary Stars

STAR	TYPE	MAX	MIN II	MIN I	PERIOD (days)	ECLIPSE DURATION (hours)
TV Cas	EA/SD	7.2	7.3	8.2	1.81	8
U Cep	EA	6.8	6.9	9.4	2.49	9
β Per (Algol) *	EA	2.1	2.2	3.4	2.87	9.6

- Both **TV Cas** and **U Cep** would benefit from more intense coverage of their eclipses.
- As one of the brightest eclipsing binary stars in the sky, **β Per** (**Algol**) is well worth looking at, especially as it is so easy to find.

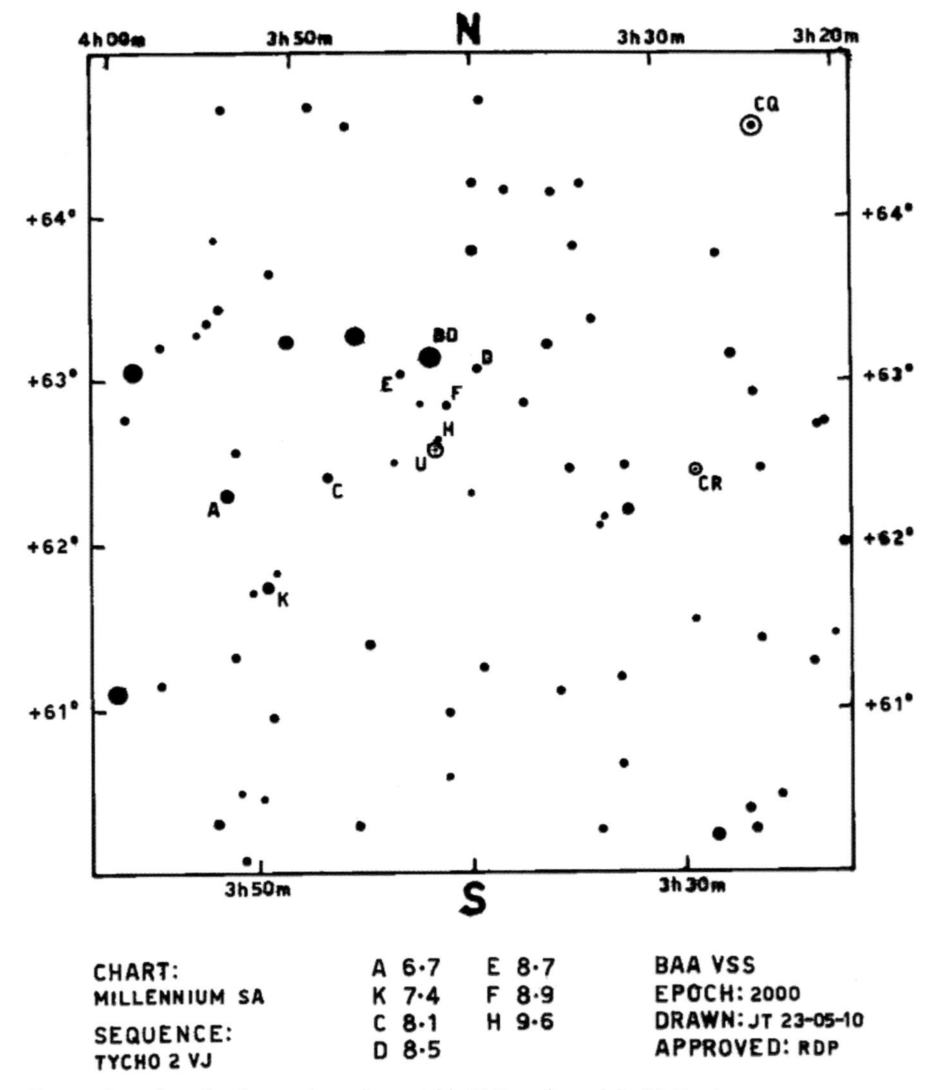

Comparison chart for the semi-regular variable U Camelopardalis (U Cam)

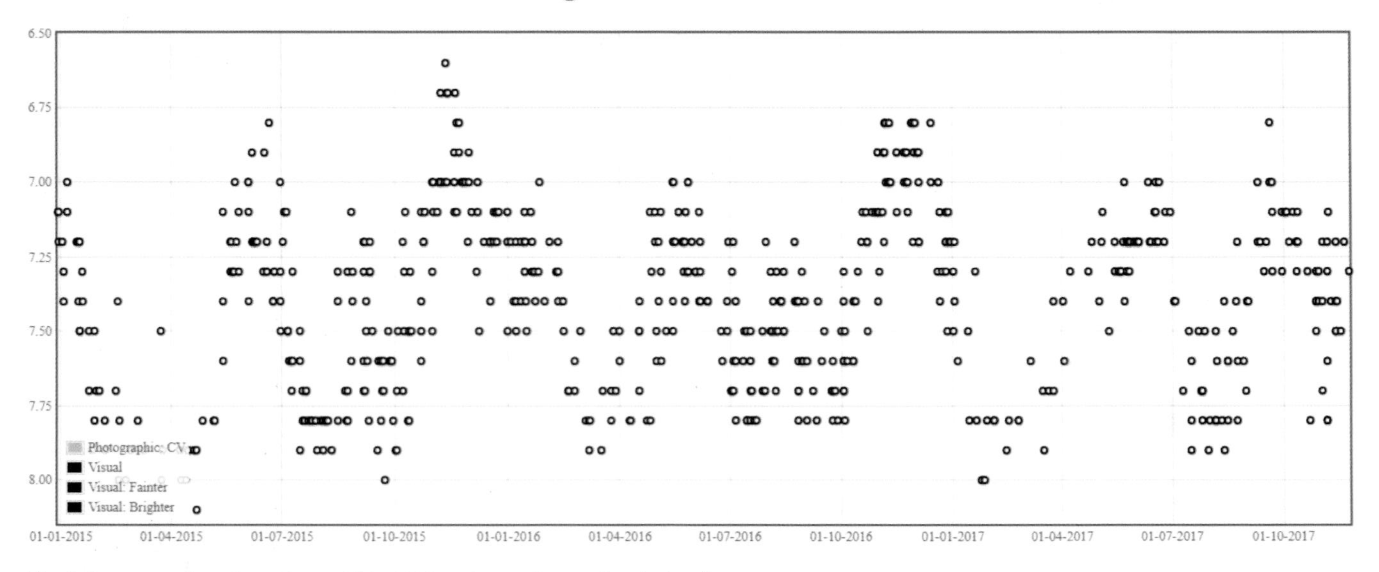

The light curve of semi-regular variable AF Cyg shows plenty of variation from 2015 to 2017.

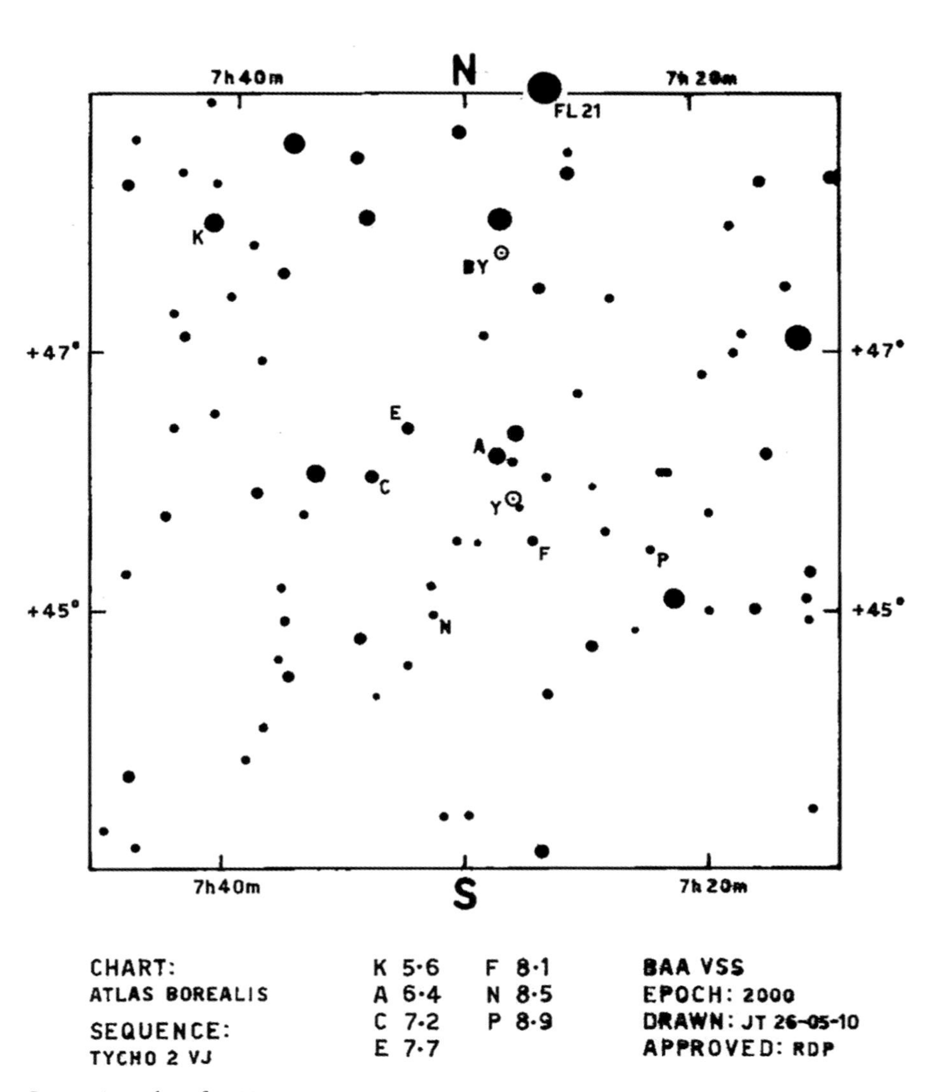

Comparison chart for Y Lyncis (Y Lyn)

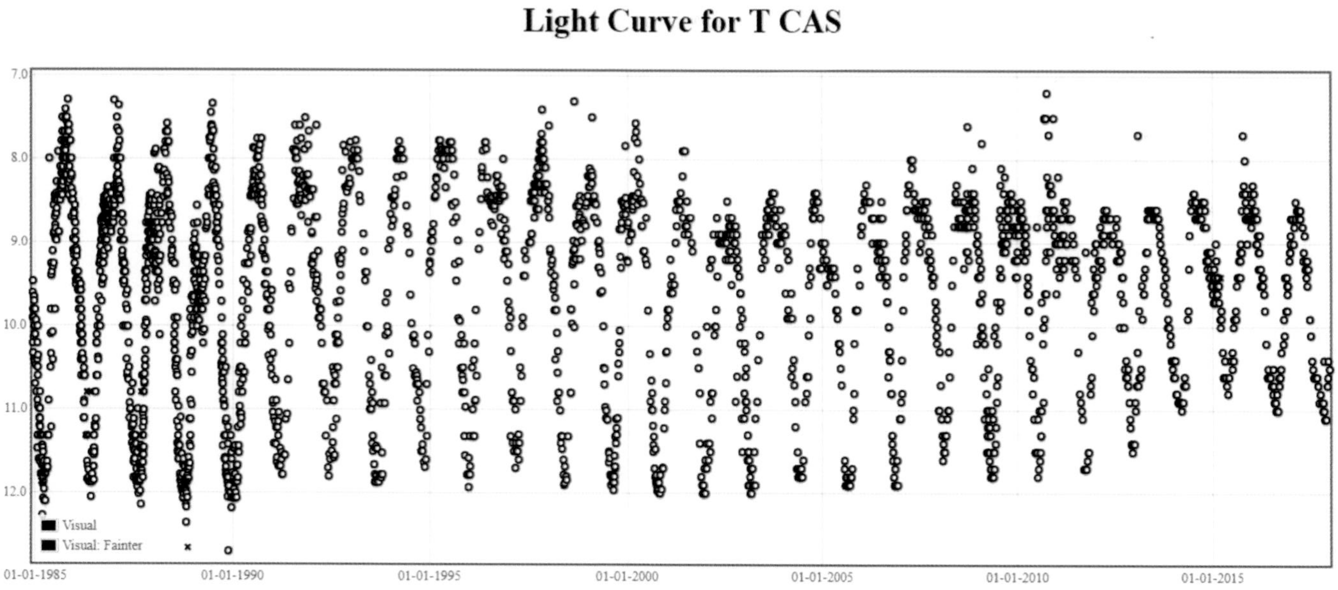

Light Curve for T CAS

7.0
8.0
9.0
10.0
11.0
12.0

Visual
Visual: Fainter

01-01-1985
01-01-1990
01-01-1995
01-01-2000
01-01-2005
01-01-2010
01-01-2015

The light curve for Mira-type variable T Cassiopeiae (T Cas) clearly shows the reduced range in more recent years.

041·04 9° FIELD DIRECT

R CORONAE BOREALIS 15h 48m 34·4s +28°09′24″ (2000)

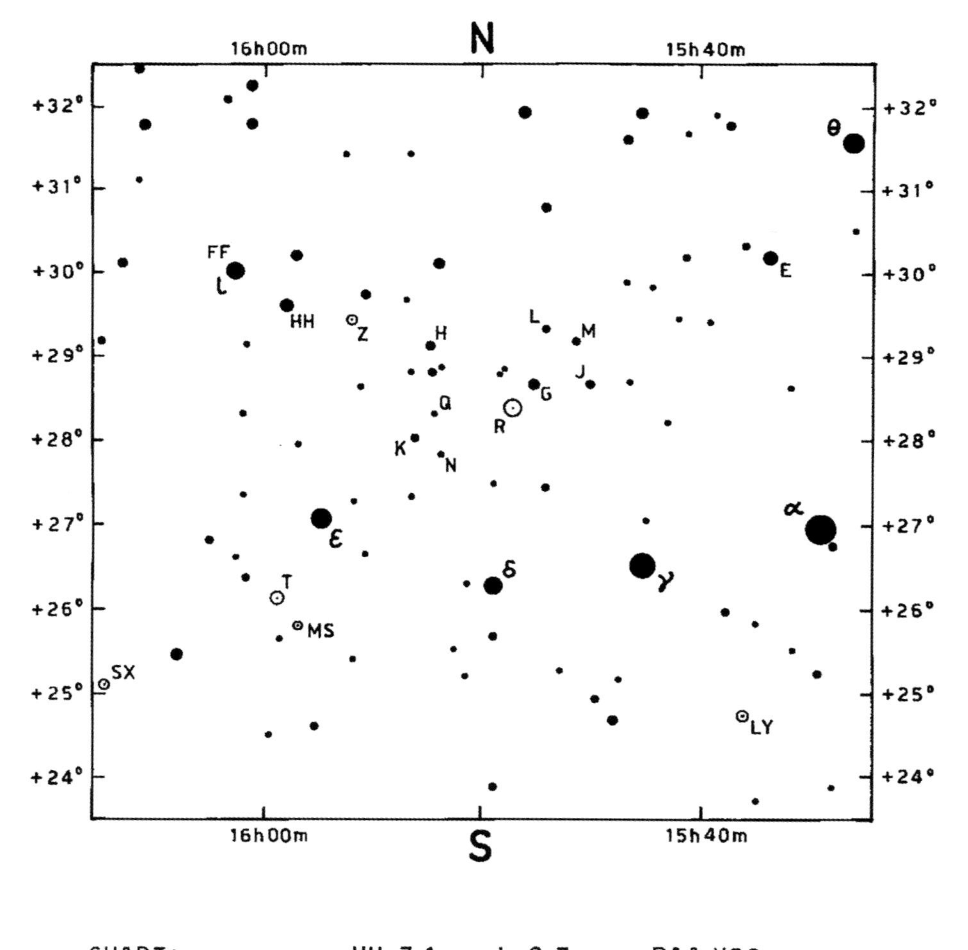

CHART:	HH 7·1	L 8·7	BAA VSS
ATLAS ECLIPTICALIS	G 7·4	M 8·9	EPOCH: 2000
SEQUENCE:	H 7·8	N 9·3	DRAWN: JT 30-01-09
TYCHO 2 VJ	J 8·1	Q 9·5	APPROVED: RDP
	K 8·3		

Comparison chart for R Coronae Borealis (R CrB)

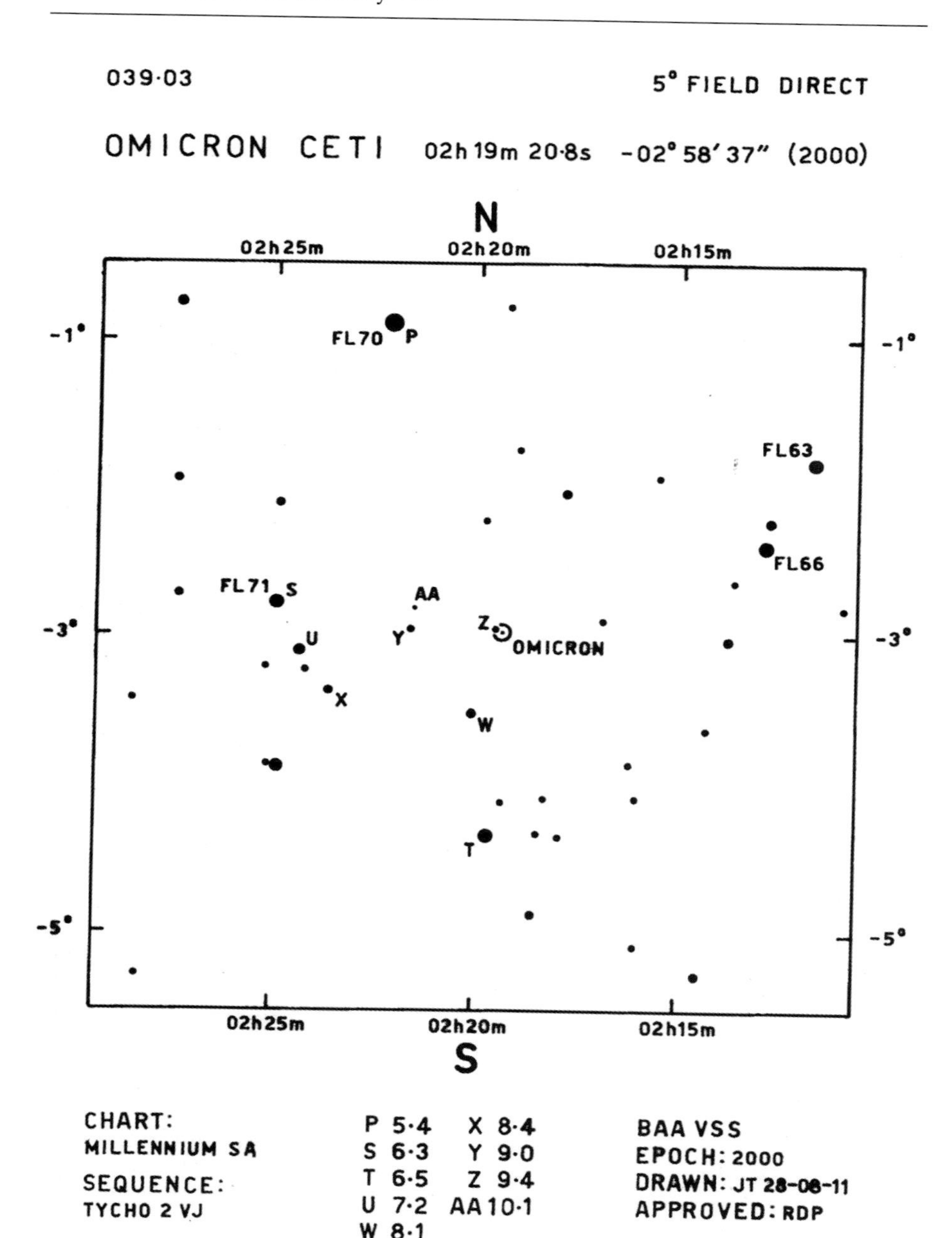

The 5 degree comparison chart for o Cet (Mira) for use when the star is at or near its faintest and needing a pair of binoculars or a small telescope to observe it.

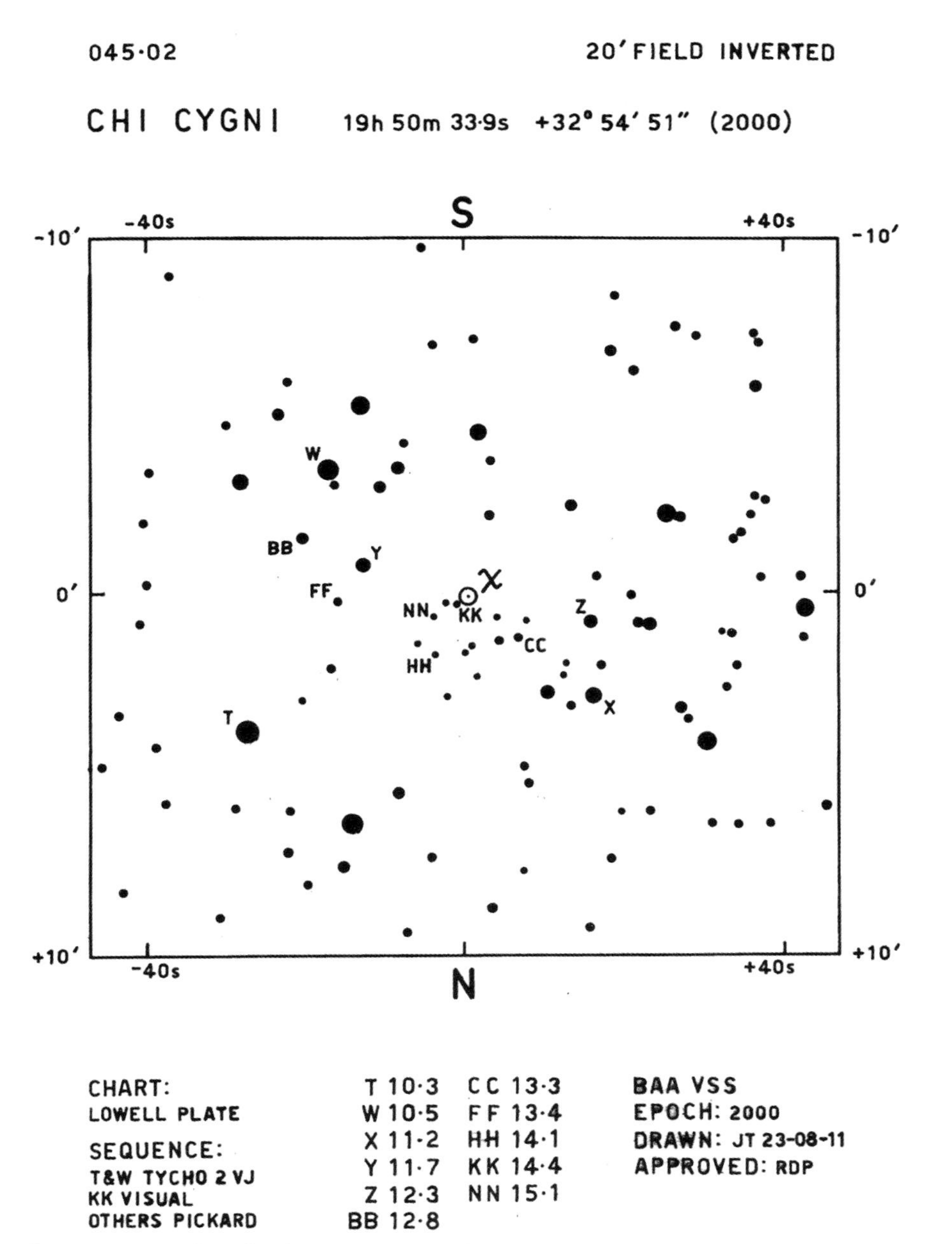

This comparison chart for the Mira-type variable Chi Cyg depicts the inverted field of view visible through a higher power eyepiece on a typical telescope when the star is faint.

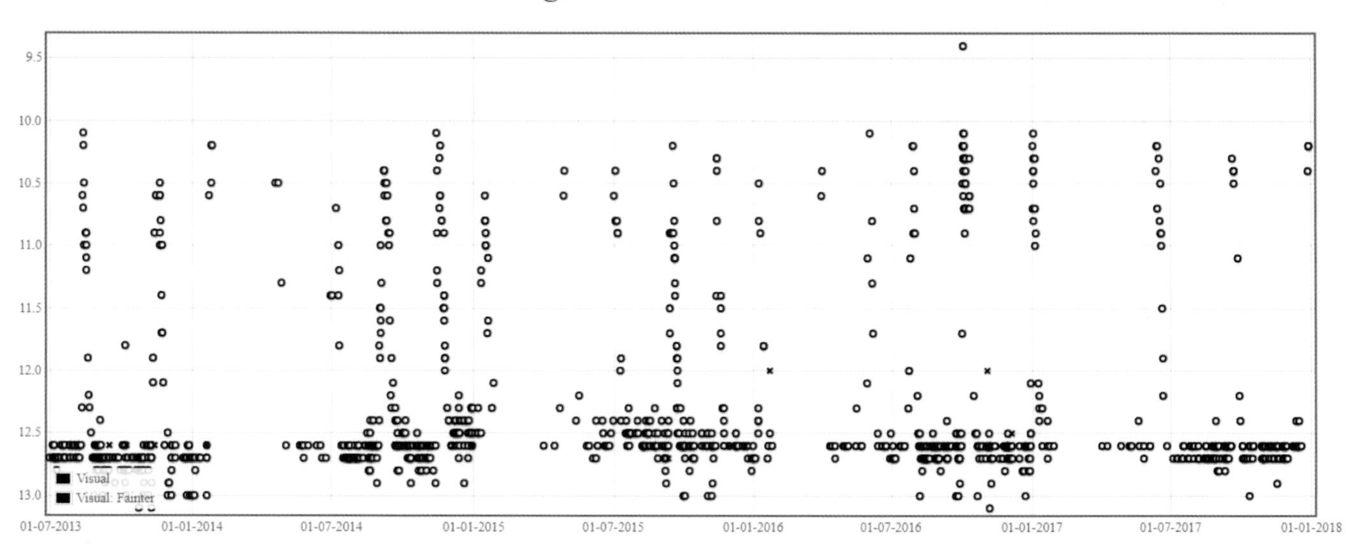

The light curve for RU Peg highlights its frequent outbursts from around magnitude 12.5 to about 10.0.

327·01 40° FIELD DIRECT

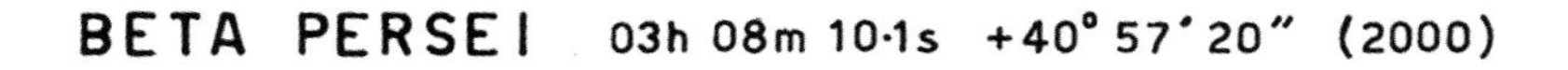

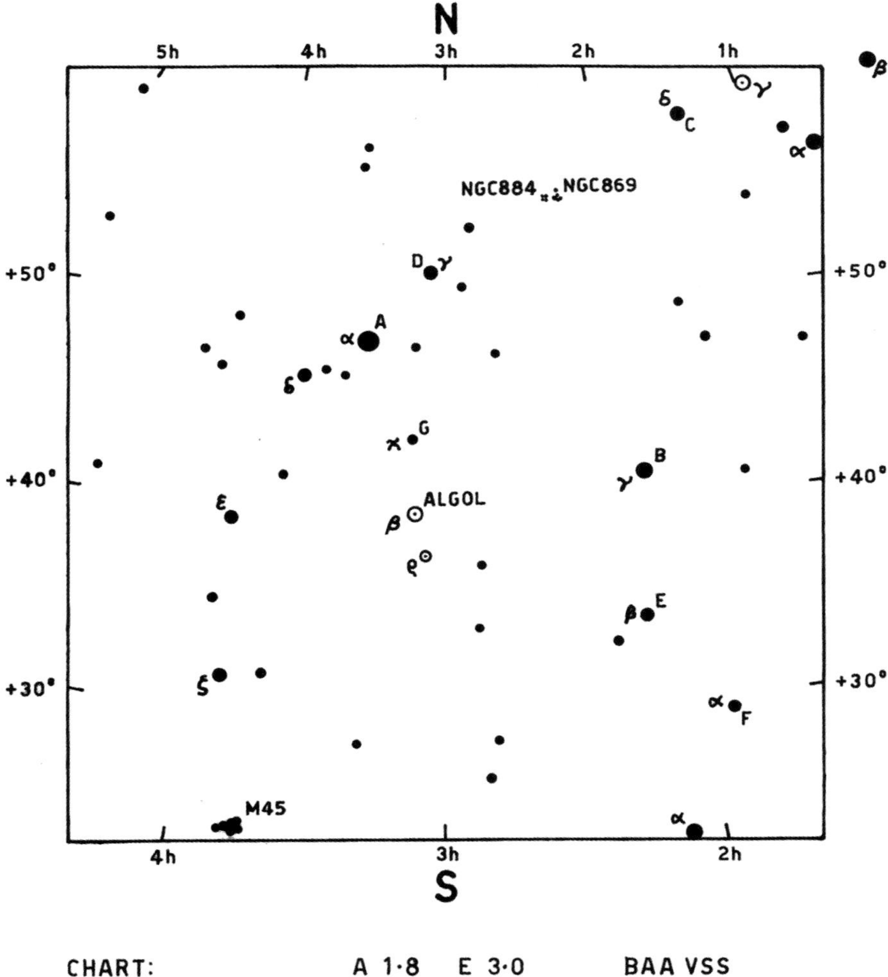

CHART:
NORTONS STAR ATLAS
SEQUENCE:
HIPPARCOS VJ

A 1·8 E 3·0
B 2·1 F 3·4
C 2·7 G 3·8
D 2·9

BAA VSS
EPOCH: 2000
DRAWN: JT 19-06-11
APPROVED: RDP

Comparison chart for β Per (Algol)

Minima of Algol in 2019

Beta (β) Persei (Algol): Magnitude 2.1 to 3.4 / Duration 9.6 hours

		h				*h*				*h*				*h*	
Jan	3	16.2		Feb	1	8.3		Mar	2	0.5	*	Apr	2	13.5	
	6	13.0			4	5.2			4	21.3	*		5	10.3	
	9	9.8			7	2.0	*		7	18.1			8	7.1	
	12	6.6			9	22.8	*		10	15.0			11	3.9	
	15	3.5	*		12	19.6			13	11.8			14	0.7	
	18	0.3	*		15	16.4			16	8.6			16	21.6	
	20	21.1	*		18	13.2			19	5.4			19	18.4	
	23	17.9			21	10.1			22	2.2	*		22	15.2	
	26	14.7			24	6.9			24	23.0	*		25	12.0	
	29	11.5			27	3.7			27	19.9			28	8.8	
									30	16.7					
May	1	5.6		Jun	1	18.6		Jul	3	7.6		Aug	3	20.6	
	4	2.5			4	15.4			6	4.4			6	17.4	
	6	23.3			7	12.3			9	1.2			9	14.2	
	9	20.1			10	9.1			11	22.0			12	11.0	
	12	16.9			13	5.9			14	18.9			15	7.8	
	15	13.7			16	2.7			17	15.7			18	4.6	
	18	10.5			18	23.5			20	12.5			21	1.5	*
	21	7.4			21	20.3			23	9.3			23	22.3	
	24	4.2			24	17.1			26	6.1			26	19.1	
	27	1.0			27	14.0			29	2.9			29	15.9	
	29	21.8			30	10.8			31	23.8					
Sep	1	12.7		Oct	3	1.7		Nov	3	14.7		Dec	2	6.8	
	4	9.6			5	22.5	*		6	11.5			5	3.7	*
	7	6.4			8	19.3			9	8.3			8	0.5	*
	10	3.2			11	16.2			12	5.1			10	21.3	
	12	0.0	*		14	13.0			15	2.0	*		13	18.1	
	15	20.8			17	9.8			17	22.8	*		16	14.9	
	18	17.6			20	6.6			20	19.6			19	11.7	
	21	14.4			23	3.4	*		23	16.4			22	8.6	
	24	11.3			26	0.2	*		26	13.2			25	5.4	
	27	8.1			28	21.1			29	10.0			28	2.2	*
	30	4.9			31	17.9							30	23.0	*

Minima marked with an asterisk (*) are favourable from the British Isles, taking into account the altitude of the variable and the distance of the Sun below the horizon (based on longitude 0° and latitude 52°N). Note that during the winter months some favourable partial eclipses may also be visible.

Some Interesting Double Stars

Brian Jones

The accompanying table describes the visual appearances of a selection of double stars. These may be optical doubles (which consist of two stars which happen to lie more or less in the same line of sight as seen from Earth and which therefore only appear to lie close to each other) or binary systems (which are made up of two stars which are gravitationally linked and which orbit their common centre of mass).

Other than the location on the celestial sphere and the magnitudes of the individual components, the list gives two other values for each of the double stars listed – the angular separation and position angle (PA). Further details of what these terms mean can be found in the article *Double and Multiple Stars* published in the 2018 edition of the Yearbook of Astronomy.

Double-star observing can be a very rewarding process, and even a small telescope will show most, if not all, the best doubles in the sky. You can enjoy looking at double stars simply for their beauty, such as Albireo (β Cygni) or Almach (γ Andromedae), although there is a challenge to be had in splitting very difficult (close) double stars, such as the demanding Sirius (α Canis Majoris) or the individual pairs forming the Epsilon (ε) Lyrae 'Double-Double' star system.

The accompanying list is a compilation of some of the prettiest double (and multiple) stars scattered across both the Northern and Southern heavens. Once you have managed to track these down, many others are out there awaiting your attention …

Star	*RA* h	m	*Declination* °	′	*Magnitudes*	*Separation (arcsec)*	*PA* °	*Comments*
Beta[1,2] (β[1,2]) Tucanae	00	31.5	−62	58	4.36 / 4.53	27.1	169	Both stars again double, but difficult
Achird (η Cassiopeiae)	00	49.1	+57	49	3.44 / 7.51	13.4	324	Easy double
Mesarthim (γ Arietis)	01	53.5	+19	18	4.58 / 4.64	7.6	1	Easy pair of white stars
Almach (γ Andromedae)	02	03.9	+42	20	2.26 / 4.84	9.6	63	Yellow and blue-green components
32 Eridani	03	54.3	−02	57	4.8 / 6.1	6.9	348	Yellowish and bluish
Alnitak (ζ Orionis)	05	40.7	−01	57	2.0 / 4.3	2.3	167	Difficult, can be resolved in 10cm telescopes
Gamma (γ) Leporis	05	44.5	−22	27	3.59 / 6.28	95.0	350	White and yellow-orange components, easy pair
Sirius (α Canis Majoris)	06	45.1	−16	43	−1.4 / 8.5			Binary, period 50 years, difficult
Castor (α Geminorum)	07	34.5	+31	53	1.93 / 2.97	7.0	55	Binary, 445 years, widening
Gamma (γ) Velorum	08	09.5	−47	20	1.83 / 4.27	41.2	220	Pretty pair in nice field of stars
Upsilon (υ) Carinae	09	47.1	−65	04	3.08 / 6.10	5.03	129	Nice object in small telescopes
Algieba (γ Leonis)	10	20.0	+19	50	2.28 / 3.51	4.6	126	Binary, 510 years, orange-red and yellow
Acrux (α Crucis)	12	26.4	−63	06	1.40 / 1.90	4.0	114	Glorious pair, third star visible in low power
Porrima (γ Virginis)	12	41.5	−01	27	3.56 / 3.65			Binary, 170 years, widening, visible in small telescopes
Cor Caroli (α Canum Venaticorum)	12	56.0	+38	19	2.90 / 5.60	19.6	229	Easy, yellow and bluish
Mizar (ζ Ursae Majoris)	13	24.0	+54	56	2.3 / 4.0	14.4	152	Easy, wide naked-eye pair with Alcor
Alpha (α) Centauri	14	39.6	−60	50	0.0 / 1.2			Binary, beautiful pair of stars
Izar (ε Boötis)	14	45.0	+27	04	2.4 / 5.1	2.9	344	Fine pair of yellow and blue stars
Omega[1,2] (ω[1,2]) Scorpii	16	06.0	−20	41	4.0 / 4.3	14.6	145	Optical pair, easy
Epsilon1 (ε[1]) Lyrae	18	44.3	+39	40	4.7 / 6.2	2.6	346	The Double-Double, quadruple system with ε[2]
Epsilon2 (ε[2]) Lyrae	18	44.3	+39	40	5.1 / 5.5	2.3	76	Both individual pairs just visible in 80mm telescopes
Theta[1,2] (θ[1,2]) Serpentis	18	56.2	+04	12	4.6 / 5.0	22.4	104	Easy pair, mag 6.7 yellow star 7 arc minutes from θ2

Star	*RA*		*Declination*		*Magnitudes*	*Separation*	*PA*	*Comments*
	h	*m*	°	′		*(arcsec)*	°	
Albireo (β Cygni)	19	30.7	+27	58	3.1 / 5.1	34.3	54	Glorious pair, yellow and blue-green
Giedi ($\alpha^{1,2}$ Capricorni)	20	18.0	−12	32	3.7 / 4.3	6.3	292	Optical pair, easy
Gamma (γ) Delphini	20	46.7	+16	07	5.14 / 4.27	9.2	265	Easy, orange and yellow-white
61 Cygni	21	06.9	+38	45	5.20 / 6.05	31.6	152	Binary, 678 years, both orange
Delta (δ) Tucanae	22	27.3	−64	58	4.49 / 8.7	7.0	281	Beautiful double, white and reddish

Some Interesting Nebulae, Star Clusters and Galaxies

Brian Jones

Object	*RA*		*Declination*		*Remarks*
	h	*m*	°	′	
47 Tucanae (in Tucana)	00	24.1	−72	05	Fine globular cluster, easy with naked eye
M31 (in Andromeda)	00	40.7	+41	05	Andromeda Galaxy, visible to unaided eye
Small Magellanic Cloud	00	52.6	−72	49	Satellite galaxy of the Milky Way
NGC 362 (in Tucana)	01	03.3	−70	51	Globular cluster, impressive sight in telescopes
M33 (in Triangulum)	01	31.8	+30	28	Triangulum Spiral Galaxy, quite faint
NGC 869 and NGC 884	02	20.0	+57	08	Sword Handle Double Cluster in Perseus
M34 (in Perseus)	02	42.1	+42	46	Open star cluster near Algol
M45 (in Taurus)	03	47.4	+24	07	Pleiades or Seven Sisters cluster, a fine object
Large Magellanic Cloud	05	23.5	−69	45	Satellite galaxy of the Milky Way
30 Doradus (in Dorado)	05	38.6	−69	06	Star-forming region in Large Magellanic Cloud
M79 (in Lepus)	05	24.2	−24	31	Globular cluster
M38 (in Auriga)	05	28.6	+35	51	Open star cluster
M42 (in Orion)	05	33.4	−05	24	Orion Nebula
M1 (in Taurus)	05	34.5	+22	01	Crab Nebula, near Zeta (ζ) Tauri
M36 (in Auriga)	05	36.2	+34	08	Open star cluster
M37 (in Auriga)	05	52.3	+32	33	Open star cluster
M35 (in Gemini)	06	06.5	+24	21	Open star cluster near Eta (η) Geminorum
M41 (in Canis Major)	06	46.0	−20	46	Open star cluster to south of Sirius
M44 (in Cancer)	08	38.0	+20	07	Praesepe, visible to naked eye
IC 2391 (in Vela)	08	40.6	−53	02	Omicron (ο)Velorum (open star) Cluster
IC 2602 (in Carina)	10	42.9	−64	24	Theta (θ) Carinae Cluster, naked eye object
Carina Nebula (in Carina)	10	45.2	−59	52	NGC 3372, large area of bright and dark nebulosity
M104 (in Virgo)	12	40.0	−11	37	Sombrero Hat Galaxy to south of Porrima
Coal Sack (in Crux)	12	50.0	−62	30	Prominent dark nebula, visible to naked eye
NGC 4755 (in Crux)	12	53.6	−60	22	Jewel Box open cluster, magnificent object
Omega (ω) Centauri	13	23.7	−47	03	Splendid globular in Centaurus, easy with naked eye
M51 (in Canes Venatici)	13	29.9	+47	12	Whirlpool Galaxy
M3 (in Canes Venatici)	13	40.6	+28	34	Bright Globular Cluster

Object	*RA*		*Declination*		*Remarks*
	h	*m*	°	'	
M4 (in Scorpius)	16	21.5	−26	26	Globular cluster, close to Antares
M12 (in Ophiuchus)	16	47.2	−01	57	Globular cluster
M10 (in Ophiuchus)	16	57.1	−04	06	Globular cluster
M13 (in Hercules)	16	40.0	+36	31	Great Globular Cluster, just visible to naked eye
M92 (in Hercules)	17	16.1	+43	11	Globular cluster
M6 (in Scorpius)	17	36.8	−32	11	Open cluster
M7 (in Scorpius)	17	50.6	−34	48	Bright open cluster
M20 (in Sagittarius)	18	02.3	−23	02	Trifid Nebula
M8 (in Sagittarius)	18	03.6	−24	23	Lagoon Nebula, just visible to naked eye
M16 (in Serpens)	18	18.8	−13	49	Eagle Nebula and star cluster
M17 (in Sagittarius)	18	20.2	−16	11	Omega Nebula
M11 (in Scutum)	18	49.0	−06	19	Wild Duck open star cluster
M27 (in Lyra)	18	52.6	+32	59	Ring Nebula, brightest planetary
M27 (in Vulpecula)	19	58.1	+22	37	Dumbbell Nebula
M29 (in Cygnus)	20	23.9	+38	31	Open cluster
M15 (in Pegasus)	21	30.1	+12	10	Bright globular cluster near Epsilon (ε) Pegasi
M39 (in Cygnus)	21	31.6	+48	27	Open cluster, good with low powers
M52 (in Cassiopeia)	23	24.2	+61	35	Open star cluster near 4 Cassiopeiae

M = Messier Catalogue Number   NGC = New General Catalogue Number

The positions in the sky of each of the objects contained in this list are given on the Monthly Star Charts printed elsewhere in this volume.

Astronomical Organizations

American Association of Variable Star Observers

49 Bay State Road, Cambridge, Massachusetts 02138, USA

www.aavso.org

The AAVSO is an international non-profit organization of variable star observers whose mission is to enable anyone, anywhere, to participate in scientific discovery through variable star astronomy. We accomplish our mission by carrying out the following activities:

- observation and analysis of variable stars
- collecting and archiving observations for worldwide access
- forging strong collaborations between amateur and professional astronomers
- promoting scientific research, education and public outreach using variable star data

American Astronomical Society

1667 K Street NW, Suite 800, Washington, DC 20006, USA

https://aas.org

Established in 1899, the American Astronomical Society (AAS) is the major organization of professional astronomers in North America. The mission of the AAS is to enhance and share humanity's scientific understanding of the universe, which it achieves through publishing, meeting organization, education and outreach, and training and professional development.

Astronomical Society of Australia

c/o A/Prof. J.W. O'Byrne, School of Physics, The University of Sydney, NSW 2006, Australia

asa.astronomy.org.au

The Astronomical Society of Australia (ASA) was formed in 1966 as the organisation of professional astronomers in Australia. Membership of the ASA is open to anyone contributing to the advancement of Australian astronomy or a closely related field. As well as publishing a refereed journal *Publications of the Astronomical Society of Australia*, the Society runs an annual conference and several workshops and schools.

Astronomical Society of the Pacific
390 Ashton Avenue, San Francisco, CA 94112, USA
www.astrosociety.org
Formed in 1889, the Astronomical Society of the Pacific (ASP) is a non-profit membership organization which is international in scope. The mission of the ASP is to increase the understanding and appreciation of astronomy through the engagement of our many constituencies to advance science and science literacy. We invite you to explore our site to learn more about us; to check out our resources and education section for the researcher, the educator, and the backyard enthusiast; to get involved by becoming an ASP member; and to consider supporting our work for the benefit of a science literate world!

Astrospeakers.org
www.astrospeakers.org
A website designed to help astronomical societies and clubs locate astronomy and space lecturers which is also designed to help people find their local astronomical society. It is completely free to register and use and, with over 50 speakers listed, is an excellent place to find lecturers for your astronomical society meetings and events. Speakers and astronomical societies are encouraged to use the online registration to be added to the lists.

British Astronomical Association
Burlington House, Piccadilly, London, W1J 0DU, England
www.britastro.org
The British Astronomical Association is the UK's leading society for amateur astronomers catering for beginners to the most advanced observers who produce scientifically useful observations. Our Observing Sections provide encouragement and advice about observing. We hold meetings around the country and publish a bi-monthly Journal plus an annual Handbook. For more details, including how to join the BAA or to contact us, please visit our website.

British Interplanetary Society

Arthur C Clarke House, 27/29 South Lambeth Road, London, SW8 1SZ, England

www.bis-space.com

The British Interplanetary Society is the world's longest-established space advocacy organisation, founded in 1933 by the pioneers of British astronautics. It is the first organisation in the world still in existence to design spaceships. Early members included Sir Arthur C Clarke and Sir Patrick Moore. The Society has created many original concepts, from a 1938 lunar lander and space suit designs, to geostationary orbits, space stations and the first engineering study of a starship, Project Daedalus. Today the BIS has a worldwide membership and welcomes all with an interest in Space, including enthusiasts, students, academics and professionals.

Canadian Astronomical Society

Société Canadienne D'astronomie (CASCA)

100 Viaduct Avenue West, Victoria, British Columbia, V9E 1J3, Canada

www.casca.ca

CASCA is the national organization of professional astronomers in Canada. It seeks to promote and advance knowledge of the universe through research and education. Founded in 1979, members include university professors, observatory scientists, postdoctoral fellows, graduate students, instrumentalists, and public outreach specialists.

Royal Astronomical Society of Canada

203-4920 Dundas St W, Etobicoke, Toronto, ON M9A 1B7, Canada

www.rasc.ca

Bringing together over 5,000 enthusiastic amateurs, educators and professionals RASC is a national, non-profit, charitable organization devoted to the advancement of astronomy and related sciences and is Canada's leading astronomy organization. Membership is open to everyone with an interest in astronomy. You may join through any one of our 29 RASC centres, located across Canada and all of which offer local programs. The majority of our events are free and open to the public.

Federation of Astronomical Societies

The Secretary, 147 Queen Street, SWINTON, Mexborough, S64 8NG

www.fedastro.org.uk

The Federation of Astronomical Societies (FAS) is an umbrella group for astronomical societies in the UK. It promotes cooperation, knowledge and information sharing and encourages best practice. The FAS aims to be a body of societies united in their attempts to help each other find the best ways of working for their common cause of creating a fully successful astronomical society. In this way it endeavours to be a true federation, rather than some remote central organization disseminating information only from its own limited experience. The FAS also provides a competitive Public Liability Insurance scheme for its members.

International Dark-Sky Association

darksky.org

The International Dark-Sky Association (IDA) is the recognized authority on light pollution and the leading organization combating light pollution worldwide. The IDA works to protect the night skies for present and future generations, our public outreach efforts providing solutions, quality education and programs that inform audiences across the United States of America and throughout the world. At the local level, our mission is furthered through the work of our U.S. and international chapters representing five continents.

The goals of the IDA are to:

- advocate for the protection of the night sky
- educate the public and policymakers about night sky conservation
- promote environmentally responsible outdoor lighting
- empower the public with the tools and resources to help bring back the night

Royal Astronomical Society of New Zealand

PO Box 3181, Wellington, New Zealand

www.rasnz.org.nz

Founded in 1920, the object of The Royal Astronomical Society of New Zealand is the promotion and extension of knowledge of astronomy and related branches of science. It encourages interest in astronomy and is an association of observers and others for mutual help and advancement of science. Membership is open to all interested in astronomy. The RASNZ has about 180 individual members including both professional and amateur astronomers and many of the astronomical research and observing programmes carried out in New Zealand involve collaboration between the two. In addition the society has a number of groups or sections which cater for people who have interests in particular areas of astronomy.

Astronomical Society of Southern Africa

Astronomical Society of Southern Africa, c/o SAAO, PO Box 9, Observatory, 7935, South Africa

assa.saao.ac.za

Formed in 1922, The Astronomical Society of Southern Africa comprises both amateur and professional astronomers. Membership is open to all interested persons. Regional Centres host regular meetings and conduct public outreach events, whilst national Sections coordinate special interest groups and observing programmes. The Society administers two Scholarships, and hosts occasional Symposia where papers are presented. For more details, or to contact us, please visit our website.

Royal Astronomical Society

Burlington House, Piccadilly, London, W1J 0BQ, England

www.ras.org.uk

The Royal Astronomical Society, with around 4,000 members, is the leading UK body representing astronomy, space science and geophysics, with a membership including professional researchers, advanced amateur astronomers, historians of science, teachers, science writers, public engagement specialists and others.

Society for Popular Astronomy

Secretary: Guy Fennimore, 36 Fairway, Keyworth, Nottingham, NG12 5DU

www.popastro.com

The Society for Popular Astronomy is a national society that aims to present astronomy in a less technical manner. The bi-monthly society magazine *Popular Astronomy* is issued free to all members.

Our Contributors

Mike Brotherton is a professor of astronomy at the University of Wyoming where he investigates the most luminous active galactic nuclei, the quasars. Powered by supermassive black holes, quasars outshine the galaxies within which they exist and shape the course of their evolution. He uses the Hubble Space Telescope, the Very Large Array in New Mexico, the Chandra X-ray Observatory, and any other telescope that will grant him observing time. He is also the author of the science fiction novels *Star Dragon* and *Spider Star*, both from Tor Books, as well as a number of short stories. He is the founder of the NASA and National Science Foundation funded Launch Pad Astronomy Workshop for Writers, which brings professional writers to Wyoming every summer in order to better educate and inspire their audiences. He has previously edited or co-edited the anthologies *Diamonds in the Sky*, *Launch Pad* (with Jody Lynn Nye) and *Science Fiction by Scientists*. His webpage is **www.mikebrotherton.com**

Steve Brown is an amateur astronomer based in Stokesley in the North Riding of Yorkshire. He has been interested in astronomy from a young age and has observed seriously since buying his first telescope (a 130mm reflector) in 2011. He regularly observes from his garden (when the Yorkshire weather allows) dividing his observing time between astrophotography and sketching. His image 'The Rainbow Star' won the Stars and Nebulae category of the Insight Astronomy Photographer of the Year competition in 2016. Steve also appeared on the Sky at Night episode 'Moore Winter Marathon', filmed at Kielder Observatory and broadcast in March 2013. As well as a number of his astro sketches being featured on the programme, Steve has also had sketches, images and articles published in *Sky at Night* and *All About Space* magazines. As a member of the British Astronomical Association, Steve regularly submits meteor shower observations to their Meteor Section. You can follow Steve on Twitter via **@sjb_astro** and see his images on Flickr at **www.flickr.com/photos/sjb_astro**

Neil Haggath has a degree in astrophysics from Leeds University, and has been a Fellow of the Royal Astronomical Society since 1993. A member of Cleveland and Darlington Astronomical Society for 37 years, he has served on its committee for 29 years. Neil is an avid umbraphile, clocking up six total eclipse expeditions so far – four of them successful – to locations as far flung as China and Australia. In Wyoming on 21 August 2017, he succeeded in observing two successive eclipses of a Saros series – having previously achieved the unenviable opposite! In 2012, he may have set a somewhat unenviable record among British astronomers – for the greatest distance travelled (6,000 miles to Thailand) to NOT see the transit of Venus. He saw nothing on the day … and got very wet!

Dr. David M. Harland gained his BSc in astronomy in 1977, lectured in computer science, worked in industry, and managed academic research. In 1995 he 'retired' in order to write on space themes.

David Harper, FRAS has had a varied career which includes teaching mathematics, astronomy and computing at Queen Mary University of London, astronomical software development at the Royal Greenwich Observatory, bioinformatics support at the Wellcome Trust Sanger Institute, and a research interest in the dynamics of planetary satellites, which began during his Ph.D. at Liverpool University in the 1980s and continues in an occasional collaboration with colleagues in China. He is married to fellow contributor Lynne Stockman.

Rod Hine was aged around ten when he was given a copy of *The Boys Book of Space* by Patrick Moore. Already interested in anything to do with science and engineering he devoured the book from cover to cover. The launch of Sputnik 1 shortly afterwards clinched his interest in physics and space travel.

He took physics, chemistry and mathematics at A-level and then studied Natural Sciences at Churchill College, Cambridge. He later switched to Electrical Sciences and subsequently joined Marconi at Chelmsford working on satellite communications installations which involved a lot of travelling. Whilst working on the new earth station in Kenya he met and subsequently married a Yorkshire lass. Enjoying the expatriate life-style he then worked in meteorological communications in Nairobi and later took a teaching post at the Kenya Polytechnic. The couple moved back to the UK in 1976 and since then he has had a variety of jobs in electronics and industrial controls, and has recently been lecturing part-time at the University of Bradford. Rod got fully back into astronomy in around 1992 when his wife bought him an astronomy book, at which time he joined Bradford Astronomical Society.

Brian Jones hails from Bradford in the West Riding of Yorkshire and was a founder member of the Bradford Astronomical Society. His fascination with astronomy began at the age of five when he first saw the stars through a pair of binoculars, although he spent the first part of his working life developing a career in mechanical engineering. However, his true passion lay in the stars and his interest in astronomy took him into the realms of writing sky guides for local newspapers, appearing on local radio and television, teaching astronomy and space in schools and, in 1985, leaving engineering to become a full time astronomy and space writer. His books have covered a range of astronomy and space-related topics for both children and adults and his journalistic work includes writing articles and book reviews for several astronomy magazines as well as for many general interest magazines, newspapers and periodicals. His passion for bringing an appreciation of the universe to his readers is reflected in his writing.

You can follow Brian on Twitter via **@StarsBrian** and check out the sky by visiting his blog at **www.starlight-nights.co.uk** from where you can also access his Facebook group Starlight Nights.

John McCue graduated in astronomy from the University of St Andrews and began teaching. He gained a Ph.D. from Teesside University studying the unusual rotation of Venus. In 1979 he and his colleague John Nichol founded the Cleveland and Darlington Astronomical Society, which then worked in partnership with the local authority to build the Wynyard Planetarium and Observatory in Stockton-on-Tees. John is currently double star advisor for the British Astronomical Association.

Carl Murray is Professor of Mathematics and Astronomy at Queen Mary University of London. His research involves studying the motion of objects in the solar system, from dust to planets. In 1990 he was selected as a member of the Cassini Imaging Team and he uses images of the Saturn system to understand the gravitational interaction between rings and moons. With Stan Dermott he has co-authored the standard textbook *Solar System Dynamics*.

Neil Norman, FRAS first became fascinated with the night sky when he was five years of age and saw Patrick Moore on the television for the first time. It was the Sky at Night programme, broadcast in March 1986 and dedicated to the Giotto probe reaching Halley's Comet, which was to ignite his passion for these icy interlopers. As the years passed, he began writing astronomy articles for local news magazines before moving into internet radio where he initially guested on the Astronomyfm show 'Under British Skies', before becoming a co-host for a short time. In 2013 he created Comet Watch, a Facebook group dedicated to comets of the past, present and future. His involvement with Astronomyfm led to the creation of the monthly radio show 'Cometwatch', which is now in its fourth year. On 7 November 2017 the main belt asteroid 314650/2006 OC1 was named in his honour. Neil lives in Suffolk with his partner and three children.

Richard Pearson, FRAS, FRGS was born and raised in Nottingham and has worked on local newspapers as a journalist for over 20 years. A member of the British Astronomical Association and a Fellow of the Royal Astronomical Society, he has written several books on astronomy and is the presenter of the monthly internet TV program Astronomy & Space, now in its fifth year, and is a celebrity among members of astronomical societies worldwide.

Roger Pickard has been observing variable stars for more years than he cares to remember. Initially this was as a visual observer, but he then dabbled in photoelectric photometry (PEP) just before the advent of affordable CCDs. He now operates his telescope and CCD camera, which is located at the end of his garden, from within the warmth of his house. Roger has been the Director of the British Astronomical Association Variable Star Section since 1999 and can be contacted at **roger.pickard@sky.com** by anyone who needs help with variable star observing.

Peter Rea has had a keen interest in lunar and planetary exploration since the early 1960s and frequently lectures on the subject. He helped found the

Cleethorpes and District Astronomical Society in 1969. In April of 1972 he was at the Kennedy Space Centre in Florida to see the launch of Apollo 16 to the moon and in October 1997 was at the southern end of Cape Canaveral to see the launch of Cassini to Saturn. He would still like to see a total solar eclipse as the expedition he was on to see the 1973 eclipse in Mali had vehicle trouble and the meteorologists decided he was not going to see the 1999 eclipse from Devon. He lives in Lincolnshire with his wife Anne and has a daughter who resides on the Gold Coast in Australia.

Lynne Marie Stockman holds degrees in mathematics from Whitman College, the University of Washington and the University of London. She is a native of North Idaho but has lived in Britain for the past 26 years and is a visiting research student in astronomy at Queen Mary University of London. Lynne was an early pioneer of the world-wide web: with her husband David Harper, she created the web site **obliquity.com** in 1998 to share their interest in astronomy, family history and cats.

Susan Stubbs first became interested in stars as a child when she was astonished by the appearance of the Milky Way after seeing it from a dark sky site in Sussex whilst camping with the Guides. However, progressing to her Guide Stargazers badge showed her that not so many people were interested in astronomy back then, as finding an enthusiast to test her proved problematic! University, career and family then got in the way for quite a few years, but she returned to the fold via Open University astronomy courses and through joining Bradford Astronomical Society of which she is an active member. She was delighted to find that so many more people are interested in astronomy now than all those years ago in childhood. Susan recently had a new astronomical experience, travelling with her husband Robin to see the August 2017 total solar eclipse in Wyoming. She definitely wants to do it again!

Glossary

Brian Jones and David Harper

Altitude
The altitude of a star or other object is its angular distance above the horizon. For example, if a star is located at the ***zenith***, or overhead point, its altitude is 90° and if it is on the horizon, its altitude is 0°.

Angular Distance
The angular distance between two objects on the sky is the angle subtended between the directions to the two objects, either at the centre of the Earth (geocentric angular distance) or the observer's eye (apparent angular distance). It is most commonly expressed in degrees, or for smaller angular distances, minutes of arc or seconds of arc.

Aphelion
This is the point at which an object, such as a planet, comet or asteroid travelling in an elliptical ***orbit***, is at its maximum distance from the Sun.

Apogee
This is the point in its ***orbit*** around the Earth at which an object is at its furthest from the Earth.

Apparition
The period during which a planet is visible, usually starting at ***conjunction*** with the Sun, running through ***opposition*** (for a superior planet) or ***greatest elongation*** (for Mercury or Venus), and ending with the next conjunction with the Sun.

Appulse
The close approach, as seen from the Earth, between two planets, or a planet and a star, or the Moon and a star or planet. Also known as a ***conjunction***.

Asterism
An asterism is grouping or collection of stars often (but not always) located within a ***constellation*** that forms an apparent and distinctive pattern in its own right. Well known examples include the Plough (in Ursa Major); the False Cross (formed from stars in Carina and Vela); and the Summer Triangle, which is formed from the bright stars Vega (in Lyra), Deneb (in Cygnus) and Altair (in Aquila).

Asteroid
Another name for a ***minor planet***.

Autumnal Equinox
The autumnal equinox is the point at which the apparent path of the Sun, moving from north to south, crosses the ***celestial equator***. In the Earth's northern hemisphere this marks the start of autumn, whilst in the southern hemisphere it is the start of spring.

Averted Vision
Averted vision is a useful technique for observing faint objects which involves looking slightly to one side of the object under observation and, by doing so, allowing the light emitted by the object to fall on the part of the retina that is more sensitive at low light levels. Although you are not looking directly at the object, it is surprising how much more detail comes into view. This technique is also useful when observing double stars which have components of greatly contrasting brightness. Although direct vision may not reveal the glow of a faint companion star in the glare of a much brighter primary, averted vision may well bring the fainter star into view.

Azimuth
The azimuth of a star or other object is its angular position measured round the ***horizon*** from north (azimuth 0°) through east (azimuth 90°), south (azimuth 180°) and west (azimuth 270°). The azimuth and ***altitude***, taken together, define the position of the object referred to the observer's ***local horizon***.

Barycentre
The barycentre is the centre of mass of two or more bodies that are orbiting each other (such as a planet and satellite or two components of a ***binary star*** system) and which is therefore the point around which they both ***orbit***.

Binary Star
See Double Star

Black Hole
A region of space around a very compact and extremely massive collapsed star within which the gravitational field is so intense that not even light can escape.

Caldwell Catalogue
This is a catalogue of 109 star clusters, nebulae, and galaxies compiled by Patrick Moore to complement the ***Messier Catalogue***. Intended for use as an observing guide by amateur astronomers it includes a number of bright ***deep sky objects*** that did not find their way into the Messier Catalogue, which was originally compiled as a list of known objects that might be confused with comets. Moore used his other surname (Caldwell) to name the list and the objects within it (the first letter of 'Moore' having been used for the Messier Catalogue)

and entries in the Caldwell Catalogue are designated with a 'C' followed by the catalogue number (1 to 109).

Amongst the 109 objects in the Caldwell Catalogue are the Sword Handle Double Cluster NGC 869 and NGC 884 (C14) in Perseus; supernova remnant(s) the East Veil Nebula and West Veil Nebula (C33 and C34) in Cygnus; the Hyades open star cluster (C41) in Taurus; and Hubble's Variable Nebula (C46) in Monoceros. Unlike the Messier Catalogue, which was compiled from observations made by Charles Messier from Paris, the Caldwell Catalogue contains deep sky objects visible from the southern hemisphere, such as the Centaurus A galaxy (C77) and globular cluster Omega Centauri (C80) in Centaurus; the Jewel Box open star cluster (C94) in Crux and the globular cluster 47 Tucanae (C106) in Tucana.

Although none of the objects detailed elsewhere in the Yearbook of Astronomy carry a Caldwell Catalogue reference, it was felt that an entry should appear in the Glossary as the catalogue is nonetheless an important guide for the backyard astronomer.

Celestial Equator

The celestial equator is a projection of the Earth's ***equator*** onto the ***celestial sphere***, equidistant from the ***celestial poles*** and dividing the celestial sphere into two hemispheres.

Celestial Poles

The north (and south) celestial poles are points on the ***celestial sphere*** directly above the north and south terrestrial poles around which the celestial sphere appears to rotate and through which extensions of the Earth's axis of rotation would pass.

The north celestial pole, the position of which is at marked at present by the relatively bright star Polaris, lies in the constellation Ursa Minor (the Little Bear) and would be seen directly overhead when viewed from the North Pole. There is no particularly bright star marking the position of the south celestial pole, which lies in the tiny ***constellation*** Octans (the Octant) and which would be situated directly overhead when seen from the South Pole. The north celestial pole lies in the direction of north when viewed from elsewhere on the Earth's surface and the south celestial pole lies in the direction of south when viewed from other locations.

Celestial Sphere

The imaginary sphere surrounding the Earth on which the stars appear to lie.

Circumpolar Star

A circumpolar star is a star which never sets from a given ***latitude***. When viewing the sky from either the North or South Pole, all stars will be circumpolar, although no stars are circumpolar when viewed from the equator.

Comet

A comet is an object comprised of a mixture of gas, dust and ice which travels around the Sun in an orbit that can often be very eccentric.

Conjunction
This is the position at which two objects are lined up with each other (or nearly so) as seen from Earth. Superior conjunction occurs when a planet is at the opposite side of the Sun as seen from Earth and inferior conjunction when a planet lies between the Sun and Earth.

Constellation
A constellation is an arbitrary grouping of stars forming a pattern or imaginary picture on the celestial sphere. Many of these have traditional names and date back to ancient Greece or even earlier and are associated with the folklore and mythology of the time. There are also some of what may be described as 'modern' constellations, devised comparatively recently by astronomers during the last few centuries. There are 88 official constellations which together cover the entire sky, each one of which refers to and delineates that particular region of the ***celestial sphere***, the result being that every celestial object is described as being within one particular constellation or another.

Dark Nebula
See Nebula.

Declination
This is the ***angular distance*** between a celestial object and the celestial equator. Declination is expressed in degrees, minutes and seconds either north (N) or south (S) of the ***celestial equator***.

Deep Sky Object
Deep sky objects are objects (other than individual stars) which lie beyond the confines of our ***Solar System***. They may be either galactic or extra-galactic and include such things as ***star clusters***, ***nebulae*** and ***galaxies***.

Direct Motion
A planet is in direct (or prograde) motion when its ***right ascension*** or ecliptic ***longitude*** is increasing with the passing of time. This means that it is moving eastwards with respect to the background stars.

Double Stars
Double stars are two stars which appear to be close together in space. Although some double stars (known as *optical* doubles) are made up of two stars that only happen to lie in the same line of sight as seen from Earth and are nothing more than chance alignments, most are comprised of stars that are gravitationally linked and orbit each other, forming a genuine double-star system (also known as a *binary* star).

Eclipse
An eclipse is the obscuration of one celestial object by another, such as the Sun by the Moon during a solar eclipse or one component of an eclipsing ***binary star*** by the companion star.

- A **solar eclipse** occurs when the Moon passes directly between the Earth and the Sun. There are three types of solar eclipse. A total solar eclipse takes place when the Moon completely obscures the Sun, during which event the Sun's corona, or outer atmosphere, is revealed; a partial solar eclipse occurs when the lining up of the Earth, Moon and Sun is not exact and the Moon covers only a part of the Sun; an annular solar eclipse takes place when the Moon is at or near its farthest from Earth, at which time the lunar disc appears smaller and does not completely cover the solar disc, the Sun's visible outer edges forming a 'ring of light' or 'annulus' around the Moon. Some eclipses which begin as annular may become total along part of their path; these are known as hybrid eclipses, and are quite rare.
- A **lunar eclipse** occurs when the Earth passes between the Sun and the Moon, and the Earth's shadow is thrown onto the lunar surface. There are three types of lunar eclipse. A total lunar eclipse takes place when the Moon passes completely through the ***umbra*** of the Earth's shadow, during which process the Moon will gradually darken and take on a reddish/rusty hue; a partial lunar eclipse occurs when the Moon passes through the ***penumbra*** of the Earth's shadow and only part of it enters the umbra; a penumbral lunar eclipse takes place when the Moon only enters the penumbra of the Earth's shadow without touching or entering the umbra.

Ecliptic
As the Earth orbits the Sun, its position against the background stars changes slightly from day to day, the overall effect of this being that the Sun appears to travel completely around the ***celestial sphere*** over the course of a year. The apparent path of the Sun is known as the ecliptic and is superimposed against the band of ***constellations*** we call the ***Zodiac*** through which the Sun appears to move.

Ellipse
The closed, oval-shaped form obtained by cutting through a cone at an angle to the main axis of the cone. The orbits of the planets around the Sun are all elliptical.

Elongation (and Greatest Elongation)
In its most general sense, elongation refers to the angular separation between two celestial objects as seen from a third object. It is most often used to refer to the ***angular distance*** between the Sun and a planet or the Moon, as seen from the Earth.

The greatest elongation of Mercury or Venus is the maximum angular distance between the planet and the Sun as seen from the Earth, during a particular ***apparition***.

Emission Nebula
See Nebula.

Ephemeris (plural: Ephemerides)
Table showing the predicted positions of celestial objects such as comets or planets.

Equator

The equator of a planet or other spheroidal celestial body is the great circle on the surface of the body whose latitude is zero, as defined by the axis of rotation. The ***celestial equator*** is the projection of the plane of the Earth's equator onto the sky.

Equinox

The equinoxes are the two points at which the ecliptic crosses the ***celestial equator*** (see also ***Autumnal Equinox*** and ***Vernal Equinox***). The term is also used to denote the dates on which the Sun passes these points on the ***ecliptic***.

Exoplanet

An exoplanet (or extrasolar planet) is a planet orbiting a star outside of our ***Solar System***.

Galaxy

A galaxy is a vast collection of stars, gas and dust bound together by gravity and measuring many thousands of light years across. Galaxies occur in a wide variety of shapes and sizes including spiral, elliptical and irregular and most are so far away that their light has taken many millions of years to reach us. Our ***Solar System*** is situated in the Milky Way Galaxy, a spiral galaxy containing several billion stars. Located within the ***Local Group of Galaxies***, the ***Milky Way*** Galaxy is often referred to simply as the Galaxy.

Horizon

The horizon is a great circle that is theoretically defined by a zenith distance of 90 degrees. In practice, the observer's ***local horizon*** will differ from this.

Index Catalogue (IC)

References such as that for IC 2391 (in Vela) and IC 2602 (in Carina) are derived from their numbers in the Index Catalogue (IC), published in 1895 as the first of two supplements (the second was published in 1908) to his ***New General Catalogue*** of Nebulae and Clusters of Stars (NGC) by the Danish astronomer John Louis Emil Dreyer (1852-1926). Between them, the two Index Catalogues contained details of an additional 5,386 objects.

Inferior Planet

An inferior planet is a planet that travels around the Sun inside the ***orbit*** of the Earth.

International Astronomical Union (IAU)

Formed in 1919 and based at the Institut d'Astrophysique de Paris, this is the main coordinating body of world astronomy. Its main function is to promote, through international cooperation, all aspects of the science of astronomy. It is also the only authority responsible for the naming of celestial objects and the features on their surfaces.

Latitude

The latitude of the Sun, Moon or planet is its angular distance above or below the ***ecliptic***. Note that the ***angular distance*** of a celestial body north or south of the ***celestial equator*** is called ***declination***, and not latitude.

The latitude of a point on the Earth's surface is its angular distance north or south of the ***equator***.

Light Year

To express distances to the stars and other galaxies in miles would involve numbers so huge that they would be unwieldy. Astronomers therefore use the term 'light year' as a unit of distance. A light year is the distance that a beam of light, travelling at around 300,000 km (186,000 miles) per second, would travel in a year and is equivalent to just under 10 trillion km (6 trillion miles).

Local Group of Galaxies

This is a gravitationally-bound collection of galaxies which contains over 50 individual members, one of which is our own Milky Way Galaxy. Other members include the Large Magellanic Cloud, the Small Magellanic Cloud, the Andromeda Galaxy (M31), the Triangulum Spiral Galaxy (M33) and many others.

Galaxies are usually found in groups or clusters. Apart from our own Local Group, many other groups of galaxies are known, typically containing anywhere up to 50 individual members. Even larger than the groups are clusters of galaxies which can contain hundreds or even thousands of individual galaxies. Groups and clusters of galaxies are found throughout the universe.

Local Horizon

The horizon seen by an observer on land or at sea differs from the ideal theoretical horizon, defined as 90 degrees from the ***zenith***, due to several factors. This can affect astronomical observations. On land, distant features such as mountains may delay the appearance of the rising Sun, Moon or stars by minutes or even hours compared to rising times tabulated in almanacs. At sea, altitudes measured relative to the sea horizon are affected by the observer's height above sea level. At a height of 30 metres above sea level (an aircraft carrier deck, for example), this 'dip' of the sea horizon is 10 arc-minutes, and the ***altitude*** of a star observed using a nautical sextant must have this amount subtracted before it can be used to determine position at sea. The effect may seem small, but 1 arc-minute of observed altitude corresponds to one nautical mile, so ignoring the 10 arc-minute dip correction would lead to an error of 10 nautical miles in the position of the ship.

Local Hour Angle

The local hour angle of a star or other celestial object is the difference between the local ***sidereal time*** and the object's ***right ascension***. At upper ***transit***, an object's local hour angle is zero. Before transit, the local hour angle is negative, whilst after transit, it is positive.

Longitude

The longitude of the Sun, Moon or planet is its angular position, measured along the ***ecliptic*** from the First Point of Aries.

The longitude of a point on the Earth's surface is its ***angular distance*** east or west of the ***prime meridian*** through Greenwich. By convention, terrestrial longitude is positive east of Greenwich and negative west of Greenwich.

Lunar

[illegible] or appertaining to the Moon.

[illegible]lipse

[illegible] is purely and simply a measurement of its brightness. In around [illegible]ner Hipparchus divided the stars up into six classes of brightness, [illegible]ars being ranked as first class and the faintest as sixth. This system [illegible] and other celestial objects according to how bright they actually appear [illegible]ver. In 1856 the English astronomer Norman Robert Pogson refined the system [illegible]ed by Hipparchus by classing a 1st magnitude star as being 100 times as bright as one of 6th magnitude, giving a difference between successive magnitudes of $\sqrt[5]{100}$ or 2.512. In other words, a star of magnitude 1.00 is 2.512 times as bright as one of magnitude 2.00, 6.31 (2.512 x 2.512) times as bright as a star of magnitude 3.00 and so on. The same basic system is used today, although modern telescopes enable us to determine values to within 0.01 of a magnitude or better. Negative values are used for the brightest objects including the Sun (-26.8), Venus (−4.4 at its brightest) and Sirius (−1.46). Generally speaking, the faintest objects that can be seen with the naked eye under good viewing conditions are around 6th magnitude, with binoculars allowing you to see stars and other objects down to around 9th magnitude.

Meridian

This is a great circle crossing the ***celestial sphere*** and which passes through both ***celestial poles*** and the ***zenith***.

Messier Catalogue and References

References such as that for Messier 1 (M1) in Taurus, Messier 31 (M31) in Andromeda and Messier 57 (M57) in Lyra relate to a range of deep sky objects derived from the *Catalogue des Nébuleuses et des Amas d'Étoiles* (Catalogue of Nebulae and Star Clusters) drawn up by the French astronomer Charles Messier during the latter part of the eighteenth century.

Meteor

This is a streak of light in the sky seen as the result of the destruction through atmospheric friction of a ***meteoroid*** in the Earth's atmosphere.

Meteorite
A meteorite is a ***meteoroid*** which is sufficiently large to at least partially survive the fall through Earth's atmosphere.

Meteoroid
This is a term applied to particles of interplanetary meteoritic debris

Milky Way
This is the name given to the faint pearly band of light that we sometimes see crossing
sky and which is formed from the collective glow of the combined light from the
of stars that lie along the main plane of our Galaxy as seen from Earth. The vast
these stars are too faint to be seen individually without some form of optical
provided the sky is really dark and clear, the Milky Way itself is easily
eye, and any form of optical aid will show that it is indeed made up
individual stars. Our ***Solar System*** lies within the main plane of the
located inside one of its spiral arms. The Milky Way is actually our
along the main galactic plane. The glow we see is the combined light
and is visible as a continuous band of light stretching completely around

Nadir
This is the point on the ***celestial sphere*** directly opposite the ***zenith***.

Nebula
Nebulae are huge interstellar clouds of gas and dust. Observed in other galaxies as well as our own, their collective name is from the Latin '*nebula*' meaning 'mist' or 'vapour', and there are three basic types:

- **Emission nebulae** contain young, hot stars that emit copious amounts of ultra-violet radiation which reacts with the gas in the nebula causing the nebula to shine at visible wavelengths and with a reddish colour characteristic of this type of nebula. In other words, emission nebulae *emit* their own light. A famous example is the Orion Nebula (M42) in the constellation Orion which is visible as a shimmering patch of light a little to the south of the three stars forming the Belt of Orion.
- The stars that exist in and around **reflection nebulae** are not hot enough to actually cause the nebula to give off its own light. Instead, the dust particles within them simply *reflect* the light from these stars. The stars in the Pleiades star cluster (M45) in Taurus are surrounded by reflection nebulosity. Photographs of the Pleiades cluster show the nebulosity as a blue haze, this being the characteristic colour of reflection nebulae.
- **Dark nebulae** are clouds of interstellar matter which contain no stars and whose dust particles simply blot out the light from objects beyond. They neither emit or reflect light and appear as dark patches against the brighter backdrop of stars or nebulosity, taking on the appearance of regions devoid of stars. A good example is the Coal Sack in the constellation Crux, a huge blot of matter obscuring the star clouds of the southern Milky Way.

Neutron Star
This is the remnant of a massive star which has exploded as a ***supernova***.

New General Catalogue (NGC)
References such as that for NGC 869 and NGC 884 (in Perseus) and NGC 4755 (in Crux) are derived from their numbers in the New General Catalogue of Nebulae and Clusters of Stars (NGC) first published in 1888 by the Danish astronomer John Louis Emil Dreyer (1852-1929) nd which contains details of 7,840 star clusters, nebulae and galaxies.

ltation
temporary covering up of one celestial object, such as a star, by another, such as planet.

the orbit of a ***superior planet*** when it is located directly opposite

path of one object around another under the influence of gravity.

Parallax
Parallax describes the change in the apparent direction to a distant object caused by a change in the observer's location. In astronomy, it refers specifically to the very small change in the position of a star when observed from opposite sides of the Earth's orbit. This change, when measured, can be used to infer the distance to the star. The parallax of the nearest star, Proxima Centauri, is 0.768 seconds of arc.

Parsec
A unit of distance, often used by professional astronomers in preference to light years. A star at a distance of one parsec has a ***parallax*** of one second of arc. It is equal to 3.26 light years. The nearest star, Proxima Centauri, is 1.3 parsecs from the Sun. Distances within our Galaxy are generally expressed in kiloparsecs (1,000 parsecs; abbreviation kpc), whilst distances between galaxies are expressed in megaparsecs (1,000,000 parsecs; abbreviation Mpc).

Penumbra
This is the area of partial shadow around the main cone of shadow cast by the Moon during a solar ***eclipse*** or the Earth during a lunar ***eclipse***. The term penumbra is also applied to the lighter and less cool region of a sunspot.

Perigee
This is the point in its ***orbit*** around the Earth at which an object is at its closest to the Earth.

Perihelion

This is the point in its ***orbit*** around the Sun at which an object, such as a planet, comet or asteroid, is at its closest to the Sun.

Planetary Nebula

Planetary nebulae consist of material ejected by a star during the latter stages of its evolution. The material thrown off forms a shell of gas surrounding the star whose newly-exposed surface is typically very hot. Planetary nebulae have nothing whatsoever to do wit planets. They derive their name from the fact that, when seen through a telescope, so planetary nebulae take on the appearance of luminous discs, resembling a gaseous such as Uranus or Neptune. Probably the best known example is the famous Ri (M57) in Lyra.

Precession

The Earth's axis of rotation is an imaginary line which passes thr Poles of the planet. Extended into space, this line defines the North in the sky. The North ***Celestial Pole*** currently lies close to Polaris in Bear), so the daily rotation of our planet on its axis makes the rest of the stars to travel around Polaris, their paths through the sky being centred on the Pole Star.

However, the position of the north celestial pole is slowly changing, this because of gradual change in the Earth's axis of rotation. This motion is known as 'precession' and is identical to the behaviour of a spinning top whose axis slowly moves in a cone. Precession is caused by the combined gravitational influences of the Sun and Moon on our planet. Each resulting cycle of the Earth's axis takes around 25,800 years to complete, the net effect of precession being that, over this period, the north (and south) celestial poles trace out large circles around the northern (and southern) sky. This results in slow changes in the apparent locations of the celestial poles. Polaris will be closest to the North Celestial Pole in the year 2102, but it will then begin to move slowly away and eventually relinquish its position as the Pole Star. Vega will take on the role some 11,500 years from now.

Prime Meridian

The celestial prime ***meridian*** is the meridian on the sky that passes through the ***vernal equinox***. It marks the zero point for measuring ***right ascension***.

On the surface of the Earth, the prime meridian is the line of constant ***longitude*** which passes through the centre of the Airy transit telescope at the Royal Observatory at Greenwich in London. It was adopted by international agreement in 1884 as the origin for measuring longitude. Unlike the celestial prime meridian, it has no physical significance.

Prograde Motion

See Direct Motion.

Pulsar

This is a rapidly-spinning neutron star which gives off regular bursts of radiation.

Quadrature
This refers to the geometric configuration of the Sun, Earth and a ***superior planet*** when the elongation of the planet from the Sun, as seen from the Earth, is 90°.

Quasar
These are small, extremely remote and highly luminous objects which at the cores of active galaxies. They are comprised of a super-massive black hole surrounded by an accretion disk of gas which is falling into the black hole.

[illegible]ction Nebula
[illegible]la.

[illegible]ion
[illegible]de motion when its ***right ascension*** or ecliptic ***longitude*** is decreasing [illegible]e. This means that it is moving westwards with respect to the [illegible]rior planets*** undergo a period of retrograde motion around the [illegible]

[illegible]sion
[illegible] angular distance, measured eastwards, of a celestial object from the ***vernal equinox***. Right ascension is expressed in hours, minutes and seconds.

Satellite
A satellite is a small object orbiting a larger one.

Shooting Star
The popular name for a ***meteor***.

Sidereal Period
The time taken for an object to complete one ***orbit*** around another, measured with respect to a fixed direction in space.

Solar
Of or appertaining to the Sun.

Solar Eclipse
See Eclipse.

Solar System
The Solar System is the collective description given to the system dominated by the Sun and which embraces all objects that come within its gravitational influence. These include the planets and their satellites and ring systems, minor planets, comets, meteoroids and other interplanetary debris, all of which travel in orbits around our parent star.

Solstice

These are the points on the ***ecliptic*** at which the Sun is at its maximum angular distance (***declination***) from the ***celestial equator***. The term is also used to denote the dates when the Sun passes these points on the ecliptic.

Spectroscope

An instrument used to split the light from a star into its different wavelengths or colours.

Spectroscopic Binary

This is a ***binary star*** whose components are so close to each other that they can[illegible] resolved visually and can only be studied through ***spectroscopy***.

Spectroscopy

This is the study of the spectra of astronomical objects.

Star

A star is a self-luminous object shining through the release of energy [illegible] reactions at its core.

Star Colours

When we look up into the night sky the stars appear much the same. Some stars appear brighter than others but, with a few exceptions, they all look white. However, if the stars are looked at more closely, even through a pair of binoculars or a small telescope, some appear to be different colours. A prominent example is the bright orange-red Arcturus in the ***constellation*** of Boötes, which contrasts sharply with the nearby brilliant white Spica in Virgo. Our own Sun is yellow, as is Capella in Auriga. Procyon, the brightest star in Canis Minor, also has a yellowish tint. To the west of Canis Minor is the constellation of Orion the Hunter, which boasts two of the most conspicuous stars in the whole sky; the bright red Betelgeuse and Rigel, the brilliant blue-white star that marks the Hunter's foot.

The colour of a star is a good guide to its temperature, the hottest stars being blue and blue-white with surface temperatures of 20,000 degrees K or more. Classed as a yellow dwarf, the Sun is a fairly average star with a temperature of around 6,000 degrees K. Red stars are much cooler still, with surface temperatures of only a few thousand degrees K. Betelgeuse in Orion and Antares in Scorpius are both red giant stars that fall into this category.

Star Clusters

Although most of the stars that we see in the night sky are scattered randomly throughout the spiral arms of the Galaxy, many are found to be concentrated in relatively compact groups, referred to by astronomers as star clusters. There are two main types of star cluster – open and globular. Open clusters, also known as galactic clusters, are found within the main disc of the Galaxy and have no particularly well-defined shape. Usually made up of young hot stars, over a thousand open clusters are known, their diameters generally being no more than a few tens of light years. They are believed to have formed from vast interstellar gas and dust clouds within our Galaxy and indeed occupy the same regions of the Galaxy as the nebulae. A number of open clusters are visible to the naked eye including Praesepe (M44) in Cancer, the Hyades in Taurus and perhaps the most famous open cluster of all the Pleiades (M45), also in Taurus.

Globular clusters, as their name suggests, are huge spherical collections of stars. Located in the area of space surrounding the Galaxy, they can have diameters of anything up to several hundred light years and typically contain many thousands of old stars with little or none of the nebulosity seen in open clusters. When seen through a small telescope or binoculars, they take on the appearance of faint, misty balls of greyish light superimposed against the background sky. Although some form of optical aid is usually needed to see globular clusters, there are three famous examples which can be spotted with the naked eye. These are 47 Tucanae in Tucana, Omega Centauri in Centaurus and the Great Hercules Cluster (M13) in Hercules.

Stationary Point
A planet is at a stationary point when its motion with respect to the background stars changes from ***direct*** (motion) to ***retrograde*** (motion) or vice versa. All ***superior planets*** pass through two stationary points at each ***apparition***, once before ***opposition*** and again after opposition.

Superior Planet
A superior planet is a planet that travels around the Sun outside the ***orbit*** of the Earth.

Supernova
Supernovae are huge stellar explosions involving the destruction of massive stars and resulting in sudden and tremendous brightening of the stars involved.

Synodic Period
The synodic period of a planet is the interval between successive ***oppositions*** or ***conjunctions*** of that planet.

Transit
1 – The instant when an object crosses the local ***meridian***. When the object's ***local hour angle*** is zero, this is known as upper transit, and marks the maximum ***altitude*** of the object above the observer's ***horizon***. When the object's local hour angle is 12 hours, it is known as lower transit.

2 – The passage of Mercury or Venus across the disk of the Sun, as seen from the Earth, or of a planetary satellite across the disk of the parent planet.

Umbra
This is the main cone of shadow cast by the Moon during a solar ***eclipse*** or the Earth during a lunar ***eclipse***. The term umbra is also applied to the darkest, coolest region of a sunspot.

Variable Stars

A variable star is a star whose brightness varies over a period of time. There are many different types of variable star, although the variations in brightness are basically due either to changes taking place within the star itself or the periodic obscuration, or eclipsing, of one member of a ***binary star*** by its companion.

Vernal Equinox

The vernal equinox is the point at which the apparent path of the Sun, moving from south to north, crosses the ***celestial equator***. In the Earth's northern hemisphere this marks the start of spring, whilst in the southern hemisphere it is the start of autumn.

Zenith

This is the point on the ***celestial sphere*** directly above the observer.

Zodiac

The band of ***constellations*** along which the Sun appears to travel over the course of a year. The Zodiac straddles the ***ecliptic*** and comprises the 12 constellations Aries, Taurus, Gemini, Cancer, Leo, Virgo, Libra, Scorpius, Sagittarius, Capricornus, Aquarius and Pisces. The ecliptic also passes through part of the constellation of Ophiuchus, as delimited by the boundaries defined by the ***International Astronomical Union***, but Ophiuchus is not traditionally considered a constellation of the Zodiac.